Advances in

Geosciences

Volume 17: Hydrological Science (HS)

ADVANCES IN GEOSCIENCES

Editor-in-Chief: Wing-Huen Ip *(National Central University, Taiwan)*

A 5-Volume Set

Volume 1: Solid Earth (SE)
ISBN-10 981-256-985-5

Volume 2: Solar Terrestrial (ST)
ISBN-10 981-256-984-7

Volume 3: Planetary Science (PS)
ISBN-10 981-256-983-9

Volume 4: Hydrological Science (HS)
ISBN-10 981-256-982-0

Volume 5: Oceans and Atmospheres (OA)
ISBN-10 981-256-981-2

A 4-Volume Set

Volume 6: Hydrological Science (HS)
ISBN-13 978-981-270-985-1
ISBN-10 981-270-985-1

Volume 7: Planetary Science (PS)
ISBN-13 978-981-270-986-8
ISBN-10 981-270-986-X

Volume 8: Solar Terrestrial (ST)
ISBN-13 978-981-270-987-5
ISBN-10 981-270-987-8

Volume 9: Solid Earth (SE), Ocean Science (OS) & Atmospheric Science (AS)
ISBN-13 978-981-270-988-2
ISBN-10 981-270-988-6

A 6-Volume Set

Volume 10: Atmospheric Science (AS)
ISBN-13 978-981-283-611-3
ISBN-10 981-283-611-X

Volume 11: Hydrological Science (HS)
ISBN-13 978-981-283-613-7
ISBN-10 981-283-613-6

Volume 12: Ocean Science (OS)
ISBN-13 978-981-283-615-1
ISBN-10 981-283-615-2

Volume 13: Solid Earth (SE)
ISBN-13 978-981-283-617-5
ISBN-10 981-283-617-9

Volume 14: Solar Terrestrial (ST)
ISBN-13 978-981-283-619-9
ISBN-10 981-283-619-5

Volume 15: Planetary Science (PS)
ISBN-13 978-981-283-621-2
ISBN-10 981-283-621-7

A 6-Volume Set

Volume 16: Atmospheric Science (AS)
ISBN-13 978-981-283-809-4
ISBN-10 981-283-809-0

Volume 17: Hydrological Science (HS)
ISBN-13 978-981-283-811-7
ISBN-10 981-283-811-2

Volume 18: Ocean Science (OS)
ISBN-13 978-981-283-813-1
ISBN-10 981-283-813-9

Volume 19: Planetary Science (PS)
ISBN-13 978-981-283-815-5
ISBN-10 981-283-815-5

Volume 20: Solid Earth (SE)
ISBN-13 978-981-283-817-9
ISBN-10 981-283-817-1

Volume 21: Solar Terrestrial (ST)
ISBN-13 978-981-283-819-3
ISBN-10 981-283-819-8

Advances in Geosciences

Volume 17: Hydrological Science (HS)

Editor-in-Chief

Wing-Huen Ip

National Central University, Taiwan

Volume Editor-in-Chief

Namsik Park

Dong-A University, South Korea

World Scientific

NEW JERSEY • LONDON • SINGAPORE • BEIJING • SHANGHAI • HONG KONG • TAIPEI • CHENNAI

Published by

World Scientific Publishing Co. Pte. Ltd.
5 Toh Tuck Link, Singapore 596224
USA office: 27 Warren Street, Suite 401-402, Hackensack, NJ 07601
UK office: 57 Shelton Street, Covent Garden, London WC2H 9HE

British Library Cataloguing-in-Publication Data
A catalogue record for this book is available from the British Library.

ADVANCES IN GEOSCIENCES
A 6-Volume Set
Volume 17: Hydrological Science (HS)

ISBN-13 978-981-283-808-7 (Set)
ISBN-10 981-283-808-2 (Set)
ISBN-13 978-981-283-811-7 (Vol. 17)
ISBN-10 981-283-811-2 (Vol. 17)

Typeset by Stallion Press
Email: enquiries@stallionpress.com

Printed in Singapore.

EDITORS

Editor-in-Chief: Wing-Huen Ip

Volume 16: Atmospheric Science (AS)
Editor-in-Chief: Jai Ho Oh
Editors: G. P. Singh
C. C. Wu
K.-J. Ha

Volume 17: Hydrological Science (HS)
Editor-in-Chief: Namsik Park
Editors: Ji Chen
Joong-Hoon Kim
Jinping Liu
Young-Il Moon
Sanjay Patil
Ashok Kumar Rastogi
Simon Toze

Volume 18: Ocean Science (OS)
Editor-in-Chief: Jianping Gan
Editors: Minhan Dai
Anne Mueller
Murty Vadiamani

Volume 19: Planetary Science (PS)
Editor-in-Chief: Anil Bhardwaj

Volume 20: Solid Earth (SE)
Editor-in-Chief: Kenji Satake

Volume 21: Solar & Terrestrial Science (ST)
Editor-in-Chief: Marc Duldig

REVIEWERS

The Editors of Volume 17 (Hydrological Science) would like to acknowledge the following referees who have helped to review the papers published in this volume:

Mukund Baride
Ghosh Bobba
Chavalit Chaleeraktrakoon
D Chandrasekaram
Hongey Chen
Xi Chen
Xiaohong Chen
Xing Chen
Yongqin David Chen
Haiyun Cheng
Ke-Sheng Cheng
May Ting Fong Chui
C.G. Cui
Pakorn Ditthakit
TI Eldho
Ibrahim Gurer
Shin-ichiro Hayashi
Yutaka Ichikawa
Changsoo Im
Bongseog Jung
V Jyotiprakash
Kenji Kawaike
Dae-Hong Kim
Hung Soo Kim
Jong Suk Kim
Joong Hoon Kim
Sung Kim
Young Il Kim
Toshi Kojiri
Hyun-Han Kwon
Chang Hae Lee
Tiejian Li
Yan Li
Bingzhang Lin
Shie-Yui Liong
Zhiyu Liu
Sukomal Mandal
Akira Mano
Shashi Mathur
S Mohan
Young-Il Moon
Toshiyuki Moriyama
Ahmed Murad
Tae Suk Oh
Kyungrock Paik
Namsik Park
PI Park
Haresh Patel
JN Patel
Sanjay Patil
Prem Paudel
Dingzhi Peng
Mingkai Qian
BK Sahu
Hussain Saleh
Mohammed Rafi Sayyed
Uma Seeboonruang
Gunwant Sharma
Jatinder Sidhu
Israel Silva
Sri Srikanthan
Yasuto Tachikawa
Sampam Thaikruawan
Guangsheng Wang

Guoqing Wang
Chun-Chieh Wu
Yiping Wu
Yue-Ping Xu
Zongxue Xu
Yang Yang
Jaeeung Yi
Yuki Yusa
Jian Zhang
Jianyun Zhang
Liping Zhang
Qiang Zhang
Ping Zhao
Yongguang Zheng
Maichun Zhou
Yunmei Zhu
Michael Ziobro

PREFACE

The present volume set of Advances in Geosciences (ADGEO) contain papers from the Busan annual meeting in 2008 and some from the Singapore annual meeting in 2009. As Editor-in-Chief, I must apologize to the AOGS members and authors for this delay. Since 2006 we have published 20 volumes in total. This publication project has been supported by the AOGS Council, World Scientific Publication Company (WSPC), the team of hard working editors and the broad membership and participants of AOGS. As with the main purpose of the Society, ADGEO is meant to promote information exchange and to forge scientific cooperation in the area of Earth science and environmental study. As witnessed by the negotiation efforts at the United Nations Climate Change Conference in Copenhagen in December 2009, all these issues have become more and more important and vital in the Asia-Pacific region. It is not a matter of exaggeration in saying that the solution to global warming, if there is one, has to come from the emerging economies and developing countries covered by AOGS. By design, ADGEO has its fundamental role to play. In practical terms, it is actually a difficult task because of many factors involved in deciding the quality of manuscripts, editorial and review processes, publication procedure, scientific impacts, readership, policy of the AOGS Council, and last but not least, marketing from the point of view of the publisher. Any small mishap in this long chain of interactive steps could lead to a major discontinuity. We have encountered such a situation with the publication of the Busan manuscripts. It is only with the cooperation of the authors, the ADGEO editorial team, WSPC, and the AOGS Secretariat Office, that we are able to produce these volumes, albeit a long delay. With this lesson learned, we hope to consolidate the ADGEO management and editorial system so that it would become an essential publication in our understanding of Earth and space science and information tool books in the battle against climate change. Finally, I would like to take this opportunity to thank the Volume Editors-in-Chief who are the driving force in making ADGEO possible: A. Bhardwaj (Planetary Science), M. Duldig (Solar and Terrestrial Science), J.P. Gan (Ocean Science), J.H. Oh (Atmospheric

Science), N.S. Park (Hydrological Science), K. Satake and C.H. Lo (Solid Earth). They have to work very hard to ensure both the quantity and quality of the published papers in ADGEO. Of equal importance, the support from WSPC is essential and its foresight in identifying the academic and social values of Earth science and environmental study to be sustained and articulated by ADGEO is very much appreciated.

Wing-Huen Ip
Editor-in-Chief

PREFACE TO HS VOLUME

The papers in this volume are presented at the 5th and the 6th AOGS Annual Meeting held in Busan, Korea in 2008 and in Singapore in 2009, respectively. Participants of the two Meetings were asked to submit full manuscripts. All manuscripts were then subject to peer reviews. Subjects of the papers range from precipitation, groundwater, river flow, sedimentation, floods and water resources. Impact on precipitation due to "climate change" is a hot issue. Downscaling is one of the hot themes. Short-term forecasting is also of interest, especially for water related disaster management. In this volume, flooding is the most popular topic. Although damages due to flood can be generally reduced by investing in infrastructure system, the magnitude of flash floods also seems to grow, threatening securities of people, eco-systems, low-land area, water resources, etc. Forecasting is also the major theme of papers on flood. Several paper deal with groundwater. Some papers present site-specific problems while others describe general methodologies. Water resources are also a timely topic as both quantity and quality degenerate due to anthropogenic and natural causes. All in all, this volume is a must-have to scientists and engineers who are interested in hydrological problems in this region and beyond.

The chief editor would like to express his sincere gratitude to all authors and anonymous reviewers. Their commitments to the promulgation of scientific knowledge and to the upkeep of scientific quality are the essential driving force of all professional publications, including AdGeo. Fellow editors of HS deserve special thanks as they had to deal with the chief editor, authors, reviewers and the online publication system. Chief editor Ip's leadership has been instrumental in putting everything in order. Secretarial support from MeetMatt and WSPC are noted. Special thank is due on Ms. Zaiyi at MeetMatt. She built a nice on-line portal which made the lives of everybody involved in this publication much easier.

Namsik Park
HS Volume Editor-in-Chief

CONTENTS

Advances in Geosciences
Vol. 17: Hydrological Science (2008)
Eds. Namsik Park *et al.*

RAINGAUGE-BASED RAINFALL NOWCASTING WITH ARTIFICIAL NEURAL NETWORK

SHIE-YUI LIONG* and SHAN HE
Tropical Marine Science Institute,
National University of Singapore,
Singapore 119223
**tmslsy@nus.edu.sg*

Rainfall forecasting and nowcasting are of great importance, for instance, in real-time flood early warning systems. Long term rainfall forecasting demands global climate, land, and sea data, thus, large computing power and storage capacity are required. Rainfall nowcasting's computing requirement, on the other hand, is much less. Rainfall nowcasting may use data captured by radar and/or weather stations. This paper presents the application of Artificial Neural Network (ANN) on rainfall nowcasting using data observed at weather and/or rainfall stations. The study focuses on the North-East monsoon period (December, January and February) in Singapore. Rainfall and weather data from ten stations, between 2000 and 2006, were selected and divided into three groups for training, over-fitting test and validation of the ANN. Several neural network architectures were tried in the study. Two architectures, Backpropagation ANN and Group Method of Data Handling ANN, yielded better rainfall nowcasting, up to two hours, than the other architectures. The obtained rainfall nowcasts were then used by a catchment model to forecast catchment runoff. The results of runoff forecast are encouraging and promising. With ANN's high computational speed, the proposed approach may be deliverable for creating the real-time flood early warning system.

1. Introduction

Rainfall forecasting can be obtained by long-term forecast, and short-term forecast, namely nowcasting. Numerical weather prediction models have been used for long term forecasting. However, their ability to provide accurate rainfall forecasts at the temporal and spatial resolutions still requires more improvement for many hydrologic applications (Brath, 1999); this is particularly crucial for catchments with small/medium sizes.

A number of approaches are available for rainfall nowcasting. For examples, there are (1) linear stochastic auto-regressive moving-average models (ARMA), which express the future rainfall as a linear function of past data, (2) the use of remote sensing observations (radar data and satellite images), which nowcasts rainfall based on the extrapolation of current weather conditions, (3) adaptive-network-based fuzzy inference system (ANFIS) proposed by Jang(1992), which can construct an input-output mapping based on both human knowledge (in the form of fuzzy if-then rules) and stipulated input-output data pairs, and (4) artificial neural network (ANN), which belongs to the non-linear, data-driven approaches. ANN depends on the available data for 'learning' without any priori hypothesis about the kind of relationship. One application described by Kuligowski and Barros (1998) showed that the ANN performance was superior to that of a multi-linear regression model at least for heavy rain events. Another study (Chau *et al.*, 2005) investigated two hybrid models based on ANN and ANFIS; the models were employed for flood forecasting in a channel reach of the Yangtze River in China. The ANFIS model was found to be optimal, but it entailed a large number of parameters. The performance of the ANN-GA model was shown to be good as well, yet it required longer computation time and additional modeling parameters.

Neural networks offer a number of advantages, including requiring less formal statistical training, ability to implicitly detect complex nonlinear relationships between dependent and independent variables, ability to detect all possible interactions between predictor variables, and the availability of multiple training algorithms. Disadvantages include its "black box" nature, greater computational burden, proneness to over-fitting, and the empirical nature of model development.

The scope of this paper is to explore the nowcasting skill with artificial neural network using the observations from the rainfall/meteorological stations in Singapore. The proposed approach is then demonstrated on a relatively small urban catchment characterized by runoff with a short time of concentration.

2. Methodology

2.1. *Artificial neural network*

Artificial Neural Network (ANN) excels at problem diagnosis, decision making, and prediction. In this study, ANN is applied to train rainfall data set, learn the patterns, and develop the ability to reasonably classify new patterns or to make forecasts and predictions.

The basic building block of neural network technology is the simulated neurons. The network processes a number of inputs to produce an output,

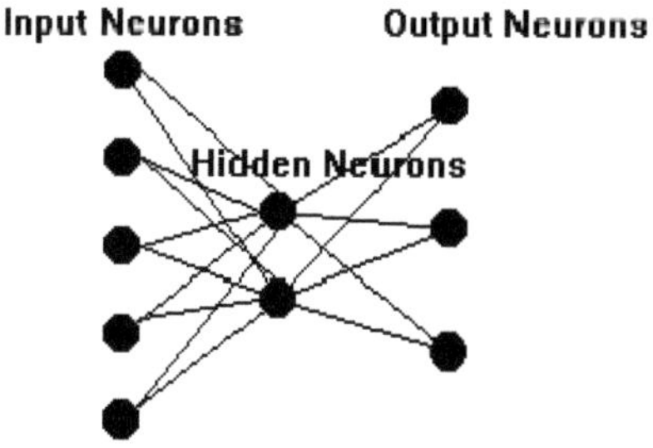

Fig. 1. Network structure.

the network's classifications or predictions. The neurons are connected by weights (depicted as lines as shown in Fig. 1) which are applied to values passed from one neuron to the next.

The network "learns" by adjusting the interconnection weights between layers. The network will be producing answer and repeatedly compared with the measured data, and each time the connecting weights are adjusted slightly in the direction of the measured data. As a result, the problem can be learnt, and a stable set of weights adaptively evolves and will produce a good prediction model.

In order to build up a successful neural network, it is important to know when to stop training. If the training is not sufficient, the net will not be able to learn the pattern; if the network is over trained, the network will memorize the training set thus it is unable to give a good answer on unseen data.

In this study a multilayer perceptron (MLP) with backpropagation (Bishop, 1995; Reed and Marks, 1999; Fausett, 1994; Rumelhart *et al.*, 1986) ANN is presented.

2.2. *Case study*

Singapore lies just north of the Equator near Latitude 1.5°N and Longitude 104°E. Because of its geographical location, maritime exposure and size, its climate is characterized by rather uniform temperature and pressure, high humidity, and abundant rainfall. The climate of Singapore can mainly be divided into two main seasons, the Northeast Monsoon and the Southwest Monsoon, separated by two relatively short inter-monsoon periods.

As shown from Fig. 2, five meteorological and five rainfall stations are considered in the study. They are spread quite throughout Singapore. The

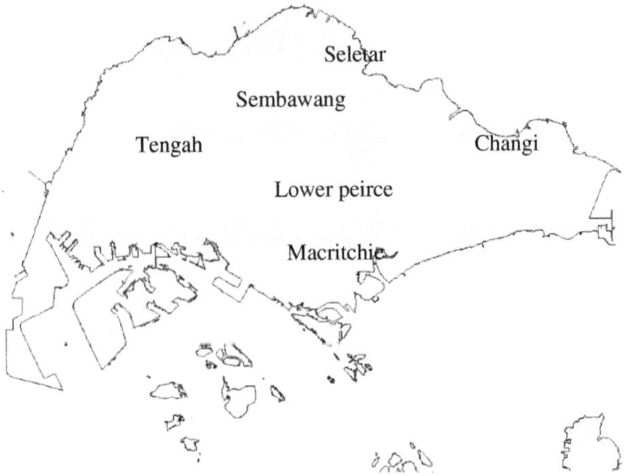

Fig. 2. Meteorological and rainfall stations considered.

five meteorological stations are Changi, Sembawang, Tengah, Seletar and Paya Lebar; hourly time series of meteorological data such as pressure, dry bulb temperature, wet bulb temperature, low cloud coverage, wind speed, WX (two figure numeric code representing the weather. E.g. 10 for mist; 60–65 for rain;), wind direction and rainfall are extracted from these five stations. The five rainguage stations are MacRitchie Reservoir, Lower Pierce Reservoir, Upper Seletar, Chestnut and Upper Pierce Reservoir; only hourly rainfall time series are considered for the current work. Table 1 summarizes the ANN input variables used for different forecast horizons.

To estimate rainfall at forecast horizons $(t + 1\Delta t)$ and $(t + 2\Delta t)$ where Δt is the incremental time step of 1 hour, the current and the past two hour weather variables (i.e. at times t, $t - 1\Delta t$, and $t - 2\Delta t$) are used. Table 1 shows the weather variables used at different time steps extracted from the five meteorological stations while at the other rainfall stations only rainfall data from times $(t - 1\Delta t)$ and $(t - 2\Delta t)$ are used as input variables.

In the study, 7 out of the 10 stations (Changi, Sembawang, Tengah, MacRitchie, Chestnut, Upper Seletar and Chestnut) are used as training stations while the remaining 3 stations are used for validation. Data considered as the training data set is from December–February of Years 2001 to 2005, and data from 17th to 19th Dec. 2006. We trained ANN with these data to later validate the trained ANN with the rainfall events from the period of 19th–21st December 2006. It should be noted that rainfall events in Dec 2006 were the third highest event ever recorded.

Table 1. Weather parameters used in the study as input variables.

parameters	$t-2\Delta t$	$t-1\Delta t$	t
Pressure	√	√	√
Dry-20	√	√	√
Cloud Coverage	√	√	√
Wind Speed	√	√	√
WX	√	√	√
Web Bulb Depression	√	√	√
Wind x-direction	√	√	√
Wind y-direction	√	√	√
Rainfall	√	√	NOT USED
Change in Pressure in the last 2 hours	—	—	√
Change in dry bulb temperature in the last 2 hours	—	—	√
change in wet bulb depression in the last 2 hours	—	—	√
Change in low cloud amount in the last 2 hours	—	—	√

Table 2. Rainfall prediction at forecast horizons $(t+1\Delta t)$, $(t+2\Delta t)$.

	Stations	Correlation Coefficient (CC) for time $(t+1\Delta t)$	Correlation Coefficient (CC) for time $(t+2\Delta t)$
Training	7 stations	0.6529	0.5639
Validation (19–21 December 2006)	Changi	0.6132	0.4630
	Sembawang	0.6877	0.4323
	Tengah	0.7386	0.5531
	MacRitchie	0.9540	0.7772
	Chestnut	0.9367	0.7330
	Upper Seletar	0.9114	0.6700
	Upper Pierce	0.9086	0.6952
	Paya Lebar	0.8003	0.6489
	Seletar	0.8541	0.5951
	Lower Pierce	0.9663	0.7290

The results are shown in Table 2. It can be seen that the performance indicator is reasonably good since the average goodness-of-fit for even $t+2\Delta t$ is around 0.6. It should be noted that the task of rainfall nowcasting is very difficult since the growth and decay of the rain storm is a very complicated process.

Figure 2 clearly shows the location situation of all the stations used. As MacRitchie station is the closest station to the Singapore River which is used later for the runoff simulation, the result of Neural Network for MacRitchie is presented in Fig. 3 and imported into rainfall-runoff model as rainfall input. It can be shown that the performance indicator is reasonably good since the correlation coefficient for even $(t + 2)$ is 0.7772.

3. Runoff Simulation

The nowcasted rainfall for times $(t + 1\Delta t)$ and $(t + 2\Delta t)$ are then used as input to a catchment runoff model, SOBEK (Introduction to SOBEK, 2004), to simulate the runoff from an urban catchment of about 12 km^2 area and characterized with a very short time of concentration about 20–30 minutes.

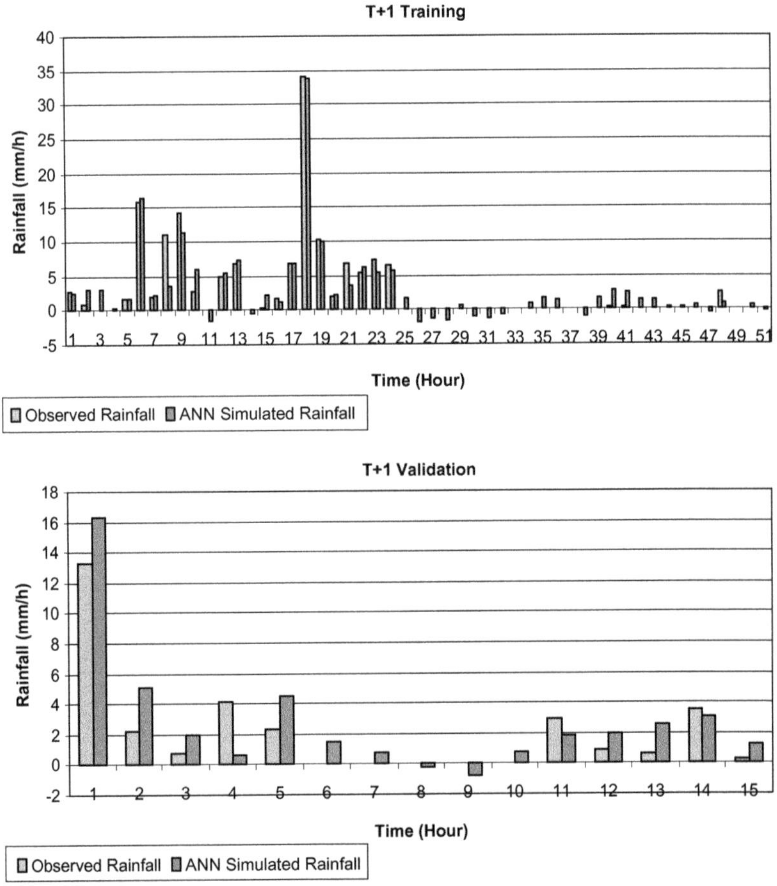

Fig. 3. Comparison of Macritchie rainfalls resulting from ANN and raingauges.

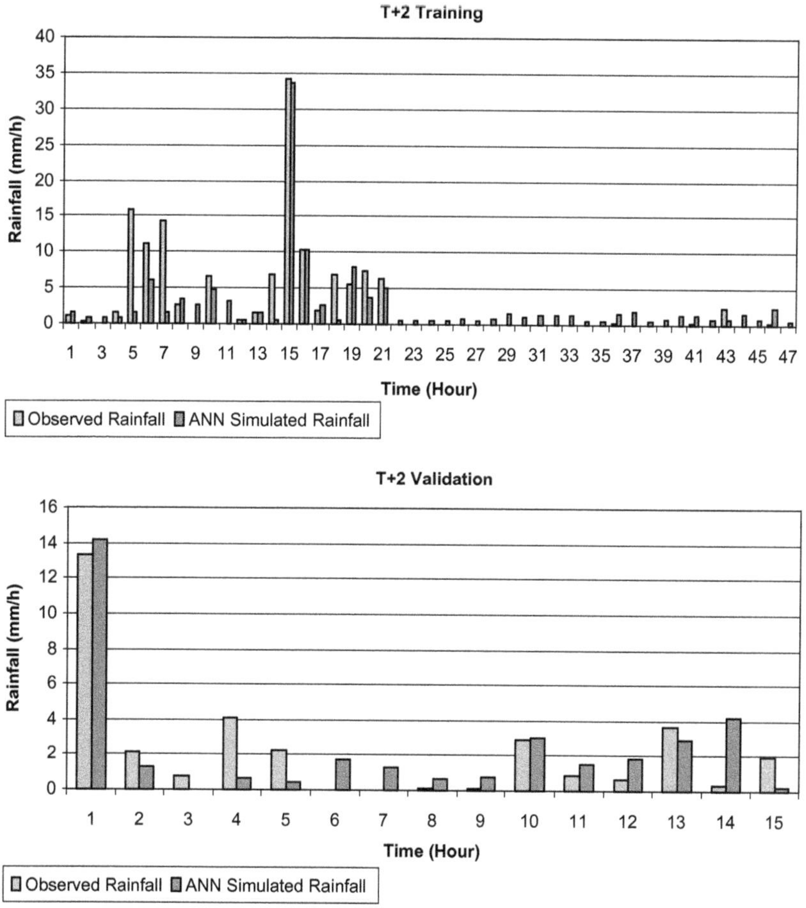

Fig. 3. (*Continued*)

Since the final target is the flow at the catchment outlet, the simulated flow obtained from ANN rainfall input is in reasonably good agreement with the flow obtained from measured rainfall as can be seen in Fig. 4 and Table 3. (Negative flow in Fig. 4 refers to tide impact.)

4. Conclusions

This paper presented a promising rainfall observation based rainfall nowcasting approach using Artificial Neural Network. Hourly meteorological and rainfall data were used to forecast rainfall quantities,

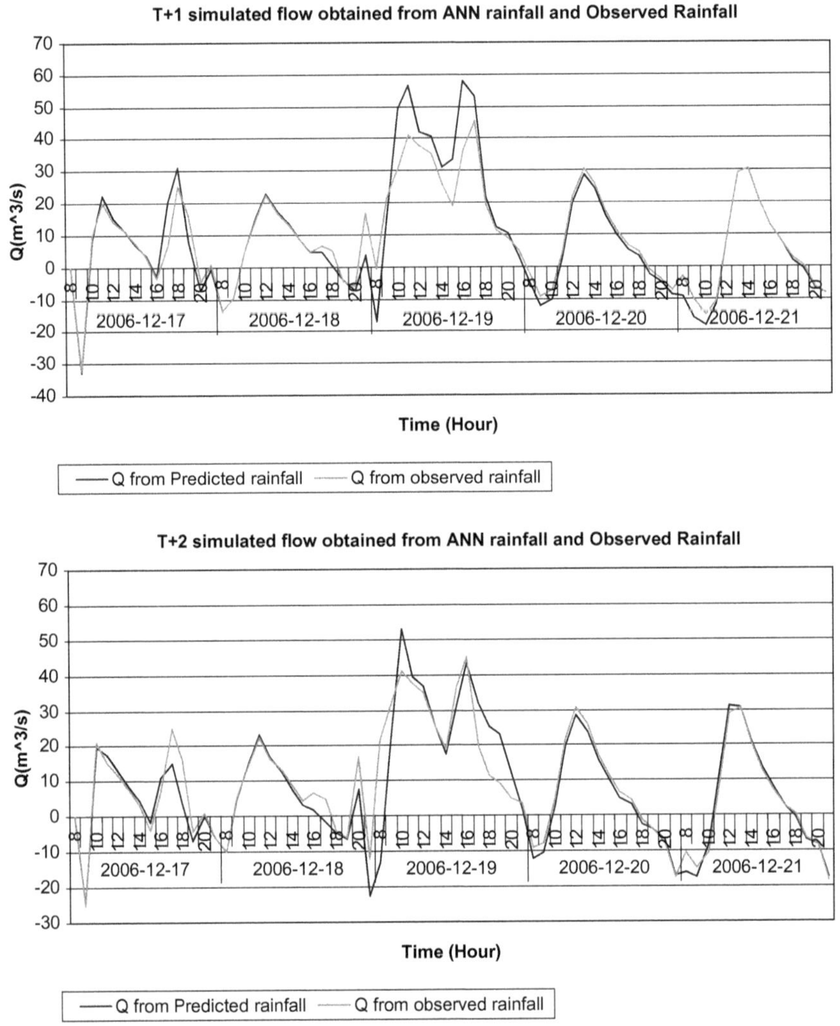

Fig. 4. Forecast flow results: (a) $t + 1\Delta t$; (b) $t + 2\Delta t$.

in this study, up to forecast horizons of 2 hours. The final rainfall forecast data were then demonstrated to simulate runoff from an urban catchment in Singapore.

The results, presented in terms of visual comparison of observed and simulated hydrographs and goodness-of-fit measures, indicate that the proposed approach is able to capture the peak discharges quite accurately.

Table 3. Result for flow forecast.

Correlation Coefficient	Lead time	
	$t+1\Delta t$	$t+2\Delta t$
Period (8:00 17/12/06–21:00 21/12/06)	0.9609	0.8828

The ability of the proposed approach to forecast rainfall and runoff may be considered necessarily needed for its application in the real time flood early warning systems.

Acknowledgment

The authors thank Executive Meteorological Officer Mr. K.H. CHEW from Meteorological Services Division for suggesting important weather parameters for ANN input.

References

1. C. M. Bishop, *Neural Networks for Pattern Recognition* (Oxford: Oxford University Press, 1995).
2. A. Brath, On the role of numerical weather prediction models in real-time flood forecasting, *Proceedings of the International Workshop on River Basin Modeling: Management and Flood Mitigation*, September 25–26, 1997, Monselice, Italy, 1999, pp. 249–259.
3. K. W. Chau, C. L. Wu and Y. S. Li, Comparison of several Flood Forecasting Models in Yangtze River. *J. Hydrologic Engrg.* **10**(6) (2005) 485–491.
4. M. N. French, W. F. Krajewski and R. R. Cuykendall, Rainfall forecasting in space and time using a neural network. *J. Hydrol.* **137** (1992) 1–31.
5. J. S. R. Jang, ANFIS: Adaptive-network-based fuzzy inference systems, *IEEE Trans. On Systems, Man, and Cybernetics*, 1992.
6. R. Krzysztofowicz, Recent advances associated with flood forecast and warning systems, US National Report to IUGG, 1991–1994, *Rev. Geophys.* 33 (Part 2, Suppl. S), (1995) 1139–1147.
7. R. Kuligowski and A. P. Barros, Localized precipitation forecasts from a numerical weather prediction model using artificial neural networks, *Weather and Forecasting* **13**, (1998) 1195–1205.
8. NHWC, Use and benefits of the National Weather Service river and flood forecasts. National Hydrologic Warning Council, USA. 21 pp (2002).
9. W. Yeh, L. Becker and R. Zettlemaoyer, Worth of inflow forecast for reservoir operation, *Journal of Water Resources Planning and Management* **108** (1982) 257–269.
10. Introduction to SOBEK, SOBEK manual with SOBEK version 2.09, WL | Delft Hydraulics, The Netherlands (2004).

Advances in Geosciences
Vol. 17: Hydrological Science (2008)
Eds. Namsik Park *et al.*

SHORT-TERM RAINFALL FORECASTING USING A BAYESIAN STOCHASTIC RAINFALL BURST MODEL

SRI SRIKANTHAN
Water Division, Bureau of Meteorology,
GPO Box 1289 Melbourne,
Victoria 3001, Australia
r.srikanthan@bom.gov.au

BENJAMIN RENARD, LIJIE CUI and GEORGE KUCZERA
School of Civil Engineering, University of Newcastle,
University Drive Callaghan, NSW 2308, Australia

There is a growing interest in knowing the uncertainty in flood forecasting and the resulting flood warnings. Flood forecasting depends essentially on rainfall measurement and forecasting during a flood event. Uncertainty generally increases with the lead-time. The sources of uncertainty and their propagation throughout the flood forecasting process are manifold, principally in the estimation of rainfall, followed by the uncertainty attributable to the internal states of hydrological and hydraulic models, the initial conditions and process parameterization. The predictive uncertainty must be quantified adequately so that methods for systematically reducing uncertainty by means of observation can be employed in a probabilistic analysis. This study focuses on rainfall forecasting by generating ensembles of future rainfall conditioned on current information about rainfall. A stochastic rainfall burst model was developed to account for both the persistence and stochasticity in the hourly rainfall during a storm event. A Bayesian Markov Chain Monte Carlo method was used to calibrate the model. The calibrated model was then used to forecast rainfall up to 12 hours lead time. Application of this model to the Georges River Catchment located near Sydney, resulted in satisfactory rainfall forecasts.

1. Introduction

There is a growing interest in knowing the uncertainty in flood forecasting and the resulting flood warnings. This is borne out of the fact that the processes involved in flood forecasting have inherent uncertainties in them. The procedure used in flood forecasting consists of a number of steps. The first step is the measurement and forecasting of rainfall during a flood event.

The rainfall is then transformed into flow using a combined water balance and runoff-routing model. There are uncertainties associated with rainfall measurement/forecasting, the model and flow measurements. All these uncertainties contribute to the uncertainty in the resulting flood forecasts. This study considers uncertainty in the first step, namely, forecasting rainfall by generating ensembles of future rainfall conditioned on current information about rainfall and atmospheric conditions.

Advances in numerical weather prediction (NWP) have made it possible to generate high resolution rainfall forecasts at the catchment scale. This makes it of particular relevance to short-term flood forecasting (up to 6 hour lead time) where forecast skill is highly dependent on real time forecasts of rainfall temporal and spatial variability. However, recent assessment of NWP forecasts reveals large bias and uncertainties in short-term forecasts (Cui *et al.*, 2007). In response to this shortcoming, this study developed a simple stochastic rain burst model to generate future rainfall ensembles.

2. Forecast Skill of Numerical Weather Prediction Model

The Australian Bureau of Meteorology operates a suite of NWP models ranging from a global system (GASP), the regional models for mid-latitudes (LAPS) and the tropics (TLAPS), and the high resolution meso-LAPS system (Mills, 1999, Puri *et al.*, 1998). The regional systems have their lateral boundary conditions specified by the global model forecast, while the meso-LAPS system is nested inside the regional system. The meso-LAPS system covers all of Australia with a grid resolution of 0.125°. Fields are specified on 600 × 480 grid points at 29 vertical levels. The computational domain extends well beyond the Australian continent, from 55° South to some 5° North, and 95° East to about 170° East. Vertically, the domain extends from 289 cm below the surface, up to approximately 20 km above the ground. Most model fields are updated every 10 seconds during the integration process. The meso-LAPS system is run twice a day up to 36 hours ahead. Hereafter, reference to NWP implies the meso-LAPS system.

Hourly NWP rainfall forecasts covering the Georges River catchment (Fig. 1) for the May 2003 event were obtained from the Bureau archive. The hourly rainfall forecasts for the 12 grid points (165, 166, 167, 190, 191, 192, 215, 216, 217, 240, 241 and 242) in and around the catchment were obtained for comparing the forecast rainfall with observed rainfall. The

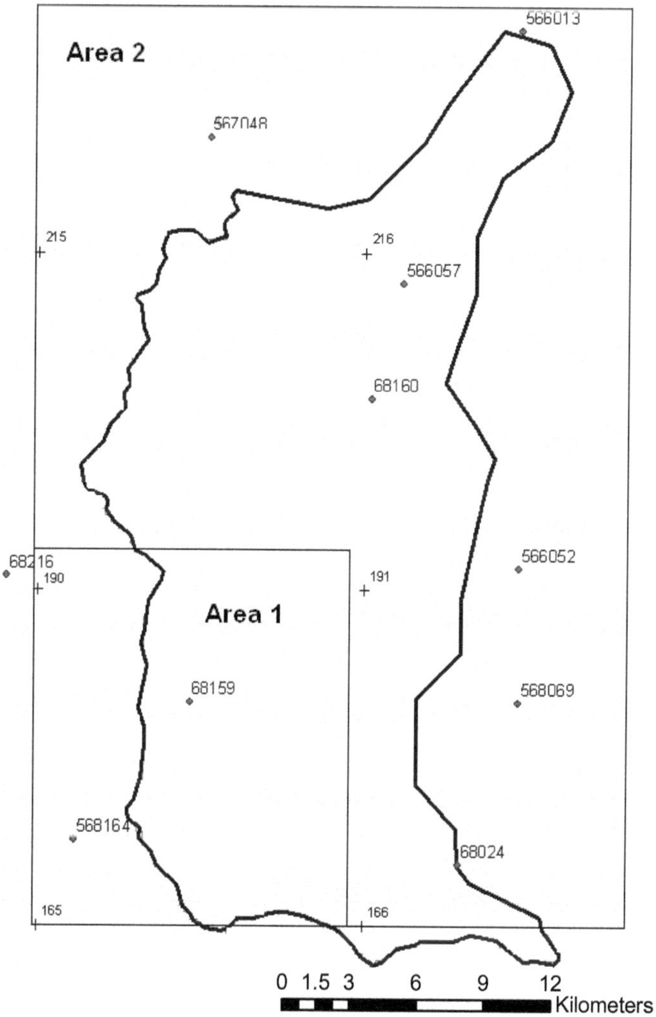

Fig. 1. Locations of the NWP grid points and rainfall gauges within the Georges River catchment.

forecast rainfall over Area 1 was compared with the observed rainfall at the rainfall stations 68159, 568164 and the average of the two gauges. For Area 2, one-hour lead forecast average rainfall from the 12 grid points was compared with the observed average rainfall from the 10 rainfall stations shown in Fig. 1. To minimise the variation in the forecasts, the average one-hour lead forecast was calculated using the forecast for the current time

Table 1. Correlation between the observed and one-hour lead forecast rainfall for May 2003 event.

Forecast time	Area 1			Area 2
Day_hours	68159	568164	Obs average	Obs average
12 12:00	0.339	0.348	0.351	0.428
13 00:00	−0.366	−0.471	−0.440	0.264
13 12:00	0.140	0.131	0.150	0.176
14 00:00	−0.103	−0.062	−0.095	−0.158
14 12:00	0.142	−0.190	−0.008	−0.233
15 00:00	0.230	0.076	0.169	0.172
15 12:00	0.144	0.380	0.287	0.023
16 00:00	0.273	0.612	0.529	0.270
Av. forecast	0.160	0.158	0.166	0.224

and the previous two forecasts. Table 1 presents the correlation between the forecast and observed rainfall.

Table 1 indicates that there is not much skill in the NWP rainfall forecasts. Also a plot (Cui *et al.*, 2007) of the NWP forecasts makes it clear that the NWP model forecasts underestimate the variability in the observed rainfall, underestimating both the extreme peaks and systematically overestimating rainfall during periods of low rain activity, and also exhibit considerable timing errors. Because the NWP forecast is only updated once every 12 hours, it is not possible to generate ensembles to represent forecast uncertainty. For these reasons it was decided to develop a stochastic model that is calibrated to real-time raingauge data and NWP forecasts and which generates rainfall ensembles to represent the variability of future rainfall.

3. Stochastic Rainfall Burst Model

The low NWP forecast skill is attributed, inter alia, to the fact that the NWP model uses a $14\,\text{km} \times 14\,\text{m}$ grid cell making it unlikely to account for the sub-grid variability affecting rainfall. We conclude that a single forecast from a NWP model is not reliable. A mechanism is required that can capture the information in the NWP forecast and also that can generate ensembles of future rainfall to account for the stochastic influence on rainfall at the subgrid scale. We propose using a stochastic model that can reproduce the stochastic characteristics of observed hourly rainfall and also make use of the NWP forecasts.

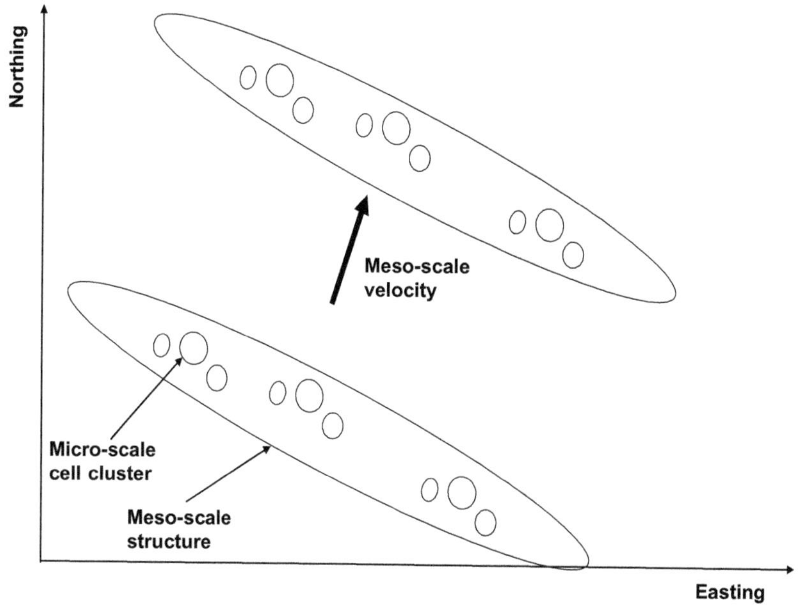

Fig. 2. Waymire *et al.* (1984) conceptualisation of rainfall field structure.

3.1. *Conceptualisation of stochastic rain burst model*

Waymire *et al.* (1984) developed a cluster-based conceptualisation of storm rainfall as illustrated in Fig. 2. It consists of mesoscale rainband structures (size 20–200 km) that advect with the prevailing storm system velocity. Embedded in the mesoscale structure are high rainfall microscale cells that cluster and persist over time scales usually less than one hour in the general direction of advection.

This conceptualisation suggests that a NWP model using 14 km computational cells is unlikely to capture the fine scale dynamics of the microscale cells. One would expect the NWP model to have some skill in identifying the characteristics of the mesoscale structure and diminishing skill at smaller length and time scales.

To be consistent with the conceptual model of Fig. 3, a stochastic rainfall model should possess the following characteristics:

- Persistence: The passage of a rainband over a gauge is likely to produce a sustained period of rainfall, whose overall depth is likely to be determined by the properties of the air mass within the band.

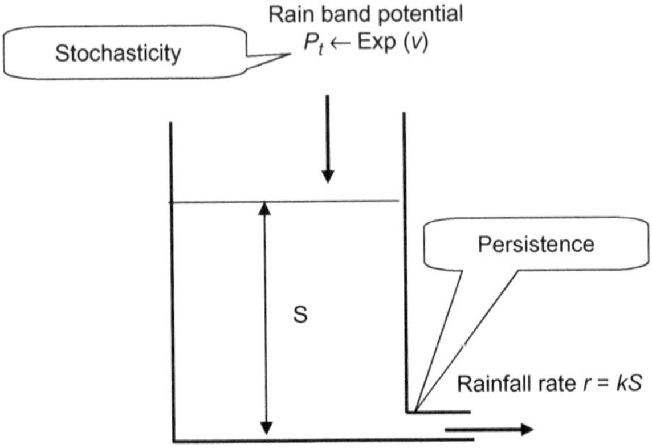

Fig. 3. Stochastic rain burst model.

- Stochasticity: The path and age of individual micro cells within the rainband affect the rainfall at a specific location. This within-grid variability is unlikely to be adequately simulated by the NWP model and thus must be treated as stochastic.

In the light of these considerations, the following stochastic rain burst model (hereafter referred to as the SRB model) is proposed. The model consists of a linear reservoir with exponentially distributed input.

The storm potential P_t is conceptualised as a random sample from an exponential distribution with parameter ν. The parameter ν represents the activity of the mesoscale structure, while the sampling from the exponential distribution represents the activity of the microscale structures embedded within the macroscale structure. The potential P_t represents the inflow (mm/hr) into a linear reservoir that conceptually allows for persistence. The dynamics of the reservoir are described by the mass conserving differential equation

$$\frac{dS}{dt} = P_t - kS \tag{1}$$

where P_t is the inflow to the reservoir starting at time t and constant for the interval Δt (equal to 1 hour). The reservoir outflow kS denotes the instantaneous rainfall rate. The parameter k represents the efficiency of the micro cell during the hour indexed by t.

Solution of Eq. (1) yields the following reservoir storage update

$$S_{t+1} = \frac{P_t - e^{-k\Delta t}(P_t - kS_t)}{k} \tag{2}$$

where S_t is the reservoir depth at the start of the interval indexed by t. It follows that the total rainfall over the time interval is

$$r_t = P_t\Delta t + S_t - S_{t+1}. \tag{3}$$

3.2. *Calibration of SRB model*

The calibration of the SRB model requires estimation of three sets of parameters: model parameters k and vg, and latent variables P_t, $t = 1, \ldots, n$, corresponding to the input to the reservoir. From a Bayesian perspective calibration involves computing the properties of the posterior density function $p(k, v, P|\mathbf{R})$ where $\mathbf{R}$ is the rainfall history, either observed or simulated by the NWP model. The BATEA approach described by Kuczera *et al.* (2006) was used to perform the calibration.

3.3. *Calibration to georges river may 2003 storm*

The SRB model was calibrated to the catchment average hourly data for three gauges 568164, 68216 and 566057 using the Bayesian Markov Chain Monte Carlo (MCMC) method. The posterior distribution of the parameter k is shown in Fig. 4. The modal value of k is about 0.3 which indicates there

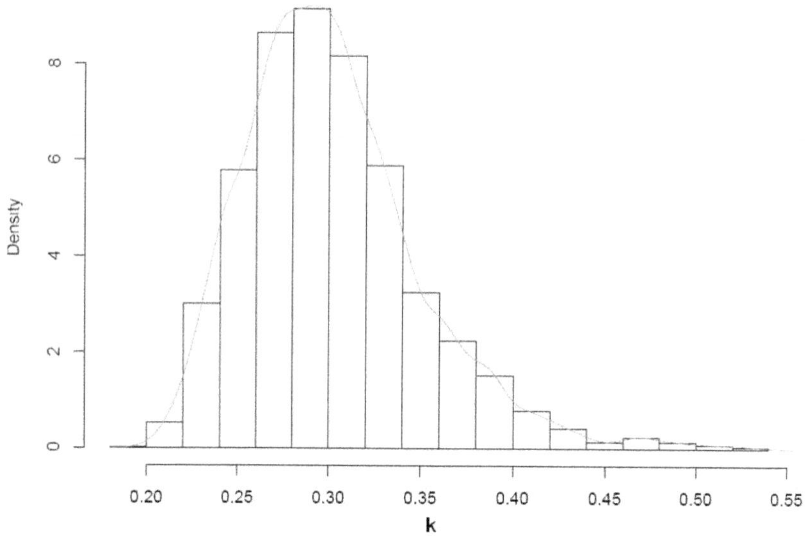

Fig. 4. Posterior distribution of k.

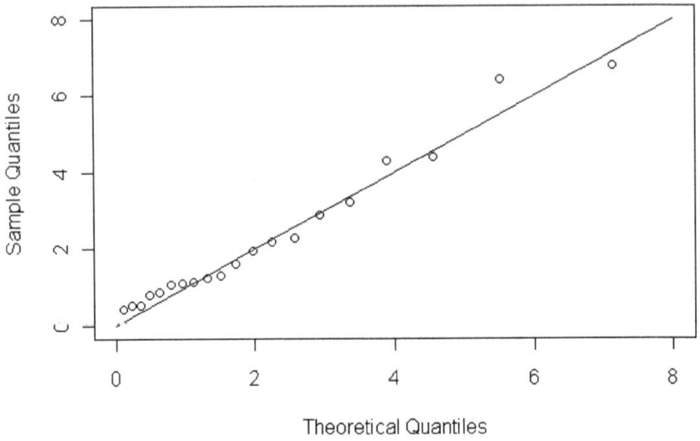

Fig. 5. QQ plot of the latent variables.

is some persistence at the hourly time scale. A quantile-quantile plot of the latent variables shown in Fig. 5 supports the assumption that the latent variables are sampled from an exponential distribution.

The coefficient of efficiency for the calibration is 0.98 and the observed and modelled rainfalls are shown in Fig. 6. These indicate a good calibration

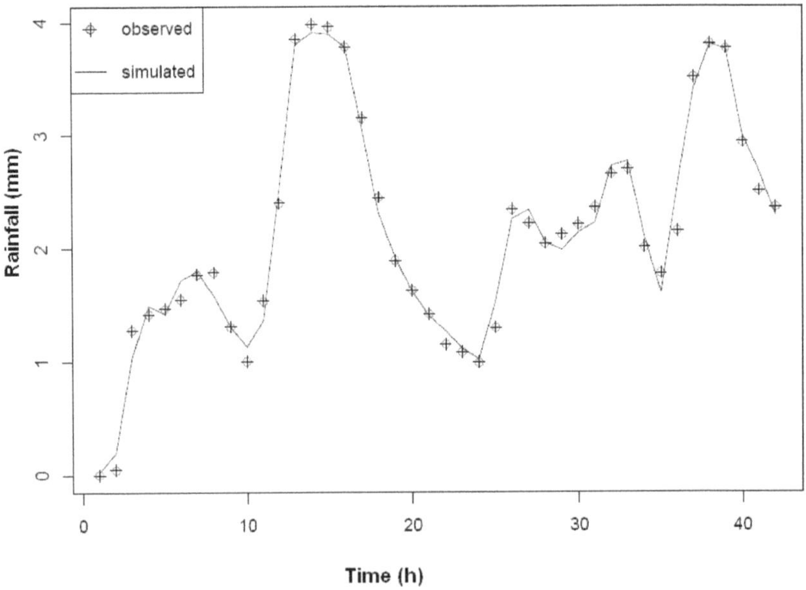

Fig. 6. Plot of the observed and the modeled rainfall during the calibration period (T = 40 h).

of the SRB model. The high value of the coefficient of efficiency is due to latent variables being fitted to individual time steps. This is not overfitting as the objective is to estimate the hyperdistribution of the latent variables. In the forecasting step, the latent variables are sampled from the hyperdistribution. The parameters of the model were reestimated at every forecast step using all available rainfall data.

3.4. *Rainfall ensemble forecasts*

In a real time flood forecasting situation, the only data available to make a forecast are the NWP forecasts along with observed raingauge data. To evaluate the value of this data for forecasting, 12-hour lead time rainfall ensembles were produced for catchment average rainfall at different times during the May 2003 event. The procedure for producing the forecast ensemble was:

1. Calibrate the SRB model using the catchment average rainfall data up to the time of forecast.
2. Using the current calibrated reservoir state, the next 12 hours of rainfall are generated 100 times by randomly sampling the two SRB model parameters (k and v) from the MCMC simulations.

The rainfall forecast ensembles were generated for 4 forecast times, namely, 20, 30, 40 and 50 hours after commencement of the storm. Figure 7 shows the forecast ensembles for the four different forecast times. Each plot compares 2.5, 50 and 97.5 percentile ensemble rainfalls against the observed rainfall at different times during the May 2003 storm.

The median rainfall forecast follows the observed rainfall reasonably well most of the time. The 95 % prediction limits encompass the observed rainfall all the time. A notable drawback with the SRB model is that it is unable to forecast the sudden increase or decrease in rainfall magnitude.

4. Conclusions

This study reports on the development of a simple stochastic rainfall model to generate ensembles of future rainfall. The developed model was applied to the May 2003 Georges River event and the results were encouraging as it overcame most of the problems associated with the NWP forecasts. However, the median ensemble value fails to track the sudden increase

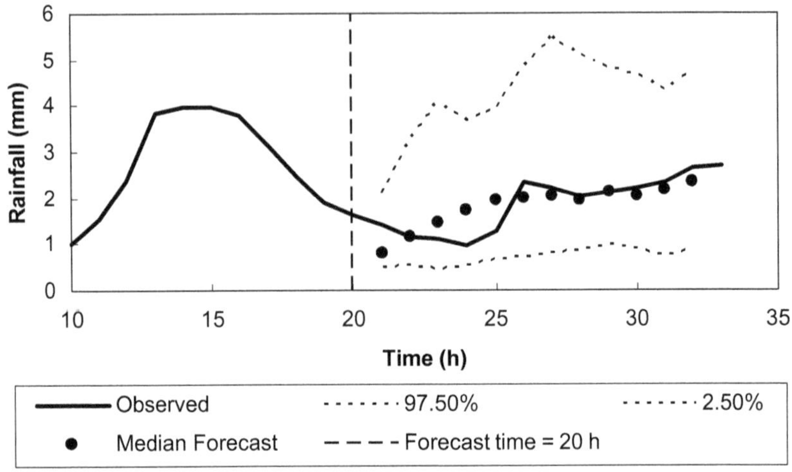

(a) Forecast time = 20 h

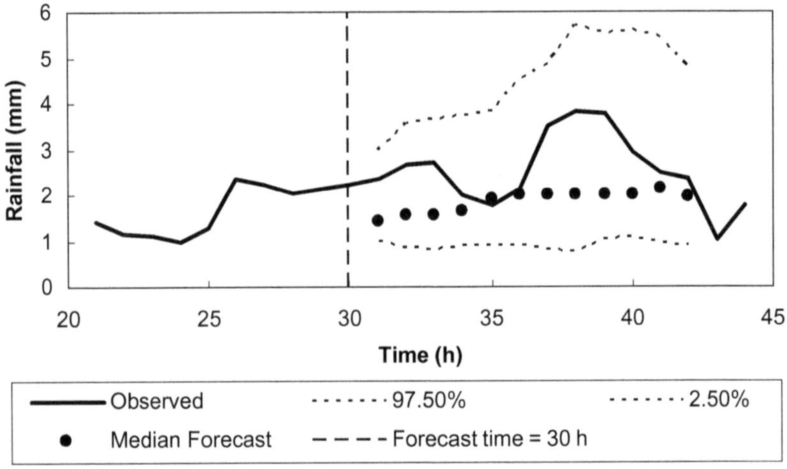

(b) Forecast time = 30 h

Fig. 7. Comparison of observed and forecast rainfall with 95 percent prediction limits for four different forecast times.

or decrease in the rainfall intensity. This is because the SRB model is stationary. One way to overcome this problem is to use the NWP forecasts as an input to the linear reservoir. Recognizing that the NWP forecasts are noisy it would be necessary to use a (latent) error multiplier to

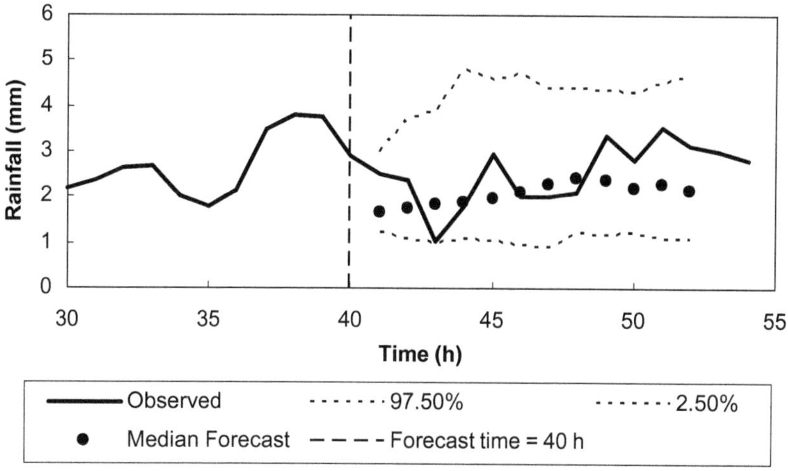

(c) Forecast time = 40 h

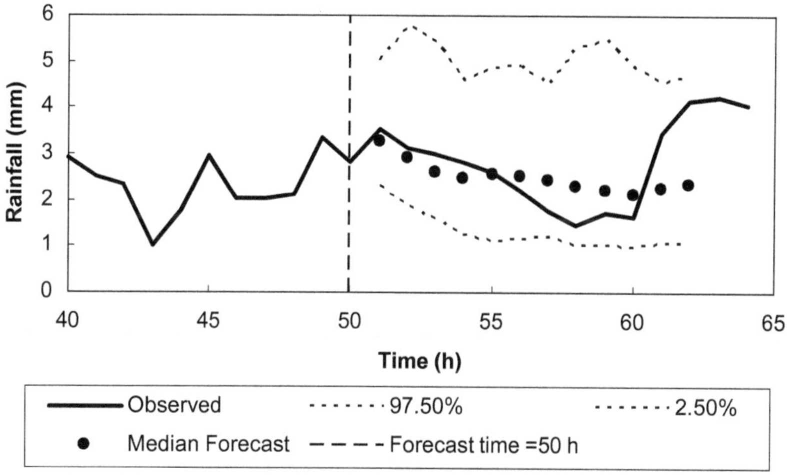

(d) Forecast time = 50 h

Fig. 7. (*continued*)

ensure consistency with the observed raingauge data. As the forecast skill of NWP increases, this scheme would better track actual rainfall and reduce ensemble variability. However, significant improvements to NWP technology will be needed.

References

1. L. Cui, G. Kuczera, R. Srikanthan, G. E. Amirthanathan, M. Thyer and T. Micevski, Short-term rainfall forecasting using a Bayesian Hierarchical Rainfall Burst Model coupled with numerical weather prediction, *International Symposium on Flood Forecasting and Management with GIS and Remote Sensing*, September, China (2007).
2. G. Kuczera, D. Kavetski, S. W. Franks and M. T. Thyer, Towards a Bayesian total error analysis of conceptual rainfall-runoff models: Characterising model error using storm-dependent parameters, *Journal of Hydrology* **331**(1–2) (2006) 161–177.
3. G. A. Mills, Mesoscale modelling of wind flows for operational fire weather forecasts, *Australian Bushfire Conference*, Albury, July (1999).
4. K. Puri, G. S. Dietachmayer, G. A. Mills, N. E. Davidson, R. A. Bowen and L. W. Logan, The new BMRC Limited Area Prediction System, LAPS. *Australian Met. Mag.*, **47** (1998) 203–223.
5. E. Waymire, V. Gupta and I. Rodrigues-Iturbe, A spectral theory of rainfall intensity at the meso-b-scale. *Water Resources. Research.* **20** (1984) 1453–1465.

Advances in Geosciences
Vol. 17: Hydrological Science (2008)
Eds. Namsik Park *et al.*

A SYNOPTIC WEATHER TYPING APPROACH TO ASSESS CLIMATE CHANGE IMPACTS ON METEOROLOGICAL AND HYDROLOGICAL RISKS AT LOCAL SCALE IN SOUTH-CENTRAL CANADA

CHAD SHOUQUAN CHENG[*,†], QIAN LI and GUILONG LI
Atmospheric Science and Applications Unit,
Meteorological Service of Canada Branch,
Environment Canada, 4905 Dufferin Street,
Toronto, Ontario M3H 5T4, Canada
[†]*shouquan.cheng@ec.gc.ca*

The synoptic weather typing approach has become popular in evaluating the impacts of climate change on a variety of environmental problems. One of the reasons is its ability to categorize a complex set of meteorological variables as a coherent index, which can facilitate analyses of local climate change impacts. The weather typing method has been applied in Environment Canada to analyze climatic change impacts on various meteorological/hydrological risks, such as freezing rain, heavy rainfall, high-/low-flow events, air pollution, and human health. These studies comprise of three major parts: (1) historical simulation modeling to verify the hazardous events, (2) statistical downscaling to provide station-scale future climate information, and (3) estimates of changes in frequency and magnitude of future hazardous meteorological/hydrological events in this century. To achieve these goals, in addition to synoptic weather typing, the modeling conceptualizations in meteorology and hydrology and various linear/nonlinear regression techniques were applied. Furthermore, a formal model result verification process has been built into the entire modeling exercise. The results of the verification, based on historical observations of the outcome variables predicted by the models, showed very good agreement. This paper will briefly summarize these research projects, focusing on the modeling exercise and results.

1. Introduction

It has become widely recognized that hot spells/heat waves and heavy precipitation events would increase over most land areas of the globe due to a changing climate.[1–7] The Fourth Assessment Report of the

*Corresponding author: shouquanc.cheng@ec.gc.ca.

Intergovernmental Panel on Climate Change (IPCC) has indicated that frequency and intensity of these extreme weather events were projected to *very likely* increase over the world in this century under global warming.[8] It is further projected that climate change could increase mortality/morbidity from heat waves, floods, and droughts as well as the heavy rainfall-related flooding risks.[9] In the light of these concerns in south-central Canada, Environment Canada has completed several research projects on local climate change impacts analysis. For example, a three-year research project on climate change and human mortality has been completed for four cities in south-central Canada, in partnerships with Toronto Public Health, McMaster University, and Public Health Agency of Canada. This project, funded by the Health Policy Research Program, Health Canada, attempted to evaluate differential and combined impacts of extreme temperatures and air pollution on human mortality under current and future climates. Another three-year research project focused on investigation of climate change and extreme rainfall-related flooding risks in southern Ontario, Canada. Environment Canada leads this project, funded by the Government of Canada's Climate Change Impacts and Adaptation Program (CCIAP) administered by Natural Resources Canada, in partnerships with Conservation Ontario, Ontario Ministry of Natural Resources, and CGI Insurance Business. This project attempted to estimate changes in frequency and magnitude of future heavy rainfall and high-/low-flow events. In addition, we have completed another study on estimates of possible impacts of climate change on freezing rain in south-central Canada.

These studies consist of three major parts: (1) historical analysis, (2) statistical downscaling, and (3) future estimates. The first part of the studies includes an automated synoptic weather typing approach to classify daily weather types or air masses and development/validation of daily within-weather-type simulation models in order to capture historical high-impact weather events. The second part of the studies focuses on statistical downscaling of hourly and daily future climate scenarios at station scale for various meteorological variables. The last part of the studies is to estimate changes in frequency and intensity of future meteorological/hydrological hazardous events using historical simulation models with downscaled future climate data. This part also includes projection of future weather types using downscaled future hourly climate data. This paper will briefly summarize these studies focusing on methodology and results.

2. Data Sources

To estimate climate change impacts on high-impact weather events, various observed data, including meteorological, hydrological, air pollution, human mortality, and insurance claims, are needed. The different projects need different kinds of the data; however, some of the data were commonly used for all projects. The more detailed information, regarding data sources, variables, and record length, is described in Table 1.

The study area and locations of the fifteen selected cities, observing with hourly meteorological data at international airports in south-central Canada, are shown in Fig. 1. The selected fifteen cities were used for climate change and freezing rain project. For other projects, only part of the locations was used. For example, four cities (Montreal, Ottawa, Toronto, Windsor) were selected for climate change and human mortality study; and three cities (London, Ottawa, Toronto) were used, combining with four adjacent watersheds (Grand, Humber, Rideau, and Upper Thames rivers), for the study of climate change and heavy rainfall-related flooding risks.

3. Analysis Techniques

The principal methods and steps used in the studies are summarized in Table 2. Overall, these studies are comprised of three major parts: (1) historical analysis: simulation modeling to verify the historical hazardous events, (2) statistical downscaling: projection of station-scale hourly/daily future climate information, and (3) future estimates: changes in frequency and magnitude of high-impact weather events in this century.

The synoptic weather typing approach has become popular in evaluating the impacts of climate change on a variety of environmental problems. One of the reasons is its ability to categorize a complex set of meteorological variables as a coherent index,[10,11] which can facilitate climate change impacts analysis. The synoptic weather typing attempts to have a suitable classification solution as it minimizes within-category variances and maximizes between-category variances. For different studies, the synoptic weather typing in terms of statistical methods is the same; however, the input data are different. In climate change and freezing rain project, hourly surface meteorological observations and six-hourly U.S. NCEP reanalysis data listed in Table 1 were used to classify daily

Table 1. Data used in the studies.

Project	Data			
	Description	Period	Variable	Source
All projects	Hourly surface meteorological observations	1958–2002	Air temperature (°C) Dew point (°C) Sea-level air pressure (hPa) Total cloud cover (tenths) U-wind and V-wind (m s^{-1})	Environment Canada's National Climate Data and Information Archive
	Six-hourly surface and upper-air levels (925, 850, 700, 600, and 500 hPa) reanalysis grid-point weather data	1958–2002	Air temperature (°C) Dew point (°C) Sea-level air pressure (hPa) U-wind and V-wind (m s^{-1})	U.S. National Centers for Environmental Prediction (NCEP) Website
	Daily surface and upper-air levels (925, 850, 700, 600, and 500 hPa) GCM outputs	1961–2000 2016–2035 2046–2065 2081–2100	Air temperature (°C) Sea-level air pressure (hPa) U-wind and V-wind (m s^{-1})	IPCC IS92a: CGCM1 IPCC SRES A2: CGCM2 GFDL-CM2.0 ECHAM5 IPCC SRES B2: CGCM2
Heavy rainfall-related flooding risks	Hourly/daily rainfall observations (9–13 climate stations for four selected river basins	1958–2002	Rainfall amount (mm)	Environment Canada's National Climate Data and Information Archive
	Daily stream flow observed at non-regulated monitoring stations	1958–2002	Stream flow volume (m^3 s^{-1})	Environment Canada's National Water Data Archive
	Monthly aggregated rainfall-related water damage insurance data	1992–2002	Number of insurance claims Incurred losses of water damage	CGI Insurance Business Services

(*continued*)

Table 1. (*continued*)

Project	Data: Description	Period	Variable	Source
Extreme temperature, air quality, and human mortality	Daily aggregated total non-traumatic mortality at each of four selected cities	1954–2000	Number of human mortality	Statistics Canada
	Hourly/daily air pollution observations (1–7 monitoring sites for four selected cities	Ozone: 1980–2000 Others: 1974–2000	Carbon monoxide (CO) Nitrogen dioxide (NO_2) Ozone (O_3) Sulphur dioxide (SO_2) Suspended particles (SP)	National Air Pollution Surveillance (NAPS) network
Freezing rain	Hourly freezing rain observations	1958–2001	Occurrence of freezing rain (1 for yes, 0 for no)	Environment Canada's National Climate Data and Information Archive

weather types. For heavy rainfall-related flooding study, only hourly surface observations were used. For climate change and human mortality project, six-hourly surface meteorological data, observed at 03:00, 09:00, 15:00, and 21:00 local standard time, were used. The selection of hourly or six-hourly meteorological data for a study was determined based on temporal resolution of the targeted weather events to be verified using within-weather-type simulation models.

There are two key issues that need to be taken into consideration when developing regression-based simulation models: (1) selection of appropriate regression methods and (2) selection of significant predictors. A variety of regression methods, listed in Table 2, was used to develop high-impact weather-event simulation models and statistical downscaling transfer functions. A regression method is suitable only for a certain kind of data with the specific distribution. For example, multinomial logit regression is a more suitable analysis tool to develop the simulation model

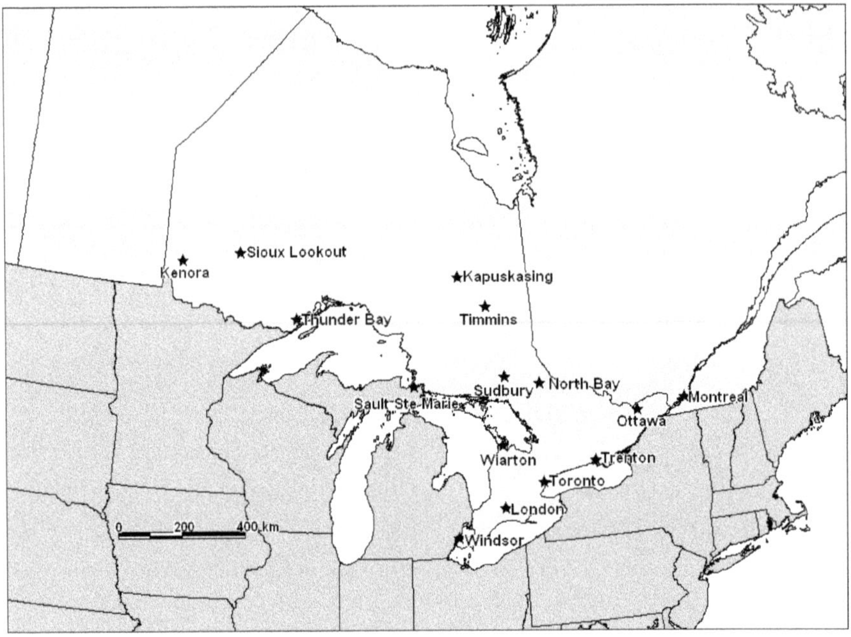

Fig. 1. Location map of study points in south-central Canada.

for an ordered categorical data (e.g. total cloud cover). To effectively select appropriate predictors, the modeling conceptualizations in meteorology and hydrology were carefully analyzed, in order to identify the predictors possessing significant relationships with the predictand. Furthermore, a formal model result verification process has been built into the entire modeling exercise. The results of the verification, based on historical observations of the outcome variables predicted by the models, showed very good agreement. The detailed information on methodologies used in the climate change impacts studies cannot be fully described in this paper due to space limitation; and refer to recent publications[12–17] for details.

4. Results

As indicated in Sec. 3 above, the climate change impacts studies summarized in this paper are comprised of three parts: (1) historical analysis, (2) statistical downscaling, and (3) future estimates. In this section, only major results for each of three parts are summarized, due to the space limitation. For detailed information, refer to the articles.[12–17]

Table 2. Research steps and methodologies used in the studies.

Study Step	Description	Method	Validation/Evaluation
Historical analysis	Synoptic weather typing	Principal components analysis Average linkage clustering procedure Discriminant function analysis	Randomly selected about 1/3 of the total years as independent validation dataset
	Identifying weather types associated with extreme events	Within-weather-type frequency of extreme event episodes Within-weather-type mean quantities (e.g. daily rainfall amount) χ^2-test	
	Development of high-impact weather simulation models	Principal component analysis Multiple stepwise regression Robust stepwise regression Logistic regression Multinomial logit regression Nonlinear regression Autocorrelation correction regression Orthogonal regression	
Hourly/daily statistical downscaling	Spatial downscaling for daily climate information		Leave-one-year-out cross-validation scheme using historical observations Comparison between downscaled GCM control runs and observations over the same period (1961–2000): • Data distributions • Diurnal/seasonal variations • Extreme weather characteristics
	Temporal downscaling for hourly climate information		

(*continued*)

Table 2. (*continued*)

Study Step	Description	Method	Validation/Evaluation
Future estimates	Estimates of future synoptic weather types	Principal components analysis Discriminant function analysis	N/A
	Estimates of future high-impact weather events	(1) Comparing future and historical identified high-impact weather type frequencies (2) Applying historical simulation models with downscaled future climate information	

The results on the first two parts: historical analysis and statistical downscaling: development and validation of simulation models/downscaling transfer functions are listed in Table 3. From Table 3, it can be seen that the simulation models of historical analysis and statistical downscaling transfer functions perform very well on verification of historical events and duplication of historical data. It is also immediately apparent, from Table 3, that the models are reliable since the results, derived from both model development and validation, are very similar.

The major results on changes in frequency and magnitude of future high-impact weather events in this century, derived from the studies, are summarized in Table 4. The values of future changes with plus and minus signs represent a percentage increase and decrease from a historical baseline (the column Historical mean in Table 4), respectively. Values of 100% and 200% indicate that future frequencies or magnitudes of the events are projected to double and triple from the current conditions. A value of −100% indicates that the future events are projected to be zero. Overall, the frequency and magnitude of future high-impact weather events in the question are projected to significantly increase in this century due to the changing climate projected by GCM scenarios.

5. Conclusions and Recommendation

The overarching purpose of these studies is to estimate possible changes in the frequency and magnitude of the high-impact weather events

Table 3. Major results on historical analysis and statistical downscaling: development and validation of simulation models/downscaling transfer functions.

Project	Study area	Description	Simulation model performance: Model development	Simulation model performance: Model validation
Heavy rainfall-related flooding risks in Southern Ontario, Canada (Apr–Nov)	Four river basins: Grand, Humber, Rideau, and Upper Thames rivers	Rainfall occurrence simulation models	Of 40 simulation models, Concordances: 0.82–0.96 27.5% models: concordances >0.92 47.5% models: concordances >0.90 82.5% models: concordances >0.87	Similar to development
		Rainfall quantity simulation models	Percentages of excellent and good simulations on daily rainfall: 69%–78% *	Corresponding percentages: 60%–68%
		Stream flow simulation models	Model R^2 Grand River: 0.71 Humber River: 0.62 Rideau River: 0.95 Upper Thames River: 0.74	Similar to development
Extreme temperature, air quality, and human mortality in south-central Canada	Four cities: Montreal, Ottawa, Toronto, and Windsor	Air quality simulation models	Of 80 simulation models, 20% models $R^2 \geq 0.7$ 50% models $R^2 \geq 0.6$ 80% models $R^2 \geq 0.5$	Similar to development with slightly smaller model R^2
		Extreme temperature- and air pollution-related mortality	Of annual elevated mortality: 20% attributable to extreme temperatures (heat/cold) 80% attributable to air pollution Daily heat-related mortality is up to twice as high as that associated with air pollution when heat occurring	Similar to development

(continued)

Table 3. (*continued*)

			Simulation model performance	
Project	Study area	Description	Model development	Model validation
Freezing rain (FR) in south-central Canada	Fifteen cities as shown in Figure 1	FR-related weather types	Four primary FR weather types for all 15 cities and two additional weather types for some cities These weather types can capture 75–100% of FR events occurred on ≥1 to ≥6 hour(s) during a day over the entire study period	Similar to development
Hourly statistical downscaling	Fifteen cities as shown in Figure 1	Hourly downscaling transfer functions	Surface temperature, dew point, and air pressure: all models' R^2 >0.95 Upper-air temperature: 0.85–0.97 Upper-air dew point: 0.77–0.87 Surface wind: 0.75–0.91 Upper-air U-wind: 0.78–0.96 Upper-air V-wind: 0.69 –0.92 Cloud: model concordance: 0.78–0.87	Similar to development

Note: *Excellent: absolute difference between observed and simulated daily rainfall ≤30% of observation;
Good: 30% of observation < absolute difference between observed and simulated daily rainfall ≤60% of observation

in this century. Automated synoptic weather typing and a variety of linear/nonlinear regression analyses were applied together with downscaled GCM data to estimate future weather types and changes in frequency and magnitude of future high-impact weather events. In these studies, a formal verification process on model results has been built into the whole exercise, including synoptic weather typing, historical simulation modeling, and the development of downscaling transfer functions. The

Table 4. Major results on changes in frequency and magnitude of future high-impact weather events in this century.

Project	Study area	Description		Historical mean	Future change (%)		
					2016–2035	2046–2065	2081–2100
Heavy rainfall-related flooding risks in Southern Ontario, Canada (Apr–Nov)	Four river basins: Grand, Humber, Rideau, and Upper Thames rivers	Seasonal mean rainfall event frequency and amount	≥ 25 mm/d	3.2 days	+30	+40	+50
			≥ 15 mm/d	10.5 days	+25	+32	+40
			≥ 0.2 mm/d	104 days	+10	+16	+20
			Rainfall totals	608 mm	+20	+25	+35
			Return values of 3-day extreme rainfall events		+20–50 in 2001–2050 +25–60 in 2051–2100		
		High-/low-flow events and seasonal mean flow	$\geq 95^{th}$ percentile of observation	11 days (May– Nov)	+45	+60	+90
			$< 5^{th}$ percentile of observation		+45	+65	+120
			Seasonal mean	varies	+15	+25	+35
			Return values of one-day extreme high-flow events		+15–35 in 2001–2050 +30–75 in 2051–2100		
	Four cities: Kitchener-Waterloo, London, Ottawa, Toronto (Apr–Sep)	Flooding risk: only consider future rainfall changes	Number of insurance claims	12K	+15	+20	+30
			Incurred economic losses	$88M			

(*continued*)

Table 4. (*continued*)

Project	Study area	Description		Historical mean	Future change (%)		
					2016–2035	2046–2065	2081–2100
Extreme temperature, air quality, and human mortality in south-central Canada	Four cities: Montreal, Ottawa, Toronto, and Windsor	Air quality–Ozone	Mean	varies		+2.7 ppb	+4.0 ppb
			$\geq$ 81 ppb	8 days		+50	+100
			51–80 ppb	45 days		+35	+40
			$\leq$ 50 ppb	312 days		−6	−9
		Air quality–others	$\geq$ 1std dev + overall mean	varies		+20	+40
			<overall mean	varies		−10	−15
		Heat-related mortality		318		+100	+200
		Cold-related mortality		332		−50	−65
		Air pollution-related mortality		2,268		+25	+40
Freezing rain (FR) in south-central Canada (Nov–Apr)	Fifteen cities as shown in Figure 1	FR Freq. (Dec–Feb)	North	1–2 days		+85	+135
			East	4–7 days		+60	+95
			South	2–5 days		+40	+45
		FR Freq. (Nov, Mar, Apr)	North	1–2 days		+10	+20
			East	3–4 days		~ 0	~ 0
			South	1–2 days		−10	−15

Note: 1. For mortality and freezing rain projects, two future time periods are available: 2040–59 and 2070–89.
2. Air quality–others: carbon monoxide, nitrogen dioxide, sulphur dioxide, and suspended particles
3. North: Kapuskasing, Kenora, Sioux Lookout, Thunder Bay, Timmins; East: Montreal, North Bay, Ottawa, Sudbury, Trenton; South: London, Sault Ste. Marie, Toronto, Wiarton, Windsor.

results of the verification, based on historical observations of the outcome variables simulated by the models, showed good agreement. As a result, a general conclusion from these studies is that a combination of synoptic weather typing, high-impact weather simulation modeling, and statistical downscaling can be useful to estimate changes in frequency and magnitude of future high-impact weather events at local scale.

The modeled results from these studies found that the frequency and magnitude of future high-impact weather events could significantly increase under the changing climate in this century. The implementation of the increases should be taken into consideration and integrated into revision of engineering infrastructure design standards and development of adaptation strategies and policies. As the IPCC recently pointed out, "More extensive adaptation than is currently occurring is required to reduce vulnerability to future climate change."[8] One of the key components of developing an effective adaptation plan for reducing meteorological and hydrological risks due to climate change is the implementation of early warning systems.

The simulation model results, derived from these studies, produced good matches to observed occurrence of the high-impact weather events for the study areas. The results and procedures from the studies have the potential to be incorporated into municipal/community emergency response plans and to provide guidance for development of the high-impact weather watch/warning systems for each of the cities. The simulation models could be used in an advance warning system for the municipalities, thus providing them with near real-time forecasting information and potentially enabling them to minimize the relevant meteorological and hydrological risks. Current Emergency Management and Civil Protection legislation requires that all municipalities, regional governments and provincial ministries assess priority risks for disasters and emergency response as well as develop emergency and disaster risk management plans.

Acknowledgments

The study of extreme temperature, air quality, and human mortality was funded through the Health Policy Research Program, Health Canada (6795-15-2001/4400011). The study of heavy rainfall-related flooding risks was funded by the Government of Canada's Climate Change Impacts and Adaptation Program (CCIAP), administered by Natural Resources Canada (A901). The authors also acknowledge in-kind and financial support for all projects provided Environment Canada. The views expressed herein are solely those of the authors and do not necessarily represent the views or official policy of Health Canada, Natural Resources Canada, or Environment Canada.

References

1. G. A. Meehl and C. Tebaldi, More intense, more frequent, and longer lasting heat waves in the 21st century. *Science* **305** (2004) 994–997.
2. J. Last and Q Chiotti, Climate Change and Health, *ISUMA* **2**(4) (2001) 62–69.
3. D. Riedel, Chapter 9: Human health and well-being, eds. D. S. Lemmen and F. J. Warren, Climate change impacts and adaptation: A Canadian Perspective Report, Government of Canada (2004) 153–169.
4. M. Medina-Ramón and J. Schwartz, Temperature, temperature extremes, and mortality: a study of acclimatisation and effect modification in 50 US cities. *Occup Environ Med*, DOI: 10.1136/oem.2007.033175 (2007).
5. U. Cubasch, J. Waszkewitz, G. Hegerl and J. Perlwitz, Regional climate changes as simulated time-slice experiments, *Climatic Change* **31** (1995) 273–304.
6. F. W. Zwiers and V. V. Kharin, Changes in the extremes of the climate simulated by CCC GCM2 under CO_2 doubling, *J. Climate* **11** (1998) 2200–2222.
7. V. V. Kharin and F. W. Zwiers, Estimating extremes in transient climate change simulations, *J. Climate* **18** (2005) 1156–1173.
8. IPCC, Summary for Policymakers, in: Climate Change 2007: The Physical Science Basis. Contribution of Working Group I to the Fourth Assessment Report of the Intergovernmental Panel on Climate Change, eds. S. Solomon, D. Qin, M. Manning, Z. Chen, M. Marquis, K. B. Averyt, M. Tignor and H. L. Miller (Cambridge University Press, Cambridge, UK and New York, NY, USA, 2007).
9. IPCC, Summary for Policymakers, in: Climate Change 2007: Impacts, Adaptation and Vulnerability. Contribution of Working Group II to the Fourth Assessment Report of the Intergovernmental Panel on Climate Change, eds. M. L. Parry, O. F. Canziani, J. P. Palutikof, P. J. van der Linden and C. E. Hanson (Cambridge University Press, Cambridge, UK, 2007), pp. 7–22.
10. B. Yarnal, *Synoptic Climatology in Environment Analysis* (London: Belhaven Press, 1993), p. 195.
11. C. S. Cheng and L. S. Kalkstein, Determination of climatological seasons for the East Coast of the U.S. using an air mass-based classification. *Climate Research* **8** (1997) 107–116.
12. C. S. Cheng, M. Campbell, Q. Li, G. Li, H. Auld, N. Day, D. Pengelly, S. Gingrich and D. Yap, A synoptic climatological approach to assess climatic impact on air quality in south-central Canada. Part I: historical analysis. *Water Air & Soil Pollution* **182** (2007) 131–148.

13. C. S. Cheng, M. Campbell, Q. Li, G. Li, H. Auld, N. Day, D. Pengelly, S. Gingrich and D. Yap, A synoptic climatological approach to assess climatic impact on air quality in south-central Canada. Part II: future estimates. *Water Air & Soil Pollution* **182** (2007) 117–130.
14. C. S. Cheng, H. Auld, G. Li, J. Klaassen and Q. Li, Possible impacts of climate change on freezing rain in south-central Canada using downscaled future climate scenarios. *Natural Hazards and Earth System Sciences* **7** (2007) 71–87.
15. C. S. Cheng, G. Li, Q. Li and H. Auld, Statistical downscaling of hourly and daily climate scenarios for various meteorological variables in south-central Canada. *Theor. Appl. Climatol.* DOI: 10.1007/s00704-007-0302-8 **91** (2008) 129–147.
16. C. S. Cheng, M. Campbell, Q. Li, G. Li, H. Auld, N. Day, D. Pengelly, S. Gingrich, J. Klaassen, D. MacIver, N. Comer, Y. Mao, W. Thompson and H. Lin, Differential and combined impacts of extreme temperatures and air pollution on human mortality in south–central Canada. Part I: historical analysis. *Air Quality, Atmosphere and Health* **1** (2008) 209–222.
17. C. S. Cheng, M. Campbell, Q. Li, G. Li, H. Auld, N. Day, D. Pengelly, S. Gingrich, J. Klaassen, D. MacIver, N. Comer, Y. Mao, W. Thompson and H. Lin, Differential and combined impacts of extreme temperatures and air pollution on human mortality in south–central Canada. Part II: future estimates. *Air Quality, Atmosphere and Health* **1** (2008) 223–235.

Advances in Geosciences
Vol. 17: Hydrological Science (2008)
Eds. Namsik Park *et al.*

OUTLINE FOR ALTERNATIVE RAINFALL-RUNOFF MODELING FRAMEWORK UNDER UNCERTAINTY

LEE GIHA* and TACHIKAWA YASUTO†
Dept. of Urban and Environmental Engineering,
Kyoto University Kyoto, 615-8540, Japan
**leegiha@flood.dpri.kyoto-u.ac.jp*
†tachikawa@mbox.kudpc.kyoto-u.ac.jp

SAYAMA TAKAHIRO and TAKARA KAORU
Disaster Prevention Research Institute,
Kyoto University Uji, 611-0011, Japan

Model identification is the most fundamental task in rainfall-runoff modeling. However, it is not easy due to various uncertainty sources involved in modeling processes. In spite of their significant effects on prediction results, a current modeling framework fails to fully incorporate them into the modeling process. This paper aims to outline an alternative rainfall-runoff modeling framework under uncertainty. The new modeling framework consists of two conditioning steps. The first step is that a set of possible models is evaluated multi-dimensionally with respect to model performance, model structural stability, and parameter identifiability. Next, the parameter space of the selected model structure in the first stage is conditioned directly or indirectly by adding physical or knowledge-based complementary constraints into the parameter identification procedure. Finally, model predictions are implemented by the filtered model structure and parameter set(s) through the two conditioning steps in the new modeling framework.

1. Introduction

The primary objective of modelers is to identify an appropriate model including its optimal parameter set which is suitable for modeling purpose, catchment characteristics and data.[1] Beven[2] reviewed the basic criteria for model choice. They are summarized as follows: (1) model availability; (2) model predictability for hydrological variables; (3) model reliability; (4) model suitability within the time and cost constraints of modeling objectives. However, he warned that all the available models are easily rejected by these criteria because of inadequate conceptualizations of the models and infeasibility of field data supporting model parameters fully.

Therefore, a more effective guideline is necessary to enable modelers to either confirm possible predictor(s) or reject unreliable one(s).

In many hydrologic model applications, the model identification has been conducted by conditioning their predictions to any available observations at the catchment of interest. This procedure is usually called calibration. Then, the calibrated model is verified against different time series or basins not used during calibration. This procedure is called model validation or verification and it commonly follows Klemes's hierarchical validation scheme.[3]

However, many studies[1,2,4] pointed out that this conventional type of model identification is basically required but is insufficient to adequately test the suitability of a model because of its lack of dealing with uncertainty involved in modeling processes. One of the significant problems in the traditional modeling framework is that often, different parameter combinations and even different model structures can become equally good representations of catchment responses.[5] This is entitled 'equifinality'[2,5] and it has become one of the noticeable issues in the hydrological modeling community such as the international working group on Uncertainty Analysis in Hydrologic Modeling, a part of the Predictions in Ungauged Basins (PUB).

Beven[5] outlined the desirable modeling framework considering uncertainty and then Wagener *et al.*[4,6] materialized it by incorporating a parsimonious model with uncertainty analysis tools such as Multi-Objective Complex Evolution Method (MOCOM, Yapo *et al.*[7]) and Dynamic Identifiability Analysis (DYNIA, Wagener *et al.*[8]) They emphasized that for building the advanced modeling framework, uniqueness to figure out the true representation of a hydrological system by calibration and validation steps has to be abandoned, instead, both extended model evaluation and enhanced parameter identification are necessary to confirm reliable predictor(s) or to reject inadequate one(s).

In this regard, this paper proposes an alternative rainfall-runoff modeling framework under uncertainty, which consists of two conditioning processes attempting to identify more reliable model structure(s) with its parameter set(s). This framework is the extended version of the frameworks proposed by Beven[5] and Wagener and Gupta.[4] It may provide a more useful guideline for model identification in practical rainfall-runoff model applications.

2. Description of a New Rainfall-Runoff Modeling Framework under Uncertainty

Figure 1 illustrates the alternative rainfall-runoff modeling framework under uncertainty. It is established by two conditioning processes: enhanced model evaluation and further parameter identification using additional information. Initially, a set of model structures is prepared for model evaluation. Here, it is assumed that all model structures can be potentially useful simulators, unless obvious evidences appear for it to be rejected.

2.1. *First conditioning process: Enhanced model evaluation*

At the first conditioning stage, three different evaluative criteria are applied to the competing models. It allows under-performing models with respect to these evaluative indices to be rejected at this stage. The first measure of model evaluation is Model Performance Index (MPI), which assesses whether the models are able to simulate the observed streamflow accurately or not. Second is Model Structural Stability Index (MSSI) for assessing whether the model structures can represent various local response modes (e.g. low and high flows) with a single parameter set. The last measure is Model Parameter Identifiability Index (MPII) for evaluating whether the model parameters are well identified or not within a predefined feasible parameter space. As a result, the anticipated criteria values will give some objective basis to search for the model with a good balance between prediction accuracy, structural stability, and parameter uncertainty.

Figure 1 includes the schematic model space with respect to the proposed three evaluative criteria. Here, each box indicates the testable model structure and the black box is the best model leading to good model performance, stable model structure and high parameter identifiability while the dark grey box on the bottom left in the three dimensional model space, is referred to as the unideal model.

The new modeling framework emphasizes that one dimensional model evaluation, which is based only on single criterion, results in many possible predictors. It implies that many models can not be rejected by only MPI while they provide different values of both MSSI and MPII. For example, both Model N (the ideal model) and Model $N-1$ can provide equally good model performance measures but Model $N-1$ is worse than Model N

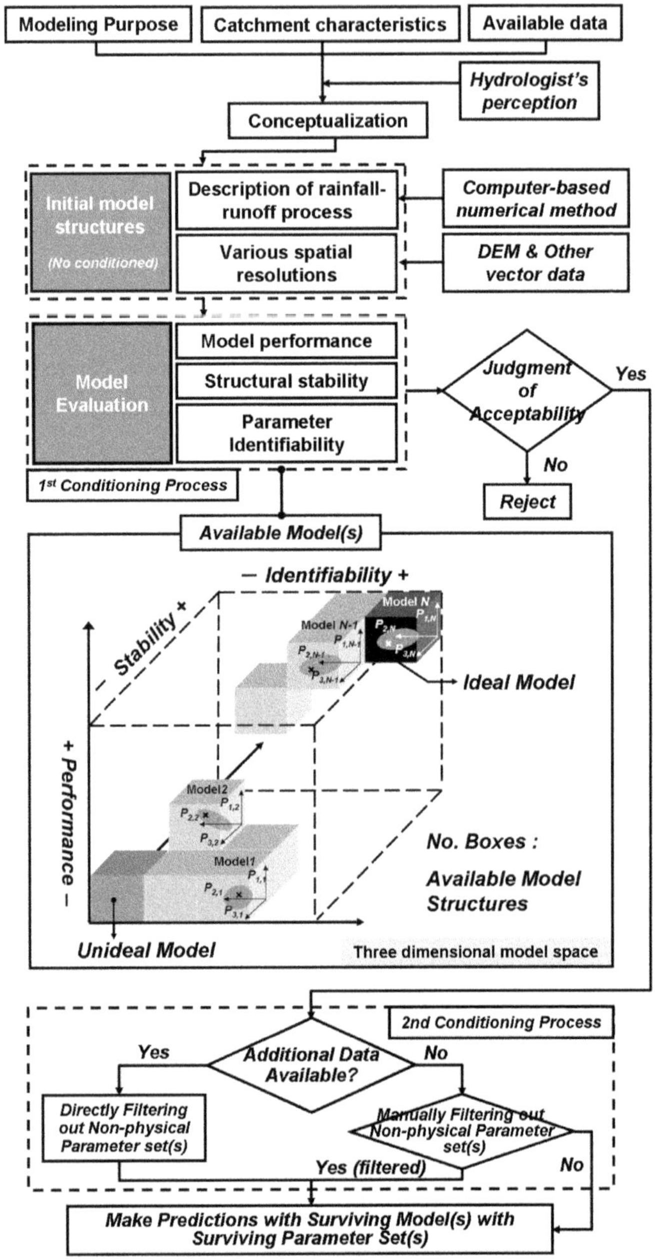

Fig. 1. Schematic stepwise procedure of the alternative rainfall-runoff modeling framework under uncertainty.

in terms of other measures for assessing structural and parameter uncertainties. As a result, additional criteria in the proposed model evaluation procedure can provide richer information enabling modelers to distinguish a balanced model, which lead to accurate and less uncertain prediction result, among a number of possible simulators.

2.2. *Second conditioning process: Further parameter identification*

However, even the best balanced model structure still suffers from the equifinality problem. Plausible parameter combinations that yield similarly good outcomes are widely distributed in the parameter space. In the three dimensional model space of Fig. 1, the light grey elliptical regions of each model (i.e. $P_{i,j}$ is a model parameter where i is the number of parameter to be calibrated and j is the number of available model structure) indicate the constrained parameter space by model calibration with streamflow data and all parameter sets within these regions provide objective function values as good as the global optimal sets marked by a symbol **x**. Note that each model structure has perhaps different parameter dimensions but parameter spaces of all models are represented schematically in three dimensions. In rainfall-runoff modeling, models are usually calibrated based only on streamflow data. However, it is just a basic requirement but insufficient information to identify model parameter(s) reliably. Kuczera and Franks[9] pointed out that a potentially more powerful approach is the use of additional data for further parameter identification.

The second conditioning process is therefore imposing complementary information on the constrained parameter space of the finally-selected model in the first conditioning stage. If additional information on hydrological responses of a catchment are available sufficiently, these constraints are used to directly reject unreliable parameter combination(s). However, despite effectiveness of complementary information, this approach is still limited to the experimental catchments[10] because of insufficient comparable measured data with multiple hydrological variables. If additional data is not available, the alternative approach in the second conditioning process is manual rejection of non-physical parameter set(s). It means that some of behavioral parameter sets, which contain conceptually unrealistic values, can be removed manually by the judgment of hydrologists or well-trained modelers. Although it is likely to receive criticism because of its subjectivity, manual calibration is still used in operational hydrology.[11] At all events, it is sure that the complementary information allows for further rejection or corroboration of model parameters, irrespective of whether it is measurable or not.

2.3. *Model prediction*

Finally, the surviving model with its behavior parameter sets should be retained until those that violate new evaluative criteria are found and then they are used for runoff prediction. In other words, the prediction result of the new modeling framework is not a single output sequence but a set of hydrographs.

3. Details of Evaluative Criteria Used in the First Conditioning Step

The purpose of the extended model evaluation is fundamentally to understand characteristics of each model structure and establish a preference between competing model structures with respect to three different criteria, model performance, structural stability and parameter identifiability. The ideal model may have a perfect (or stable) structure, which provides accurate and reliable prediction results. In addition, its response surface to parameters may be very convex or concave so that the global optimum can be easily found out using effective automatic optimization algorithms. One dimensional model evaluation based only on model performance, which is usually adopted in traditional model testing, is replaced by a three-dimensional one in order to provide a more extensive guideline with respect to selecting an adequate model.

3.1. *Model performance index (MPI)*

Model performance is a basic benchmark to support not only model selection but it is also used when discussing results with other hydrologists or stakeholders. It is typically judged by using an objective function which is to be minimized or maximized according to the modeling purposes. A wide range of statistical and hydrological objective functions is available. In general, these objective functions contain a summation of the error term and the summation of the squared errors, which is often used to avoid the canceling of errors of opposite sign. Boyle *et al.*[12] pointed out that global behavior could be described by overall measure like Root Mean Square Error (RMSE) but this aggregation of error was likely to decrease the amount of information in data. They recommended a separation of runoff time series into specific response periods to investigate the influence of individual model parameter on both global and local behaviors. Indeed, some models can reproduce the specific local behaviors (e.g. peak or rising/recession flows) very well while their overall model performances are not acceptable. For assessing model performance in the new modeling framework, the hydrograph is simply divided into two components (i.e. high and low flow periods) by the threshold, defined as the mean value of observed discharge data. The performances of each model structure are evaluated by Nash-Sutcliffe Coefficient (NSC) for two periods and then the average of the two measures is referred to as MPI, which is defined as:

$$\mathrm{NSC_{High}} = 1 - \frac{\sum_{t=1}^{N_{High}} (q_{obs}(t) - q_{sim}(t))^2}{\sum_{t=1}^{N_{high}} (q_{obs}(t) - \bar{q}_{obs}^{High})^2} \tag{1}$$

$$\mathrm{NSC_{Low}} = 1 - \frac{\sum_{t=1}^{N_{Low}} (q_{obs}(t) - q_{sim}(t))^2}{\sum_{t=1}^{N_{Low}} (q_{obs}(t) - \bar{q}_{obs}^{Low})^2} \tag{2}$$

$$\mathrm{MPI} = 0.5(\mathrm{NSC_{High}} + \mathrm{NSC_{Low}}) \tag{3}$$

where $q_{obs}(t)$ is the observed discharge at time step t; $q_{sim}(t)$ is the simulated discharge; $\bar{q}_{obs}^{High}$ and $\bar{q}_{obs}^{Low}$ are the mean observed discharge over the simulation periods, N_{High} and N_{Low}, respectively.

3.2. *Model structural stability index (MSSI)*

Structure error is an unavoidable problem in hydrological modeling since a hydrologic model is just a conversion and simplification of reality, thus those models only represent aspects of conceptualization or empiricism of modelers. In consequence, output time series of hydrologic models are as reliable as hypothesis, structure of models, quantity and quality of available data, and parameter estimates. Gupta *et al.*[13] demonstrated that one parameter set might be insufficient to represent the entire behavior of the catchment due to the imperfection of model structures. In other words, a subjective selection of objective functions for calibration of conceptual hydrologic models results in an overemphasis on different response modes. Therefore, the dependency of model performance on objective functions can be used to account for model structural stability. Two objective functions, Simple Least Squares (SLS) and Heteroscedastic Maximum Likelihood Estimator (HMLE) are recommended for evaluation of model structural stability. SLS is expressed as:

$$\mathrm{SLS} = \sum_{t=1}^{N} (q_t^{obs} - q_t(\theta))^2 \tag{4}$$

where q_t^{obs} is observed streamflow value at time t; $q_t^{(\theta)}$ is simulated streamflow value at time t using a parameter set θ and N is the number of flow values available. It has a feature that residuals between observed and simulated discharge are evenly weighed throughout a event, thus a parameter set, which matches well around peak discharge, will be obtained.

HMLE is the most successful form of the Maximum Likelihood criteria, which properly accounts for non-stationary variance in streamflow measurement errors[14]. This new measure containing weight provides more balanced performance across the entire flow range and it is calculated as:

$$\mathrm{HMLE} = \frac{1}{N} \frac{\sum_{t=1}^{N} w_t \varepsilon_t}{\left(\prod_{t=1}^{N} w_t\right)^{\frac{1}{N}}} \tag{5}$$

where $\varepsilon_t = q_t^{obs} - q_t(\theta)$ is the model residual at time t; w_t is the weight assigned to time t computed as $w_t = f_t^{2(\lambda-1)}$; $f_t = q_t^{true}$ is the expected true flow at time t; λ is the transformation parameter which stabilizes the variance. Yapo *et al.*[7] recommended the use of f_t as observed flow for more stable estimator.

Finally, the MSSI is formulated as:

$$\mathrm{MSSI} = \sqrt{\frac{1}{N} \sum_{t=1}^{N} (q_{SLS}(t) - q_{HMLE}(t))^2} \tag{6}$$

where N is total number of simulation time step; $q_{SLS}(t)$ and $q_{HMLE}(t)$ are the simulated discharges by each optimal parameter of SLS and HMLE, respectively. Here, the lower value of MSSI indicates the more stable model structure.

3.3. *Model parameter identifiability index (MPII)*

Parameter identifiability implies the level of "uniqueness" of parameters. It means that a well-identified model has a certain (or global optimal) parameter value while a poorly-identified model accepts many behavioral parameter values, which can provide model performance measures as good as the value estimated by the best-performing parameter. Wagener *et al.*[8] proposed a simple measure of parameter identifiability based on Regional Sensitivity Analysis (RSA, Spear and Hornberger[15]). The uniform random sampling method used in their research (Wagener *et al.*[8]) is able to extrapolate the parameter space easily but is computationally inefficient since a large number of samples are necessary to protect misleading results (Feyen *et al.*[16]). On the other hand, Markov Chain Monte Carlo (MCMC) methods generate samples from a Markov chain in an attempt to estimate a stationary posterior parameter distribution and it can be useful for high dimensional optimization problems (Kuczera and Parent[17]). The Shuffled Complex Evolution Metropolis (SCEM) algorithm is an effective and efficient evolutionary MCMC sampler which has enhanced search capability and operates by merging the strengths of the Metropolis algorithm, controlled random search, competitive evolution, and complex shuffling (Vrugt *et al.*[18]).

This algorithm can provide not only optimal parameter set but also its underlying stationary posterior distribution within a single optimization run. In these posterior distributions of individual parameters, the parameter value corresponding to the highest density indicates the optimal parameter value while other parameter values within these distributions are referred to as behavioral parameters. Therefore, SCEM can be used to estimate posterior distributions of individual parameters and then investigate the uniqueness of calibrated parameters. Here, the highest density value of each distribution is used as the indicator of parameter identifiability. Then, the mean of these maximum identifiability values of each parameter is regarded as MPII. Moreover, the uncertainty associated with parameters from the estimated posterior distributions is quantified and it gives the basis for making probabilistic predictions associated with parameter uncertainty. Figure 2(a) shows the example of posterior parameter distribution. In this figure, the value, marked by open circle, is the indicator of parameter

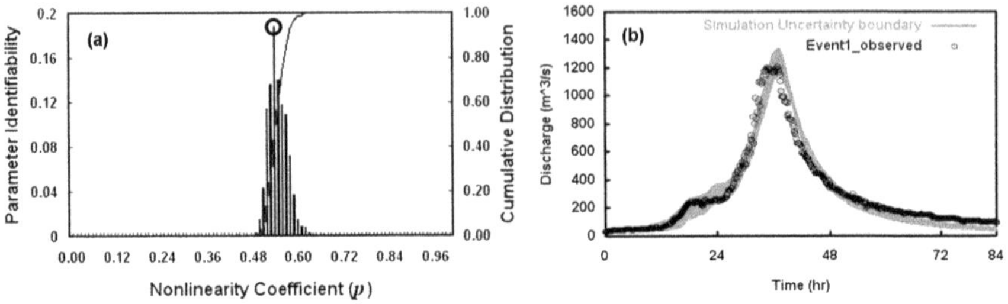

Fig. 2. (a) Example of estimation of parameter identifiability and (b) simulation uncertainty boundary associated with the behavior parameter sets within the posterior distribution.

identifiability and the parameter value corresponding to this is the best-performing parameter value. The distribution containing the sharper peak indicates that the model parameter is identified better within the feasible parameter space. Figure 2(b) presents the hydrograph simulation uncertainty estimated by the parameter sets within posterior parameter distributions.

4. Need of Complementary Constraints for Further Parameter Identification (Direct Rejection of Non-Physical Parameter Set)

A complex rainfall-runoff model due to over-parameterization can permit multiple alternative flow pathways leading to equally good hydrological outputs. Such a model usually has the poor parameter identifiability showing the uniform posterior parameter distribution. Furthermore, the commonly used streamflow data to evaluate such a model does not contain sufficient information on these possible flow pathways and thus, in spite of our deep faith in the computer-based automatic optimization tools, the model still has numerous plausible parameter combinations (Franks *et al.*[19]). Therefore, additional information is needed to offer a step forward to identifying the minimum reliable parameter sets. It can be done by further constraining the feasible parameter space.

Figures 3(a) and (b) show the parameter identification processes based on streamflow and multi-response data, respectively (Mroczkowski *et al.*[20]). The strategy shown in Fig. 3(a) only uses hydrological response A for both calibration and validation while response B is totally disregarded. This kind of framework represents the traditional split-sample test using streamflow

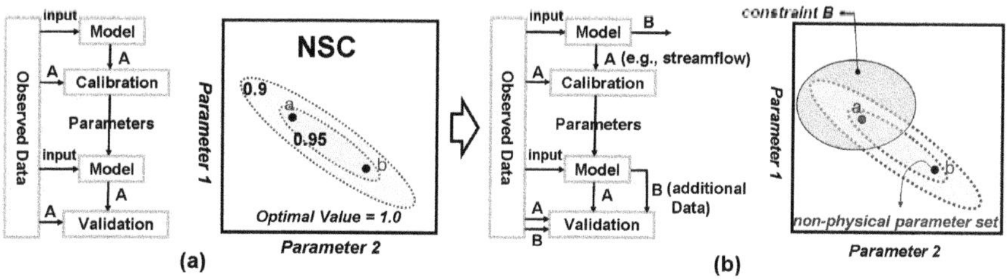

Fig. 3. Schematic diagrams of parameter identification based on (a) streamflow data and (b) multi-response data.

data alone. It is conceivable that many models cannot be rejected with regard to response A. The parameter space constrained by only response A contains numerous plausible parameter sets. For example, two parameter sets, a and b have the same model performance measure value, 0.95 because they are inside the same contourline of NSC. On the other hand, the alternative strategy illustrated in Fig. 3(b) uses response A for calibration but both responses are used for validation. This strategy provides much greater opportunity to select reliable parameter set(s) or reject unreliable one(s) because it imposes greater constraint on the parameter space. The previous parameter space based only on response A has the intersection, which comes from the additional response B. It means that parameter set b can be regarded as a non-physical parameter set not satisfying both physical constraints (i.e. outside the intersection) such that this parameter set will be rejected and not used for model prediction (i.e. reduction of parameter uncertainty).

Experimental methods such as isotope tracers, isotopic hydrograph separations and stream water residence time have been used as additional physical constraints for reducing parameter uncertainty in water quantity and quality modeling. However, these approaches that attempt to yield multiple output variables usually require the reformulation of models by adding another conceptual parameters. Ultimately, it again results in the decline of parameter identifiability. Therefore, to avoid over-parameterization, the desirable features of the enhanced parameter identification using multi-response data are as follows: (1) provide the evidence capable of rejecting unreliable parameter sets; (2) do not cause over-complexity due to numerous parameters that has to be calibrated.

Finally, statistical error estimation will be used to evaluate the improvement of simulated variables through the enhanced conditioning processes and demonstrate superiority of the proposed method to currently-available model identification methods. Therefore, filtered behavioral parameters in the second conditioning process are more physically-sound and also they can provide more reliable prediction results than the modeling case without additional constraints.

5. Conclusions

This paper has discussed the alternative rainfall-runoff modeling framework under uncertainty. Two conditioning processes incorporated into the new modeling framework was introduced. In the first step of the framework, a set of possible model structures was evaluated multi-dimensionally with respect to model performance, model structural stability, and parameter identifiability. The mean value of NSCs for low and high flow periods was proposed as MPI to assess the model predictability. The discrepancy between hydrographs reproduced by two different objective functions, SLS and HMLE was referred to as MSSI to assess the model structural uncertainty. In addition, the highest value of posterior parameter distribution was estimated by the stochastic optimization algorithm, SCEM and then it was adopted as MPII to evaluate the parameter identifiability. In the second conditioning process of the new modeling framework, the parameter space of the selected model structure in the first stage was conditioned directly or indirectly by adding physical or knowledge-based complementary constraints into parameter estimation procedure. The further parameter identification using observable multi-response data (i.e. direct filtering of non-physical parameter set) basically requires the capability providing apparent evidences to reject the unreliable parameter combination(s). Also, it should not cause over-parameterization because of over-complexity by either revision or reformulation of the model with additional parameters that is likely to worsen parameter identifiability and in turn, cause the equifinality problem.

References

1. T. Wagener, H. S. Wheater and H. V. Gupta, *Rainfall-runoff Modeling in Gauged and Ungauged Catchment* (Imperial College Press, London, 2004).
2. K. Beven, *Rainfall-runoff Modeling: The Primer* (John Wiley and Sons, Chichester, 2001).
3. V. Klemes, *Hydro. Sci. Journal* **31**(1) (1986) 13–24.

4. T. Wagener and H. V. Gupta, *Stoch. Environ. Res. and Risk Assess.* **19** (2005) 378–387.
5. K. Beven, *Hydro. Processes* **16**, 2 (2002) 189–206.
6. T. Wagener, D. P. Boyle, M. J. Leeds, H. S. Wheater, H. V. Gupta and S. Sorooshian, *Hydrol. and Earth Sys. Sci.* **5** (2001) 13–26.
7. P. O. Yapo, H. V. Gupta and S. Sorooshian, *J. Hydrol.* **204** (1998) 83–97.
8. T. Wagener, M. J. Leeds, H. S. Wheater and H. V. Gupta, *Hydrol. Processes* **17**(2) (2003) 455–476.
9. G. Kuczera and S. W. Franks, in *Mathematical models of large watershed hydrology*, eds. V. P. Singh and D. Frevert (Water Resources Publisher, Highland, 2002).
10. J. Seibert and J. J. McDonnell, *Water Resour. Res.* **38**(11) (2002) 1241.
11. T. S. Hogue, S. Sorooshian, V. K. Gupta, A. Holz and D. Braatz, *J. Hydrometeorology* **1** (2000) 524–542.
12. D. P. Boyle, H. V. Gupta and S. Sorooshian, *Water Resour. Res.* **36**(12) (2000) 3663–3674.
13. H. V. Gupta, S. Sorooshian and P. O. Yapo, *Water Resour. Res.* **134**(4) (1998) 751–763.
14. S. Sorooshian and J. A. Dracup, *Water Resour. Res.* **16**(2) (1980) 430–442.
15. R. C. Spear and G. M. Hornberger, *Water Research* **14** (1980) 43–49.
16. L. Feyen, J. A. Vrugt, B.Ó. Nualláin, J. Knijff and A. D. Roo, *J. Hydrol.* **332** (2007) 276–289.
17. G. Kuczera and E. Parent, *J. Hydrol.* **211** (1998) 69–85.
18. J. A. Vrugt, H. V. Gupta, W. Bouten and S. Sorooshian, *Water Resour. Res.* **39**(8) (2003) 1–14.
19. S. W. Franks, S. Uhlenbrook and P. Etchevers, in *Hydrology 2020: An integrating Science to Meet World Water Challenges*, eds. T. Oki, C. Valeo and K. Heal (IAHS Press, Wallingford, 2006).
20. M. Mroczkowski, G. P. Raper and G. Kuczera, *Water Resour. Res.* **33**(10) (1997) 2325–2336.

Advances in Geosciences
Vol. 17: Hydrological Science (2008)
Eds. Namsik Park *et al.*

A SIMPLE METHOD FOR INCORPORATING PARAMETER UNCERTAINTY IN STOCHASTIC DATA GENERATION

SRI SRIKANTHAN
Water Division, Bureau of Meteorology,
GPO Box 1289 Melbourne,
Victoria 3001, Australia
r.srikanthan@bom.gov.au

GEORGE KUCZERA
School of Civil Engineering,
University of Newcastle, University Drive
Callaghan, NSW 2308, Australia

MARK THYER
School of Civil Engineering, University of Newcastle,
University Drive Callaghan, NSW 2308, Australia

Stochastically generated data are used to assess the impact of climate variability on water resources systems. There are only a few models available in the literature that take parameter uncertainty into account. A commonly used approach to the traditional way of incorporating parameter uncertainty is via Bayesian analysis using Markov chain Monte Carlo simulation. This technique is complex and cannot be applied to models that have a large number of parameters like the transition probability matrix method for generating daily rainfall. Therefore, a bootstrap method is developed to incorporate parameter uncertainty in stochastic data generation. The developed method is evaluated by using an AR1 model for which an exact method of incorporating parameter uncertainty is available. Annual rainfall data from ten Australian rainfall stations are used in the evaluation and the results show that the developed model is effective in incorporating parameter uncertainty.

1. Introduction

Climate data, particularly rainfall data, are a major input to water resource and agricultural modelling systems. As the historical record only provides a single realisation of the underlying climate variability, stochastically generated data can be used to assess the impact of climate variability on water resources and agricultural systems. In most of the models, the estimated parameter values are assumed to be the "true" values and the sampling errors are ignored. The source of parameter uncertainty is insufficient sample size in calibration and is equally applicable to parametric and nonparametric models. Klemes and Bulu[7] showed that this traditional approach to generating stochastic data underestimates the range of mean and variance of the likely future sequences. Furthermore, Frost *et al.*[5] showed that ignoring parameter uncertainty underestimates extreme drought risks of at least half an order of magnitude. Several researchers have used Bayesian methods to overcome this deficiency.

Vicens *et al.*[14] incorporated parameter uncertainty into an annual data generation model for a single site using Bayesian techniques. Valdes *et al.*[13] extended the above model for a multivariate case. McLeod and Hipel[8] claim that the above methods for the incorporation of parameter uncertainty do not result in generated data that resemble the historical data. They proposed a correct method for incorporating parameter uncertainty based on large sample theory. Thyer and Kuczera[12] undertook a full Bayesian analysis to derive the posterior distribution of the hidden state Markov model and AR(1) model parameters using Markov Chain Monte Carlo (MCMC) method. To generate monthly data which reflect the uncertainty in the mean and variance correctly, Stedinger *et al.*[9] developed a condensed disaggregation model. Chaouche and Parent[3] used a Bayesian approach to estimate the uncertainty in the parameters of a two-part daily rainfall model utilising a Markov chain and a Gamma distribution. Thyer *et al.*[12] developed a procedure based on Bayesian technique to generate annual data taking into account parameter uncertainty.

In this paper, a simple bootstrap method is evaluated for incorporating parameter uncertainty in stochastic models and compared against the Bayesian method. The paper is organized as follows. The Bayesian method is described briefly in Sec. 2 and the developed method is described in Sec. 3. Section 4 describes the statistics used in the evaluation process and the application of the methods to annual rainfall data from ten Australian stations and the results are discussed in Sec. 5. Finally the conclusions are given in Sec. 6.

2. Bayesian Method

The Bayesian method is explained in terms of the AR(1) model which will be used as the case study. Thyer *et al.*[12] developed a procedure for fully evaluating the uncertainty of the parameters in the AR(1) model when used with the Box–Cox transformation to normalise the data. The AR(1) model can be written in the form:

$$z_t = \mu + \varphi_1(z_{t-1} - \mu) + \varepsilon_t \tag{1}$$

where z_t is a normally distributed time series, μ is the mean, ϕ_1 is the autoregressive parameter and ε_t is a Gaussian random variable with zero mean and variance σ_ε^2.

In the case of non-normal data, the data is normalised using the Box–Cox transformation (Box and Cox, 1964):

$$z_t = \begin{cases} \frac{x_t^\lambda - 1}{\lambda} & \lambda \neq 0 \\ \log x_t & \lambda = 0 \end{cases} \tag{2}$$

where λ is the transformation parameter which transforms the skewed data to approximately Gaussian. With the transformation parameter included, the vector of unknown AR(1) model parameters is given by:

$$\boldsymbol{\theta}^T = (\mu, \sigma_\varepsilon, \phi_1, \lambda). \tag{3}$$

The Metropolis algorithm is used to simulate the posterior distribution of the model parameters. To improve the acceptance rate of the Metropolis algorithm, the parameter space is transformed to remove the dependencies of μ and σ_ε on λ.

$$\begin{aligned} \mu &= \frac{\mu_x^\lambda - 1}{\lambda} \\ \sigma_\varepsilon &= \mu_x^{\lambda-1}\sigma_x\sqrt{1-\phi_1^2} \end{aligned} \tag{4}$$

where μ_x and σ_x represent a first order approximation to the expected value and the standard deviation of the untransformed rainfall data.

3. Bootstrap Method

An alternative to the Bayesian method, which is exact, is the bootstrap method.[2] In this method, the parameters of the stochastic model, denoted as θ, are estimated from the historic data denoted as $\mathrm{D_H}$. Using these estimated parameters $\theta(\mathrm{D_H})$, a replicate of length equal to that of the historical data, denoted as $\mathrm{D_R}$, is generated. From the generated replicate, the parameters of the stochastic model, $\theta(\mathrm{D_R})$, are again estimated. Using this second set of parameters $\theta(\mathrm{D_R})$, a replicate is generated. This replicate will incorporate the effect of parameter uncertainty. Moreover, $\theta(\mathrm{D_R})$ represents a sample from the sampling distribution of θ. In summary, the parametric bootstrap is implemented by using the following steps. Denote the probability model of the parametric model as $p(\theta)$.

Step 1: Estimate the model parameters from the historical data $\theta(D_H)$
Step 2: Generate a replicate by using the parameters estimated from the historical data:
$D_R \leftarrow p(\theta(D_H))$ where "$\leftarrow$" denotes taking a sample from a distribution.
Step 3: Estimate the model parameters from the generated replicate $\theta(D_R)$.
Step 4: Generate a replicate by using the parameters estimated in Step 3:
$D_k \leftarrow p(\theta(D_R))$
Step 5: Repeat steps 2 to 4 until the required number of replicates are generated.

The parametric bootstrap does not exactly represent parameter uncertainty. The reason for this is that Step 3 returns the sampling distribution for the model assuming the true parameter is $\theta(D_H)$, when in fact it is an estimate of the true parameter θ.

Returning to our case study model, in terms of standardised variables, an AR1 model can written as

$$X_t = rX_{t-1} + \varepsilon_t \tag{5}$$

where X_t is the standardised rainfall in year t, ε_t a normally distributed random component with zero mean and unit variance and r lag-one autocorrelation coefficient.

If the annual data are skewed, the skewness in the data can be modelled through the Wilson-Hilferty transformation (Fiering 1967).

$$\varepsilon_t = \frac{2}{g_\varepsilon}\left\{\left(1 + \frac{g_\varepsilon \eta_t}{6} - \frac{g_\varepsilon^2}{36}\right)^3 - 1\right\} \tag{6}$$

where g_ε is the skewness of ε_t which is related to the skewness (g) of annual data through

$$g_\varepsilon = \frac{(1 - r^3)}{(1 - r^2)^{3/2}} g. \tag{7}$$

For ease of application, the above model is used in preference to the Box–Cox transformation.

4. Model Evaluation

The following statistics are used to evaluate the data generation model for annual rainfall:

- Mean
- Standard deviation
- Coefficient of skewness
- Lag one autocorrelation coefficient
- Maximum rainfall
- Minimum rainfall
- 2- and 10-year low rainfall totals

For convenience, the last three items in the above list are standardised by dividing by the historical mean annual rainfall.

5. Application to Australian Rainfall Data

Annual rainfall data from ten stations located in different parts of Australia (Fig. 1) are used to evaluate the bootstrap and Bayesian methods. The

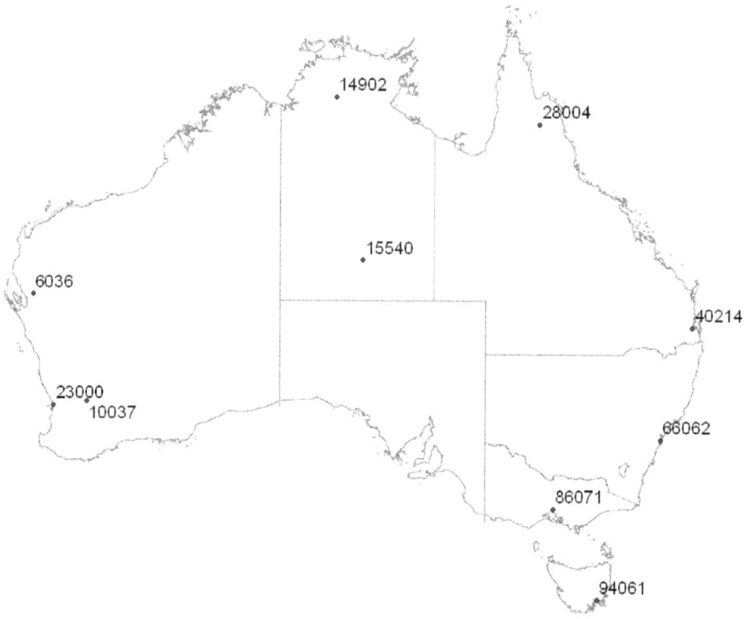

Fig. 1. Location of the rainfall stations used.

Table 1. Details of the rainfall stations used in the case study.

Station Number	Station Name	Latitude	Longitude	Years
6036	Meedo	−25.66	114.62	94
10037	Cuttening	−31.73	117.76	96
14902	Katherine	−14.46	132.26	111
15540	Alice Springs	−23.71	133.87	112
23000	Adelaide	−31.95	115.84	139
28004	Palmerville	−16.00	144.08	109
40214	Brisbane	−27.48	153.03	133
66062	Sydney	−33.86	151.20	140
86071	Melbourne	−37.81	144.97	143
94061	Sandford	−42.93	147.52	111

length of data used is given in Table 1. One thousand replicates each of length equal to the historic record were generated for each station. The proposed method was compared with the Bayesian implementation of the AR(1) model which we refer to as the exact method. For each replicate, the statistics were calculated. The average of the statistics from the 1000 replicates along with the historical values was calculated. In addition the 1000 replicate values were ranked to obtain 95% confidence limits on the observed statistic.

The ability of the model to reproduce statistics not used in calibration can be tested using the confidence limits. If the model is consistent with the data then one would expect the observed statistics to lie outside the 95% confidence limits for 5% of the stations. An unusually large or small number of stations lying outside these limits would suggest a deficiency in the model.

Historical and generated mean annual rainfalls with 95 % confidence limits are shown in Fig. 2. The parametric and bootstrap method gave similar confidence limits to those of the exact method which are used as a bench mark for model evaluation. For the standard deviation (Fig. 3) and skewness (Fig. 4) of annual rainfall, the lower limit was the same for both the models while the proposed method gave lower upper limit resulting in an underestimate of uncertainty. There is a slight negative bias in the generated lag one autocorrelation coefficient (Fig. 5) by the proposed methods. For the maximum annual rainfall (Fig. 6), the lower limit was the same for both the models while the proposed model slightly underestimated the upper limit. However, for the minimum annual rainfall, the proposed method gave slightly wider confidence limits than the exact method (Fig. 7).

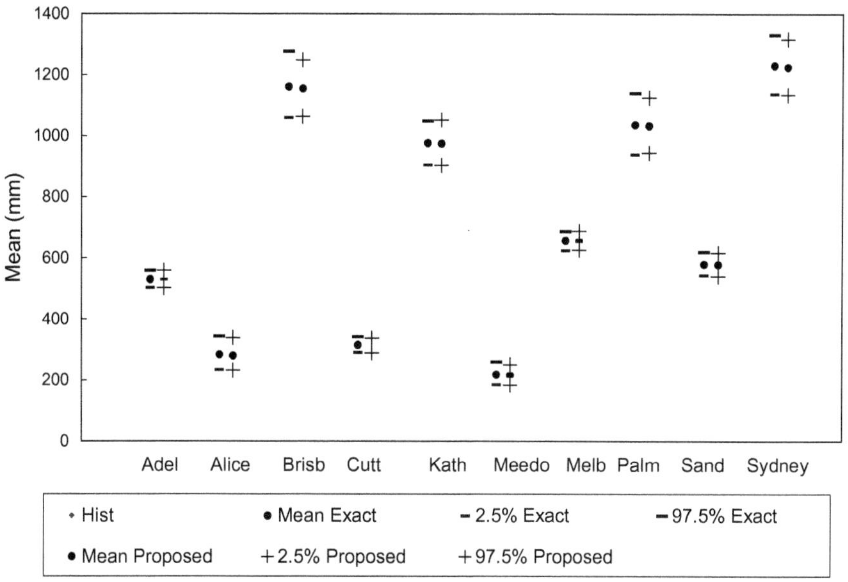

Fig. 2. Comparison of observed and generated mean annual rainfall.

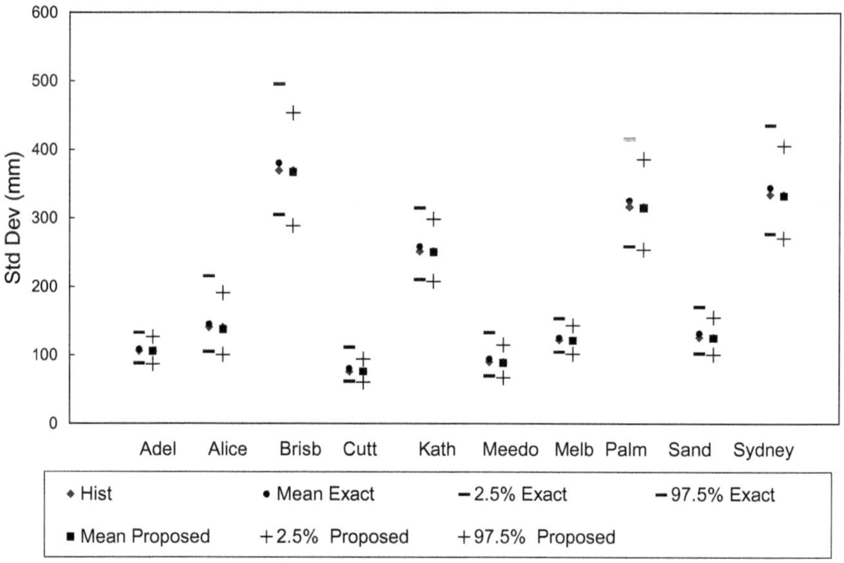

Fig. 3. Comparison of the standard deviation of the observed and generated annual rainfall.

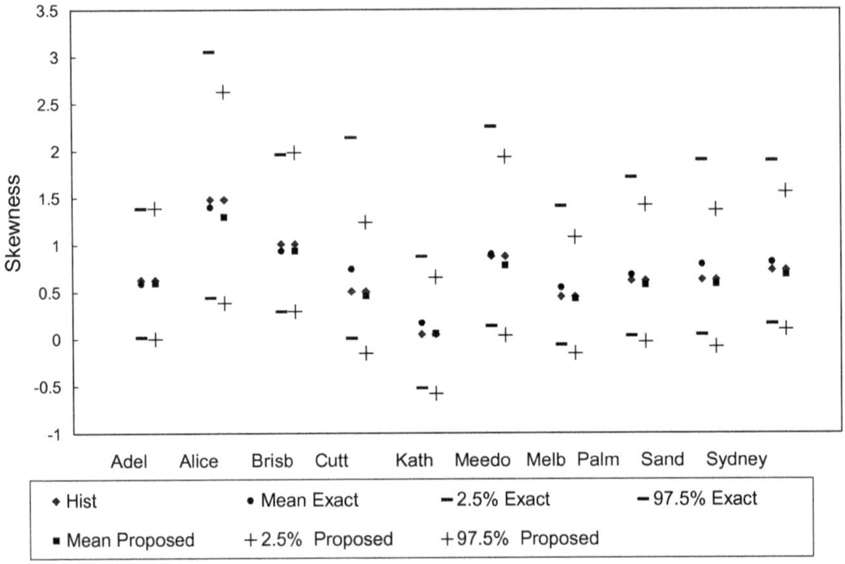

Fig. 4. Comparison of the skewness of the observed and generated annual rainfall.

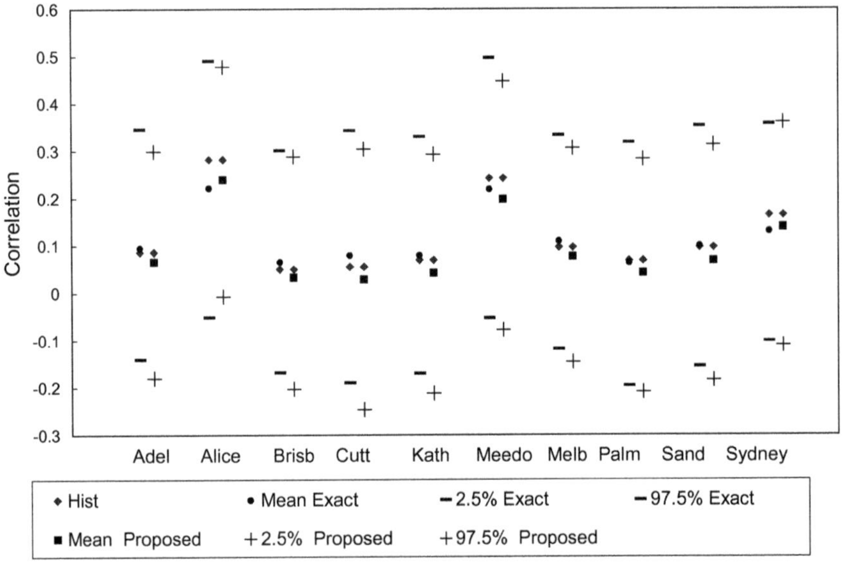

Fig. 5. Comparison of the lag one autocorrelation coefficient of the observed and generated annual rainfall.

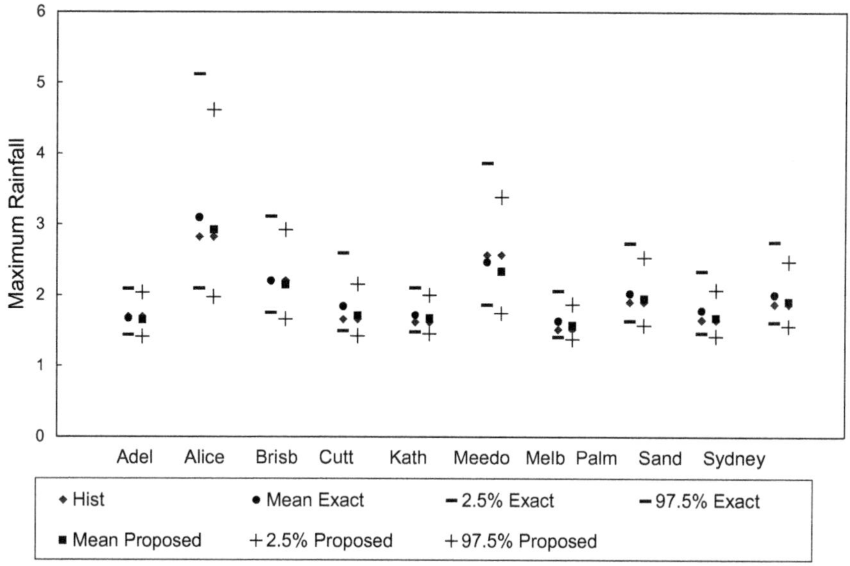

Fig. 6. Comparison of the observed and generated maximum annual rainfall.

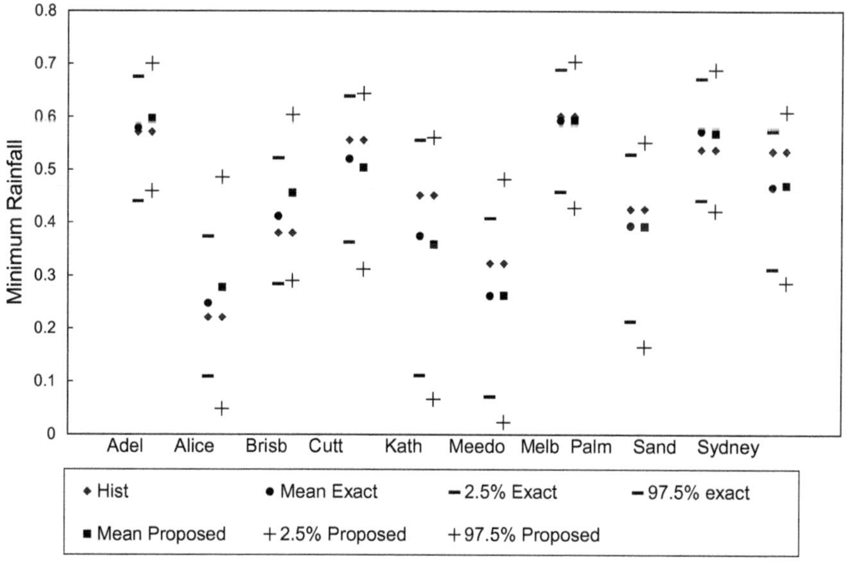

Fig. 7. Comparison of the observed and generated minimum annual rainfall.

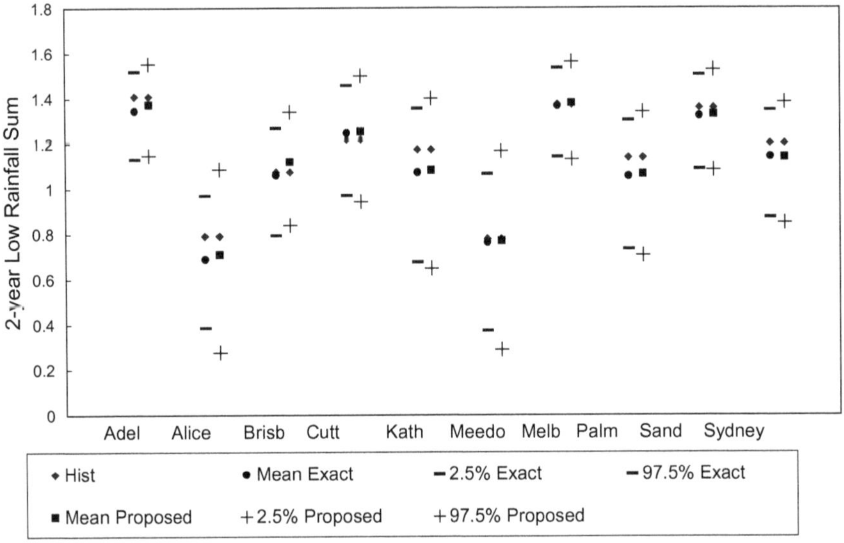

Fig. 8. Comparison of the 2-year low rainfall sum of the observed and generated annual rainfall.

The proposed method again gave a slightly wider confidence interval than the exact method for the 2-year and 10-year (Figs. 8 and 9).

It is important to stress that bootstrap methods cannot capture the full uncertainty in the data. For example, the parametric bootstrap assumes the parameter estimate derived from the historic data is the population value. Ignoring this leads to an underestimate of uncertainty which can be significant when dealing with short records and when extrapolating the model beyond the observed experience. In addition, it is implicitly assumed that the selected estimator is efficient. In the case of the AR1 model the estimators are known to have good properties. However, if a poor estimator were employed, it would lead to results considerably inferior to the Bayesian method, which always makes best use of the information in the data (because it works directly with the likelihood function). The effect of the record length of observed data on the performance of the method is investigated in another study (Srikanthan *et al.* 2009) by varying the record length from 20 to 100 years and the results show that the method works acceptably well for record lengths of 50 years or more. Provided the modeller is aware of these limitations, it is our view that resampling in parametric models can produce useful results.

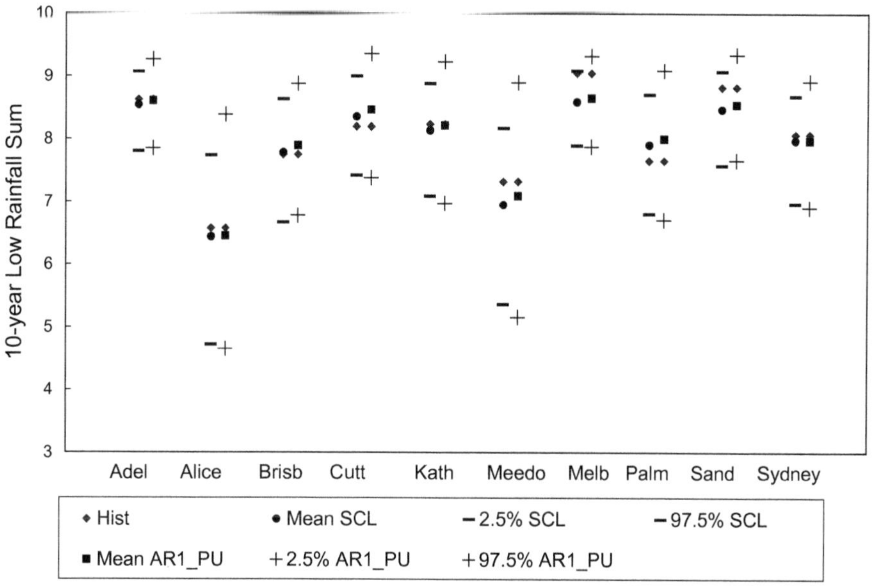

Fig. 9. Comparison of the 10-year low rainfall sum of the observed and generated annual rainfall.

6. Conclusions

Parameter uncertainty is generally incorporated in stochastic data generation via Bayesian analysis using Markov chain Monte Carlo simulation. This technique requires formulation of a likelihood function which is difficult to do for many models in use. Therefore, a simple resampling method is developed to incorporate parameter uncertainty in stochastic data generation models. The resampling model is applied to generate annual rainfall data at 10 Australian locations and the results are compared with those from a Bayesian annual rainfall model. The comparison shows that the resampling method can be a practical and effective alternative to the Bayesian method for incorporating parameter uncertainty provided the record length is sufficient. The case study demonstrates for record lengths of the order of 100 years that the resampling method produces similar results to the Bayesian method, except in some statistics such as skewness and extrema where resampling produced substantially lower uncertainty. While this conclusion is model-specific, one would expect similar results for other models employing efficient estimators.

References

1. G. E. Box and D. R. Cox, An analysis of transformations, *Jour. Royal Stat. Soc., Ser. B* **26**(2) (1964) 211–252.
2. G. Casella and R. Berger, *Statistical Inference*, Second Edition (Duxbury Press, 2001).
3. A. Chaouche and E. Parent, Bayesian identification and validation of a daily rainfall model under monsoon conditions, *Hydrol. Sci. J.* **44** (1999) 199–220 (in French).
4. M. B. Fiering, *Streamflow Synthesis* (Macmillan, New York, 1967).
5. A. J. Frost, M. A. Thyer, R. Srikanthan and G. Kuczera, A general Bayesian framework for calibrating and evaluating stochastic models of annual multi-site hydrological data, *J. Hydrol* **340**(3–4) (2007) 129–148.
6. K. W. Hipel, Stochastic research in multivariate analysis, Fourth International Hydrology Symposium, Fort Collins, Colorado, U.S.A, (1985) pp. 2–50.
7. V. Klemes and A. Bulu, Limited confidence in confidence limits derived by operational stochastic hydrologic models, *J. Hydrol.* **42** (1979) 9–22.
8. A. I. McLeod and K. W. Hipel, Simulation procedure for Box-Jenkins models, *Water Resour. Res.* **14** (1978) 969–975.
9. J. R. Stedinger, D. Pel and T. A. Cohn, A condensed disaggregation model for incorporating parameter uncertainty into monthly reservoir simulations, *Water Resour. Res.* **21** (1985) 665–675.
10. R. Srikanthan, G. Kuczera and M. Thyer, "Calibrate it Twice": A Simple Resampling Method for Incorporating Parameter Uncertainty in Stochastic Data Generation. Hydrology and Water Resources Symposium, Engineers Australia (2009).
11. M. A. Thyer and G. Kuczera, Modelling long-term persistence in hydroclimatic time series using a hidden state Markov model, *Water Resour. Res.* **36** (2000) 3301–3310.
12. M. Thyer, G. Kuczera and Q. J. Wang, Quantifying parameter uncertainty in stochastic models using the box-cox transformation, *Journal of Hydrology* **265** (2002) 246–257.
13. J. B. Valdes, I. Rodriguez-Iturbe and G. J. Vicens, Bayesian generation of synthetic streamflows 2. The multivariate case, *Water Resour. Res.* **13** (1977) 291–295.
14. G. J. Vicens, I. Rodriguez-Iturbe and J. C. Schaake, Baysian generation of synthetic streamflows, *Water Resour. Res.* **11** (1975) 827–838.

Advances in Geosciences
Vol. 17: Hydrological Science (2008)
Eds. Namsik Park *et al.*

EVALUATION OF AGRICULTURE ACTIVITIES ON GROUNDWATER OF AL-FOAH AREA, NORTH OF AL-AIN, UNITED ARAB EMIRATES (UAE)

SAMEH Y. SULEIMAN
Schlumberger, Al Khobar, Saudi Arabia

AHMED A. MURAD
Department of Geology, Faculty of Science,
UAE University, Al-Ain, UAE
ahmed.murad@uaeu.ac.ae

AHMED S. EL MAHMOUDI
Water Studies Centre, King Faisal University,
Al Hassa-31982, KSA

Forty eight samples from private and government production wells and two surface water samples were collected from Al-Foah area which is located to the north of Al-Ain area, UAE. The main objective of this study is to assess the chemical characterization of groundwater in the Quaternary aquifer of the study area. Analyses of groundwater salinity showed that about 18% of groundwater samples are considered as fresh water and about 82% are considered as brackish water. The progressive saline of groundwater might be might be ascribed to the effect of intensive agriculture activities. All groundwater samples are located above the line of seawater as observed from the relationship of Cl/Br vs. Cl, which clearly indicated that the excess of chloride could come from different sources. In addition, the positive correlation ($r^2 = 0.8$) between Cl vs. Br supported the argument that both Cl and Br might come from the same sources. One of the sources that release Br to the system might be the agriculture practices as observed from the positive relationship between K and Br. In order to test the fact of agriculture activities effect on groundwater quality, NO_3^- is measured in all groundwater samples. All the nitrate concentrations of the groundwater in the study area were less than 50 mg/l except few wells which are located within the cultivated area. A trend of increasing nitrate with potassium supported the hypothesis that the agricultural activities are one of the main sources affecting groundwater quality.

1. Introduction

Groundwater in the United Arab Emirates (UAE) is considered an important natural resource. Increasing water demand and the scarcity of water resources in the UAE is the main challenge facing the sustainable water resources management in UAE and met by wise policies and decisions. UAE is considered among the highest rates of per capita water consumption in the world; even though the county is facing severe shortages in the natural water resources due to the rapid developments of domestic, industrial, and agricultural water supplies. Due to deficit in conventional water resources, non-conventional water resources such as desalinated water and treated wastewater became important sources of water in the country and in the region to meet the water demand for different uses.

Al-Foah (previously Al Auha) area is located in the eastern region of the Abu Dhabi Emirate between longitudes of 55°20′ and 55°50′ E and latitudes of 24°10′ and 24°30′ N and close to the border of UAE and Sultanate of Oman (Fig. 1). Al Ain area, where the study area is located, has witnessed a rapid urbanization during the last few decades. The agriculture activities in the area are significant because of the land fertility. Such activity is mainly depend on groundwater of the Quaternary aquifer and fractured limestone aquifer of Simsima Formation of Upper Cretaceous.The objectives of this study are to determine the quality of the groundwater and to evaluate its degree of contamination by agriculture practices. In order to implement and achieve the goals of this study, several methods have been used. Previous

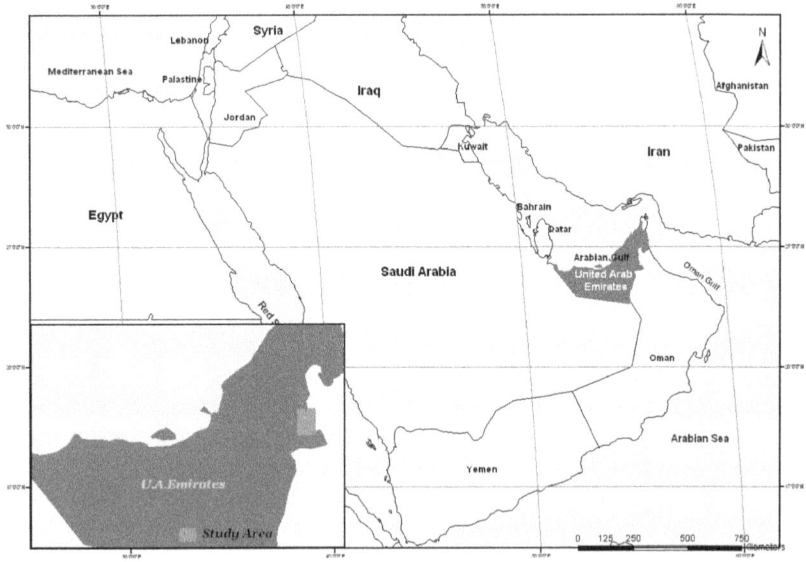

Fig. 1. Map showing the location of Al-Foah area.

studies related to the geology of the aquifer, hydrogeological parameters and climatological data have been collected, cross-checked and analyzed.

2. Geologic and Hydrogeologic Settings

There are many geomorphologic features that exist in the study area, and theses features include mountains, gravel plains and sand dunes. Jabal Auha and Jabal Huwayyah, which lie 8 km northeast of Al Ain city, are the main mountains featuring the study area. Both Jabal Mundassah and Jabal Malaqat are part of Oman Mountains but are located to the southeast of Jabal Al Auha and Jabal Huwayyah (Fig. 2). The eastern side

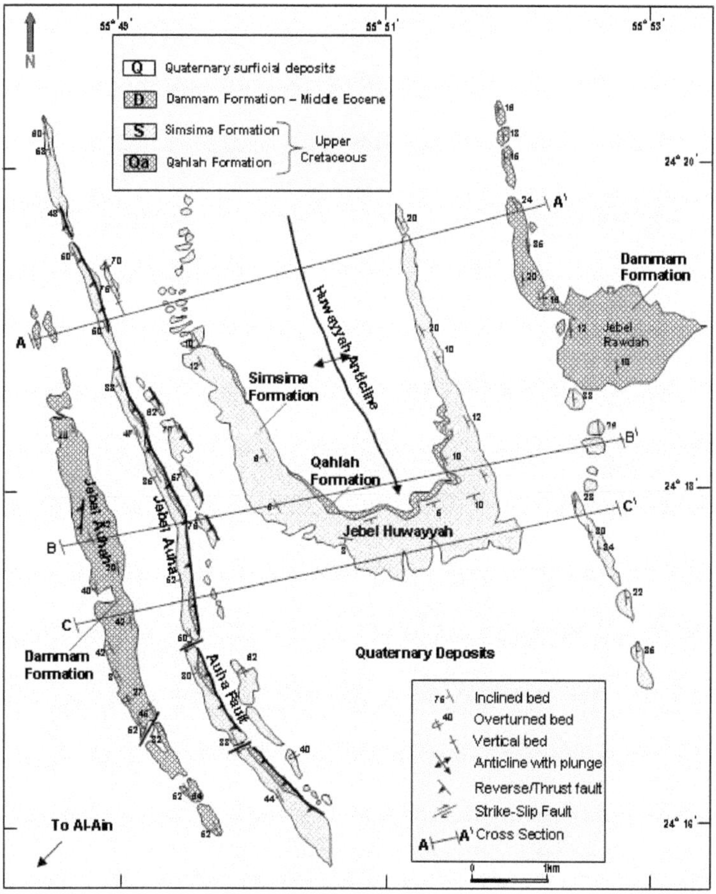

Fig. 2. Modified geologic map of the study area (Noweir and Alsharhan, 2000).

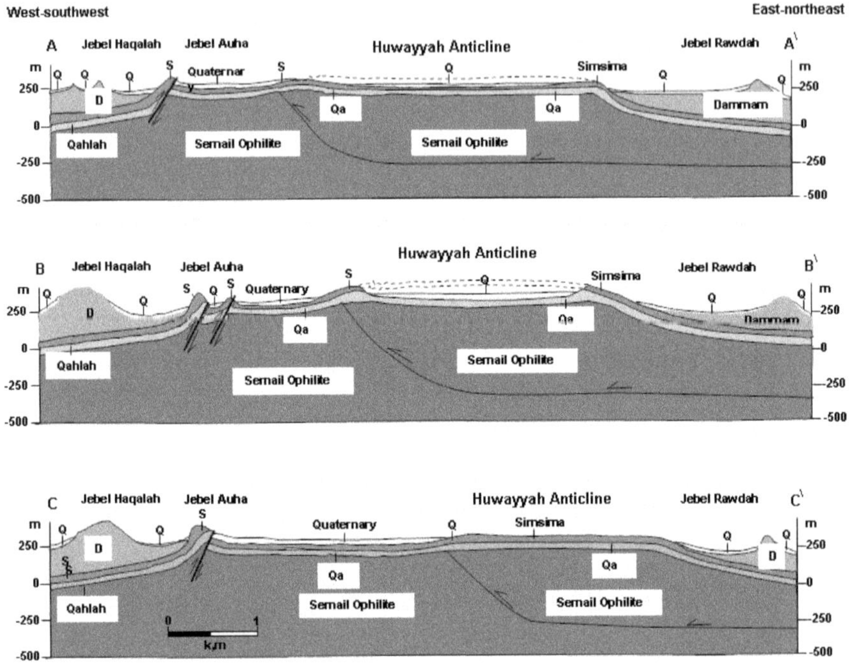

Fig. 3. Structural cross sections (A-A′, B-B′ and C-C′) through the Oha reverse faults and Huwayyah Anticline (see Fig. 2 for location and legend of section) (modified from Noweir and Alsharhan, 2000).

of Oman Mountains is bounded by gravel plains which mainly compose of alluvial sand and gravel which transported by wadis dissecting the Oman Mountains. Sand dunes and alluvial fan interrupt the gravel plains in the east of Al-Foah. The main stratigraphic units that occupy the study area are: Semail Ophiolite, Qahlah Formation, Simsima Formation, Dammam Formation and Quaternary deposits (see Figs. 2 and 3). Hamdan and Bahr (1992) stated that the age of the major units in the study area range from Upper Cretaceous to Holocene.

At the base of Jabal Auha, Jabal Huwayyah, Jabal Malaqet and Jabal Mundassah, the oldest rock units of Semail Ophiolite exist and the oldest rock units mainly composed of Pre-Maastrichtian serpentinite and serpentinized periodotite (Hamdan and Anan, 1993). The Qahlah Formation which is exposed at Jabal Auha and Jabal Huwayyah, is capped by a red chert-pebble conglomerate. The lowest part of the formation consists of weathered khaki colored shales, overlain by cross-bedded pebbly brown sandstones and chert conglomerates. The lower 5 m to 8 m of the

Maastrichtian Simsima Formation is composed of shelly, foraminiferal packstone of shallow water origin overlain by 15 m to 18 m thick of fine-grained deep-water packstone. The Simsima Formation is unconformably overlain by limestone intercalated with thin bedded of marl in Jabal Oha and Jabal Huwayyah. The carbonate facies of Damman Formation are composed of fine to coarse grains which is sign for an active shallow marine environment (Noweir and Alsharhan, 2000).

Quaternary deposits the covered most of Al-Foah area are aeolian sand, desert plain deposits and alluvial deposits. The dune sediments constitute of quartz and carbonate with minor proportions of basic and ultra basic grains. However, the deposits of desert plain are inter-layered laminated silts and carbonate cemented with dune bedding.

Alluvial deposits vary in the study area ranging from boulder gravel to conglomerate in the east and fine sand to silt further west. Pebbles in some localities are un-cemented or loosely and held by coarse-grained recrystallized calcite (Hunting, 1979).

The annual average rainfall in the study area is 91 mm and the mean evaporation in Al Ain area is ranged from 5 to 22 mm/day during the period of 1994–2000. Based on rainfall data of the study area, the annual rainfall can contribute to groundwater recharge once every 4 to 5 years (Fig. 4) (Suleiman, 2007). The Quaternary and Jabal Auha limestone aquifers are considered as the main aquifers in the study area are (Rizk *et al.*, 1997). The depth to water table in Al-Foah area is about 30 m below the ground surface and increasing towards the north and south. The continuous increasing of depth to groundwater is noticed in Suweyhan and Al Ain in which it

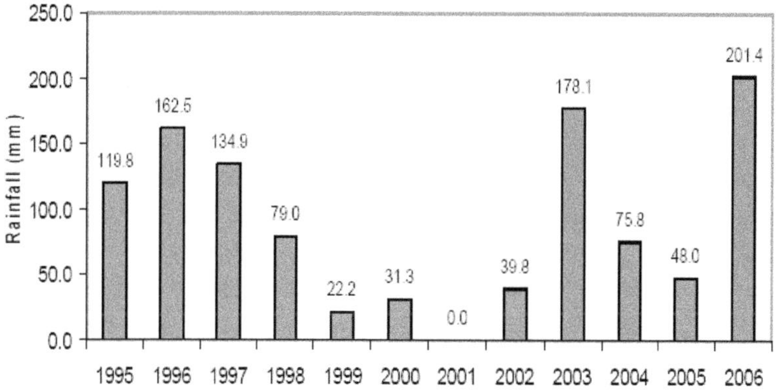

Fig. 4. Total yearly rainfall at Al-Ain (1995–2006).

reached to 50 and 90 m respectively. The depletion of groundwater depth is a result of heavy groundwater abstraction in the area for different purposes (Garamoon, 1996). The Cretaceous limestone of Jabal Mohayyer at Al-Foah area represents a good aquifer with high specific capacity wells. Water in the fractured limestone rocks originate from leakage from the overlying sands and gravel. The low dissolution of hard ophiolitic rocks forming Jabal Malaqet and Jabal Mundassah made the groundwater of eastern part of Al-Foah area of good quality (Hamdan and El-Deeb, 1990).

3. Sampling and Analysis

Forty eight samples from private and government production wells and two surface water samples (sample Nos. 25 and 43) have been collected to meet the objectives of this study (Fig. 5). The field direct measurements were conducted for electrical conductivity (EC) (μS/cm), hydrogen ion concentration (pH), and temperature (°C). The samples analyses were stored at 4°C for major anions and were preserved with a few drops of acid (HNO_3) for major cations and trace elements. The chemical analyses were conducted in the Central Laboratories Unit (CLU), United Arab Emirates (UAE) University, for major cations (K^+, Na^+, Mg^{+2} and Ca^{+2}), major anions (CO_3^{2-}, HCO_3^{2-}, SO_4^{2-}, Cl^-, PO_4^{4-} and NO_3^-) and trace elements (Mn, Cu, Fe, F, Pb, B, Ba, Sr, Cr and Zn).

4. Results and Discussions

The measured electrical conductivity of groundwater samples of the Quaternary aquifer in the Al-Foah area ranged from 1 to 12.6 mS/cm, in which the highest values of EC were observed in the north and south. However, the average TDS value in the study area is 2305 mg/l, with high concentrations observed in the north and south where agriculture activities are intense. Based on Todd's classification (1980), about 18% of groundwater samples are considered as fresh water and about 82% are considered as brackish water. It is noticed from EC and TDS data that the groundwater progressively becomes saline in groundwater at Al-Foah area. The pH of the groundwater in the study area ranges from 6.1 to 7.9. The groundwater samples are consistently meteoricl water. The temperature of groundwater samples in the Al-Foah area varied between 22.7°C and 38.4°C with an average of 31.7°C. The sequence of anions and cations dominance in groundwater of the Quaternary aquifer in the Al-Foah area has the following

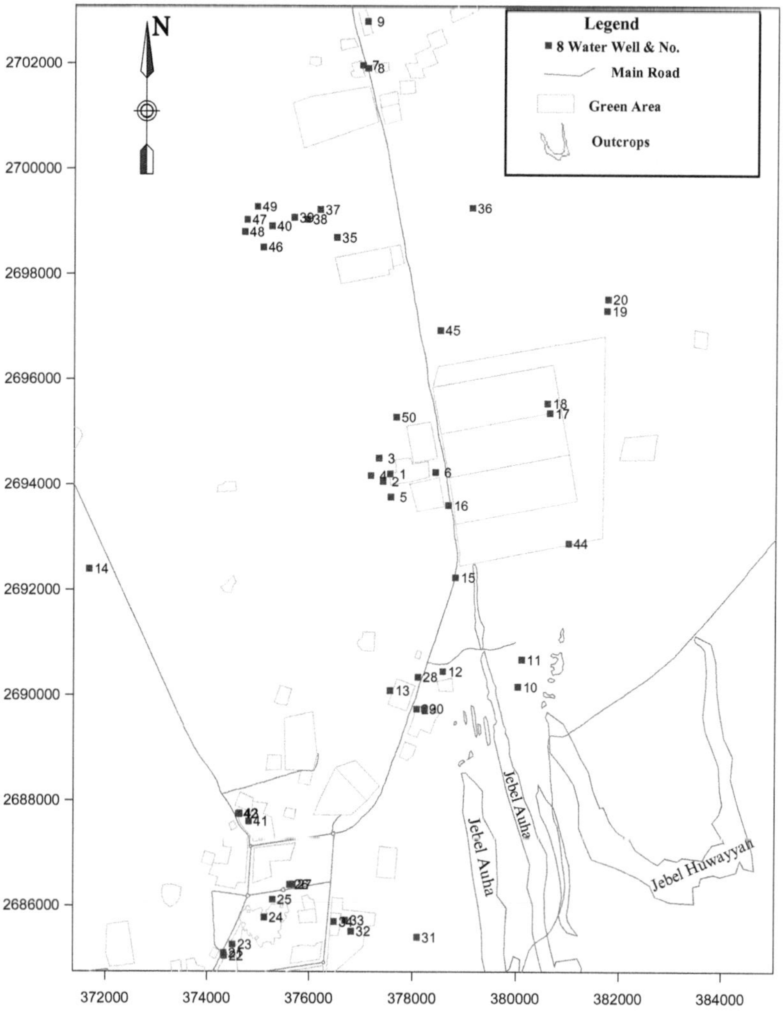

Fig. 5. Base map showing the locations of water wells at Al- Foah area, UAE.

order: $Cl^- > SO_4^{2-} > HCO_3^{2-} > CO_3^{2-}$ and $Na^+ > Mg^{+2} > Ca^{+2} > K^+$ respectively. Chloride ion concentrations in groundwater of the study area ranged from 96.7 to 3320 mg/l, except well no. 36 which has a high concentration of 6550 mg/l.

The increase of Cl^- concentration is noticed in the north of Al-Foah (Fig. 6) where the agriculture activities exist. This high concentration of chloride in groundwater might be ascribed to intense groundwater pumping,

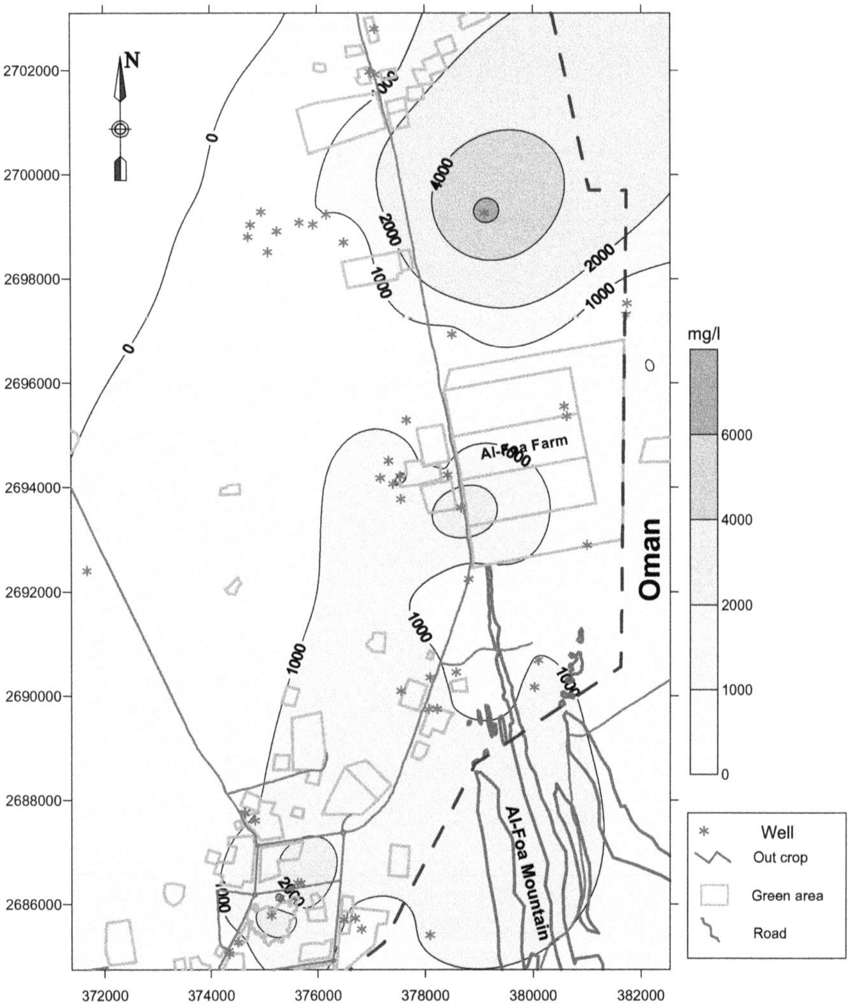

Fig. 6. The distribution of chloride in in groundwater of the study area.

agricultural activities and evaporation of return flow which is mainly used for agriculture practices in the region.

The concentration of sulfate in groundwater of Al-Foah area ranged from 40 and to 3090 ppm, in which the highest distribution of sulfate occurs in the southern part of the study area. The most extensive sulfate ions in water are occuring in sedimentary rocks as gypsum ($CaSO_4.2H_2O$) and anhydrite ($CaSO_4$) which are present in the study area. The presences of these two minerals in the study area as thick beds or streaks in limestone

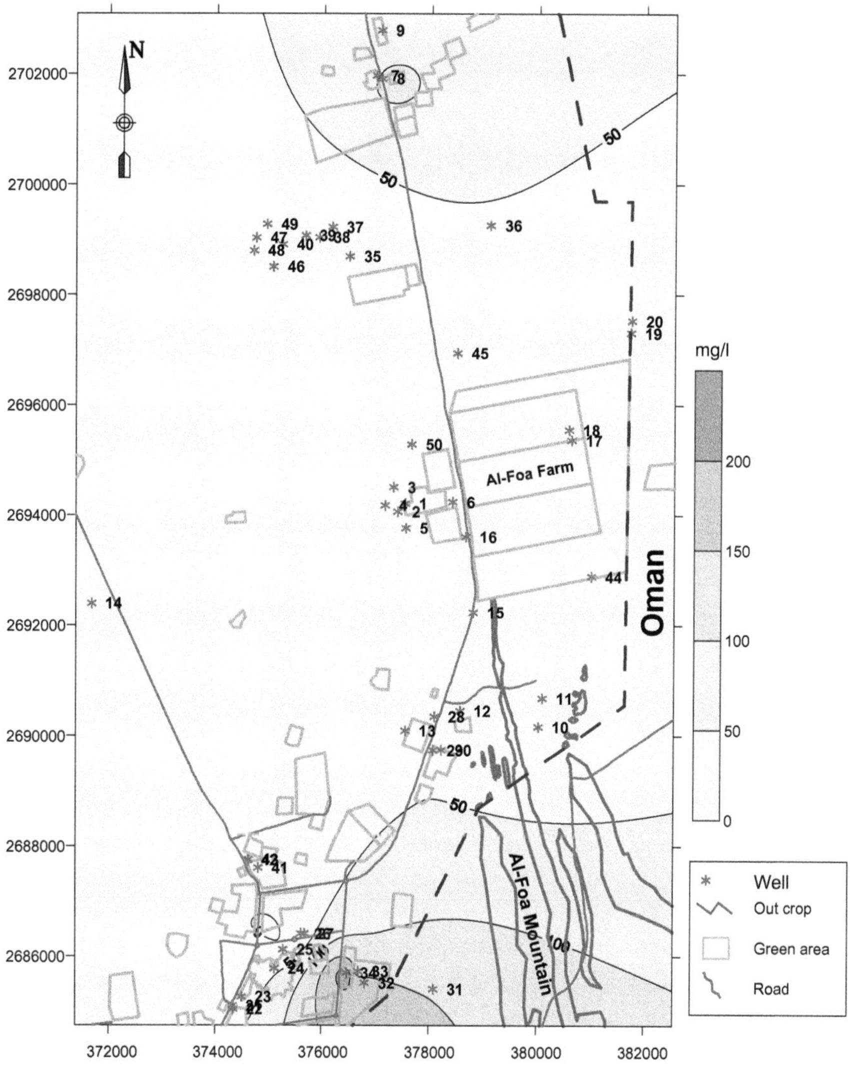

Fig. 7. The nitrate distribution in groundwater of the study area.

strata combined with their high solublity cause water in contact with them to be high in sulfate. In addition, sulfate can be added to groundwater from the breakdown of organic substances in soil from the addition of leachable sulfate in fertilizers and from other human influences (Matthess, 1982). The concentration of nitrate in collected groundwater samples of the Quaternary aquifer in Al-Foah area ranged from 2 to 230 ppm.

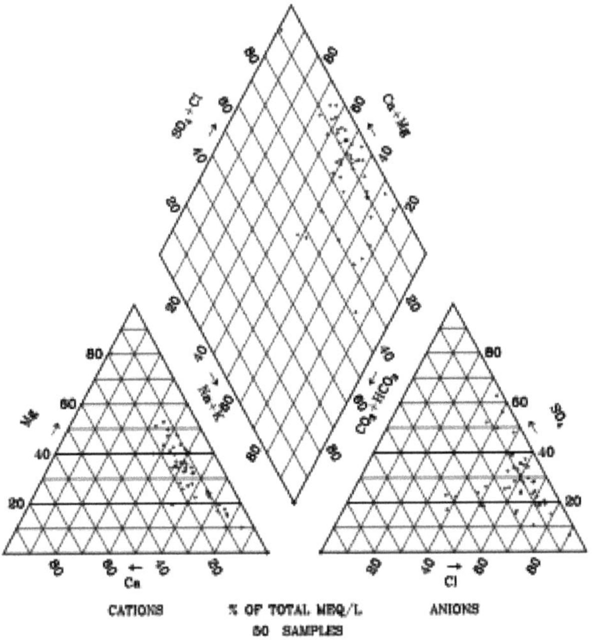

Fig. 8. Piper's trilinear diagram for classification of the groundwater samples in Al-Foah area.

The highest nitrate concentration were determined for water samples no. 7, 8, 9, 29, 31, 32, 33 and 34. All of them were taken for wells in located in the cultivated areas (Fig. 7). Also there is a trend of increasing nitrate with potassium, which mainly supports the fact that the agricultural practices are one of the main sources deteriorating the groundwater quality.

The range of sodium concentration in groundwater of the study area is from 103 to 1634 ppm. The cultivated part of the study is characterized by high concentration of Na and this indicates that irrigation water is one of the causes of increasing sodium in groundwater. The trend of increasing Na with Cl implies that the salinity is present which might come from the high evaporation rate in region, agricultural activities and urbanization. The position of groundwater samples in Piper diagram (Fig. 8) implies that Na-Cl is the dominant groundwater type in the region. The chemical analyses of groundwater samples for Mn, Cu, Fe, F, Pb, B, Ba, Sr, Al, Cr and Zn, showed that there were no significant values observed that could affect human health and the groundwater.

Based on calculated Sodium Adsorption Ratio (SAR) (U. S. Salinity Staff, 1954) for collected groundwater samples, the groundwater in the

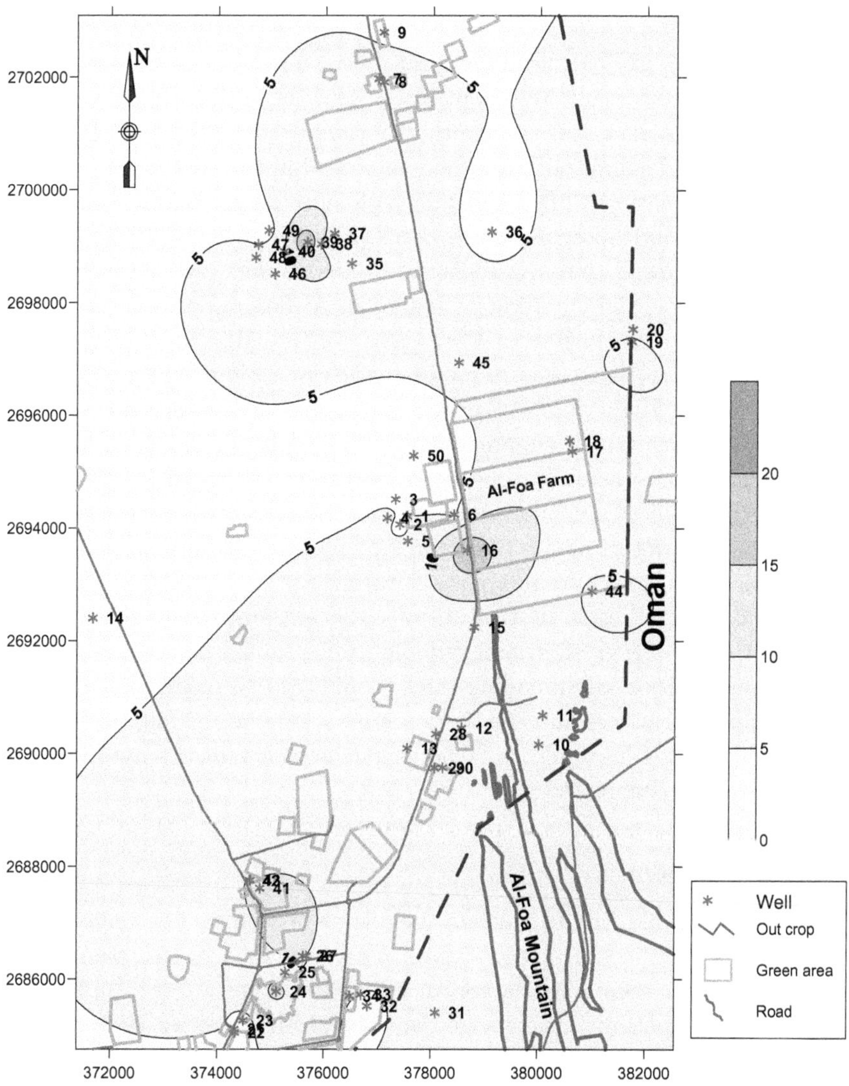

Fig. 9. The distribution map of the sodium adsorption ratio (SAR) in Al-Foah area.

study area has a low harmful effect on vegetation when used for irrigation. The SAR values for the study area are varied between 2.3 and 20 with an average of 7.23 (Fig. 9). About 88% of groundwater in the study area has low sodium hazard water, about 10% represents medium sodium hazard water and about 2% represents high sodium hazard water which lies near the cultivated area.

5. Conclusions

Groundwater in Al-Foah area has been over-pumped to meet the water demand of development and population growth. As a result of heavy groundwater abstractions, the depth to groundwater has been increased dramatically and salinity of water also increased in the study area.

Based on measured TDS, about 18% of collected groundwater samples are freshwater and about 82% are brackish water. Chemical analyses of groundwater showed that the sequence of anions and cations dominance in groundwater of the Quaternary aquifer in Al Foah area is in the order of: $Cl^- > SO_4^{-2} > HCO_3^{-2} > CO_3^{-2}$ and $Na^+ > Mg^{+2} > Ca^{+2} > K^+$ respectively. There is a trend of increasing major cations and anions from south to north. High values of nitrate were observed within the vicinity of cultivated area which support that the agricultural practices are one of the main sources deteriorating the groundwater quality in Al-Foah area. About 88% of collected groundwater samples from the study area shows low sodium hazard water, about 10% represent medium sodium hazard water and about 2% has high sodium hazard water.

Based on the result of this study , the authors recommended: the construction of new groundwater pumping wells should be fully controlled by local authorities, the pumping rates should also be monitored to avoid over pumping practices and the treated wastewater can be fully utilized in the irrigation and horticulture. This would help sustain the groundwater resources at Al Foah area.

Acknowledgments

The authors would like to express their appreciation to Mr. Jaber Latif for his assistance in the preparation of base map used in this study. The assistance of Mr. Omar Al Basher, in the collecting of water samples was valuable. The authors acknowledge the Central Laboratories Unit, UAEU for their help and facilities used in hydrochemical analysis.

References

1. H. K. Garamoon, Hydrogeological and geomorphological studies on the Abu Dhabi-Dubai-Al Ain triangle, UAE. Unpublished PhD thesis, Geol. Dep., Fac. Sci., Ain Shames University, Egypt, 1996, p. 277.
2. A. A. Hamdan and H. S. Anan, Cretaceous/Tertiary boundary in the United Arab Emirates. M.E.R.C., Ain Shams Univ., *Earth Sci. Ser.* **7** (1993) 223–231.

3. A. A. Hamdan and W. Z. El-Deeb, Stratigraphy of the Paleogene succession of Jabel Malaqet, west of the Northern Oman Mountains., Fac. Sci., UAE University **2** (1990) 30-39.
4. A. R. A. Hamdan and S. A. Bahr, Lithostratigraphy of the Paleocene succession of northern Jabel Hafit, Al-Ain, UAE. MERC, Ain Shams University, Earth science **6** (1992) 201–204.
5. Hunting Geology and Geophysics Ltd. Report on a mineral survey of UAE., Al Ain Area. Ministry of Petroleum and Mineral Resources, Abu Dhabi **9** (1979) 1–22.
6. G. Matthess, *The Properties of Groundwater* (John Wiley and Sons, NY, 1982), pp. 215–255.
7. M. A. Noweir and A. S. Alsharhan, Structural style and stratigraphy of the Huwayyah anticline: An example of an Al-Ain Tertiary Fold Northern Oman Mountains, GeoArabia: Middle East Petroleum Geosciences **5**(3) (2000) 387–400.
8. Z. S. Rizk and A. S. Alsharhan and S. Shino, Evaluation of groundwater resources of United Arab Emirates, in *Proceedings of 3rd Gulf Water Conference*, Muscat, Sultanate of Oman **1** (1997) 95–122.
9. S. Y. Suleiman, Geophysical and hydrogeological Studies of Al-Foah area, north Al Ain, United Arab Emirates, MSc Thesis, Water Resources Program, UAE University, UAE, 2007, p. 115.
10. U.S. Salinity Laboratory Staff. Diagnosis and improvement of saline and alkaline soils. US Dept. Agricultural Handbook No. 60, Washington, D.C., 1954, pp. 60–60.

Advances in Geosciences
Vol. 17: Hydrological Science (2008)
Eds. Namsik Park et al.

NUMERICAL SIMULATION OF ELECTRIC CONDUCTIVITY FLUCTUATIONS DUE TO GROUNDWATER PUMPING IN A COASTAL AQUIFER

E. D. P. PERERA*, K. JINNO and Y. HIROSHIRO
Graduate School of Engineering, Kyushu University,
744, Motooka, Nishi-ku, Fukuoka, 819-0395, Japan.
**dumindapradeep@yahoo.com*

A. TSUTSUMI
S.G. Gijutsu Consultant Co. Ltd, 4-18-25,
Kounonishi, Saga City, Japan.
a_tutumi@ybb.ne.jp

The Motooka area of Fukuoka prefecture of Japan is an agricultural area located on a coastal aquifer where the groundwater is being exploited for drinking and green house agriculture. The groundwater is the main water supply for drinking and greenhouse agriculture. With the increased water demand, groundwater is being extracted at the rate of 1,200 m^3/day. The salinity of the pumped water has been analyzed periodically and it has found that the aquifer is affected by the saltwater intrusion due to over exploitation of groundwater. The high salinity of pumped water is not suitable for both drinking and greenhouse farming. The salinity fluctuation in the pumped groundwater has become an important issue to attract much attention of the green house farmers due to its significant effect on their crops. To the authors' knowledge, research on the salinity fluctuations with groundwater pumping and their effects on seasonal recharge of groundwater in the Motooka region has not been conducted thus far. In the field of numerical modeling of saltwater intrusion, so far the simulation of salinity fluctuations is not significant. Therefore authors found a keen interest on numerical simulation of the saltwater intrusion and the salinity fluctuation phenomenon in the Motooka. To this end, a three dimensional density dependent finite difference solute transport numerical model was developed. Due to the distributed nature of pumping well locations and non-uniform geological conditions; a three-dimensional model was needed to understand the saltwater intrusion and salinity fluctuations in the Motooka. The developed model is capable of simulating the saltwater intrusion and the salinity fluctuations due to groundwater pumping for satisfactory extent.

1. Introduction

1.1. *Groundwater situation in Motooka*

The Matooka area is located in the western region of Fukuoka City of Kyushu Island in Japan as shown in Fig. 1. The elevation of the ground surface ranges from 0.3 m at the lowest point to about 100 m a.m.s.l at the highest point. The lowland area is an alluvial plain used for agriculture such as greenhouse farming and paddy fields. Under this plain, a shallow unconfined aquifer has been developed and it is partially affected by saltwater intrusion. The thickness of the unconfined aquifer under the lowland is approximately 50 m.

In the Motooka area of the Fukuoka prefecture in Japan, studied herein, water requirements are fulfilled by river water, irrigation ponds and groundwater. Since a public water supply has become available only recently in this area, groundwater has become the only reliable water supply source for drinking. Basically river water and irrigation pond water are utilized for paddy cultivation. The groundwater is then used as the main drinking water supply. In the Matooka area, groundwater is extracted for greenhouse farming at a rate of approximately 700 m^3/day, for domestic use at 400 m^3/day and at about 10 m^3/day for wineries.[3] Due to the continuous exploitation of groundwater, saltwater intrusion has become a significant cause of groundwater problems. Since the slight changes of the salt content in pumped groundwater directly affect the crops, the farmers of Motooka area are keen to know about the salinity variation with groundwater pumping. Construction of a new university in the area may also reduce the groundwater infiltration and create lower groundwater potential and invite

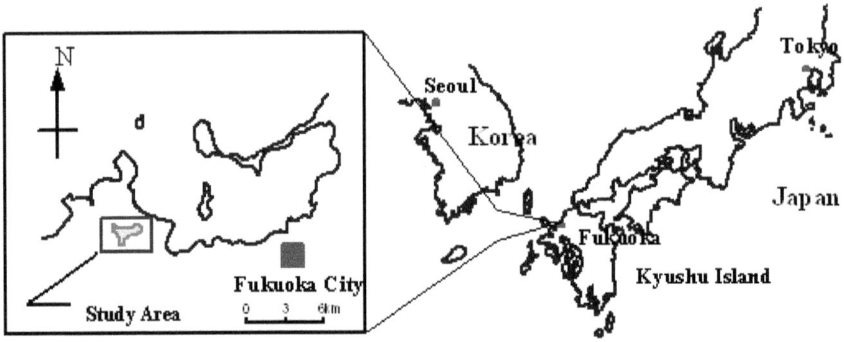

Fig. 1. Location of the Motooka and study area.

significant quantities of saltwater to flow into the freshwater aquifer. As the threat of saltwater intrusion into the groundwater aquifer is becoming more critical, the authors had a keen interest in the study of saltwater intrusion and electric conductivity fluctuation in this area. Due to the distribution of agro-wells and complex geological formations, the application of two-dimensional saltwater intrusion model is not adequate as it cannot simulate the saltwater movement to the level which would allow prediction of electric conductivity fluctuations for a satisfactory extent.

2. Methodology

To simulate the electric conductivity fluctuations due to saltwater intrusion, a three-dimensional density dependent numerical model is developed. The complexity of geology, pumping well distribution, distribution of hourly recharge rate and topography of the selected area emphasize the requirement of a three-dimensional model. In this study, the model is developed for a selected area of the Motooka coastal aquifer, as shown

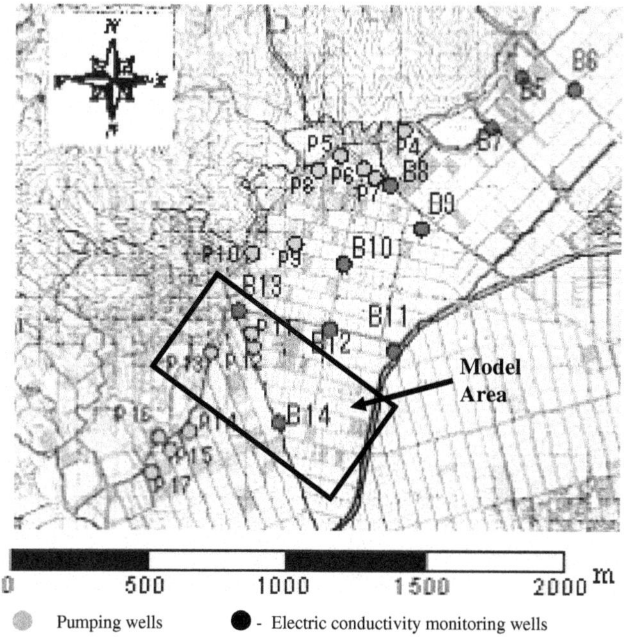

Fig. 2. Selected area for the numerical model, pumping wells' and electric conductivity monitoring wells' distribution of the Motooka area.

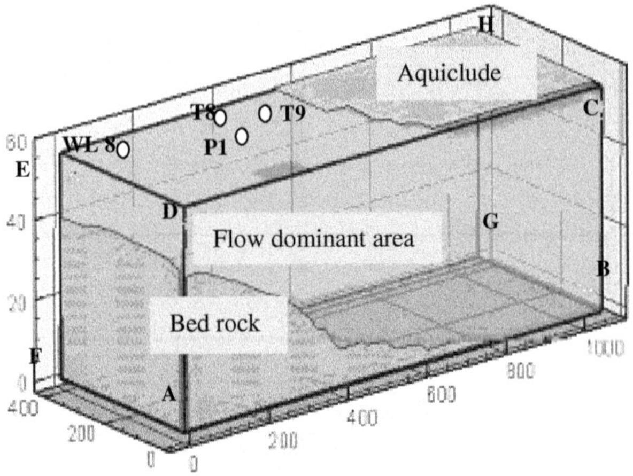

Fig. 3. Major geological regions of the numerical model, location of pumping wells (P1, T8 and T9) and location of water table observation well WL 8.

in Fig. 2. In the modeling process basically three different flow regions are considered according to the permeability values. These are bed rock, aquiclude and flow dominant areas (Fig. 3). The model is based on the finite difference method with a non uniform discretized grid system. Small grids are applied to the pumping well locations to have more precise results for the concentration of pumped water, and larger grid sizes are applied for rest. The method of characteristics is applied to solve the advection part of the mass transport equation.

2.1. *Mathematical model*

The mathematical model consists of partial differential equations that govern the groundwater flow and transport of solute in a coastal aquifer. The numerical model which is discussed here uses the groundwater flow continuity equation, Darcy's law and the mass transport equation. The X and Z axes are taken as horizontal, while Y axis is considered as vertical.

The continuity equation is given by:

$$(C_w + \alpha S_0)\frac{\partial h}{\partial t} = -\frac{\partial u}{\partial x} - \frac{\partial w}{\partial z} - \frac{\partial v}{\partial y} \tag{1}$$

where; C_w is specific moisture capacity, S_0 is specific storage coefficient, α is a dummy number which takes 0 in an unsaturated condition and 1 in a saturated condition, u is pore velocity in X direction (horizontal), w is pore velocity in Z direction (horizontal), v is pore velocity in Y direction (vertical), h is hydraulic pressure head.

Darcy's law is given by:

$$u = -k_x \frac{\partial h}{\partial x} \tag{2.1}$$

$$w = -k_z \frac{\partial h}{\partial z} \tag{2.2}$$

$$v = -k_y \left(\frac{\partial h}{\partial y} + \frac{\rho}{\rho_f} \right) \tag{2.3}$$

ρ and ρ_f are model domain water and fresh water densities, respectively.

The advection dispersion solute transport equation is written as:

$$\begin{aligned}
&\frac{\partial C}{\partial t} + \frac{1}{\theta}\left(u\frac{\partial C}{\partial x} + w\frac{\partial C}{\partial z} + v\frac{\partial C}{\partial y} \right) \\
&\quad = \frac{1}{\theta}\frac{\partial}{\partial x}\left(\theta D_{xx}\frac{\partial C}{\partial x} + \theta D_{xy}\frac{\partial C}{\partial y} + \theta D_{xz}\frac{\partial C}{\partial z} \right) \\
&\quad + \frac{1}{\theta}\frac{\partial}{\partial z}\left(\theta D_{zz}\frac{\partial C}{\partial z} + \theta D_{zy}\frac{\partial C}{\partial y} + \theta D_{zx}\frac{\partial C}{\partial z} \right) \\
&\quad + \frac{1}{\theta}\frac{\partial}{\partial y}\left(\theta D_{yy}\frac{\partial C}{\partial y} + \theta D_{yx}\frac{\partial C}{\partial x} + \theta D_{yz}\frac{\partial C}{\partial z} \right)
\end{aligned} \tag{3}$$

where; θ is the volumetric moisture content, $D_{xx}, D_{yy}, D_{zz}, D_{xy}, D_{xz}, D_{yx}$, D_{yz}, D_{zx} and D_{zy} are dispersion coefficients.

The equation of state shows the relationship between fluid density and solute concentration, and is given by:

$$C = \left(\frac{\rho - \rho_f}{\rho_s - \rho_f} \right) \times 100.00 \tag{4}$$

where C is fluid concentration and ρ_s is seawater density.

2.2. *Numerical model*

The model discussed here is based on the finite difference approach to solve the partial differential equation of flow and transport. The transient groundwater flow equation (1) is solved by an implicit finite difference method using an iterative successive over relaxation (SOR) technique. The solute transport equation (5) is solved in two step processors. Whereas advection term is computed by the method of characteristics (MOC), the dispersion is calculated by an explicit finite difference method. The method of characteristics is widely used as a high accuracy method to solve the convective-dispersion equation for the solute transport in groundwater.[2,4] The model domain is divided into unequal discretized grid system for the X and Z directions. For the Y direction, uniform grid size of 2.0 m is used. Small grid sizes are assigned to pumping area and lager grid sizes are assigned to the rest in the X and Z directions. The smallest grid sizes in the X and Z directions are 4.0 m and 5.0 m respectively. Fig. 4 shows the grid arrangement of the model: 1053.0 m, 417.0 m and 56.0 m are the length, width and height of the numerical model.

The hydro-geological parameters used in the model are obtained from borehole information, field measurements and literature.[1] The longitudinal and transverse dispersion lengths are set to 3.6 m and 0.36 m, respectively, while molecular diffusion is $1.0 \times 10^{-9}\,\mathrm{m^2/s}$. The standard seawater density value of $1025.0\,\mathrm{kg/m^3}$ is used. The density of fresh water is set to $1000.0\,\mathrm{kg/m^3}$, and $1.6 \times 10^{-8}\,\mathrm{m/s}$, $6.6 \times 10^{-7}\,\mathrm{m/s}$, and $4.6 \times 10^{-6}\,\mathrm{m/s}$ are the hydraulic conductivities of bed rock, confined layer and flow dominant region, respectively. Time increment is four hours.

3. Results and Discussion

In this numerical simulation the electric conductivity fluctuation is simulated for the selected three pumping wells of P1, T8 and T9 as shown in Fig. 3. The simulated electric conductivities with the comparison of measured data are presented. The calculated and measured water table levels of the water table observation well are also demonstrated. The variation of the electric conductivities of three pumping wells is caused by the pumping rates and the seasonal recharge rates. Figures 6, 7 and 8 show the simulated results of electric conductivity fluctuations with the pumping rates for the year 2002 to 2007. In the selected area the measured pumping rates are available only for the year 2007. In this simulation for the rest of the years the pumping rates of year 2007 were assigned assuming that the same patterns of water usage and agricultural activities have been continued from year 2002 to 2006. Simulated results shows a reasonable agreement with the measured the measured electric conductivity values.

Figure 4 shows the measured and calculated groundwater table at the well WL8. The variation of water table in the well WL8 is basically due to groundwater recharge. The numerical results show a reasonable

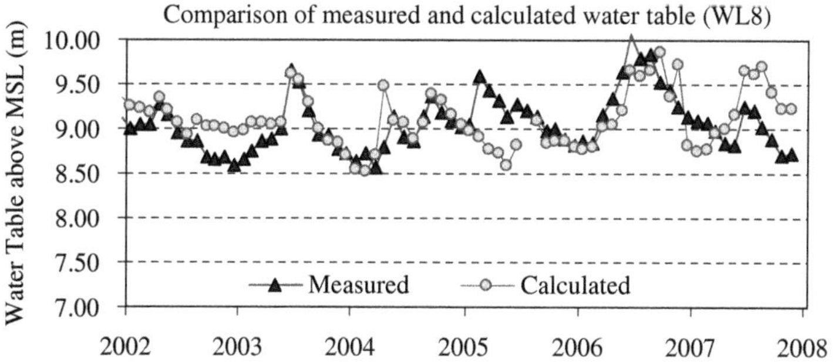

Fig. 4. Comparison of measured and calculated water tables at the observation well – WL8.

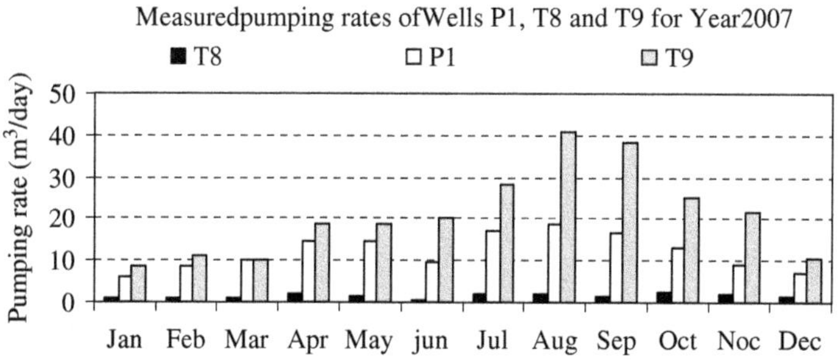

Fig. 5. Measured pumping rates for of pumping wells T8, P1 and T9 of the year 2007.

agreement with measured groundwater tables from the year 2002 to the end of year 2007.

Figure 5 shows the measured monthly pumping rates for the year 2007. During the summer period i.e. from July to September the pumping rates have increased compared to other seasons for the wells T8 and T9.

The variation of the electric conductivity of the three pumping wells is caused by the variable pumping rates and the seasonal recharge rates. In this model seasonal recharge rates were considered and time dependent boundary conditions were assigned for the six boundaries of the model as shown in the Table 1 to make the model is more realistic. Figures 6, 7 and 8 show the comparison of simulated electric conductivities of three pumping wells with the measured electric conductivities. According to the observed and calculated results, pumping well T8 shows a significant electric conductivity fluctuation then other two wells although its pumping rates are smaller than the other two wells. The reason for this is the screen depth

Table 1. Boundary conditions of the numerical model (boundaries are according to the Fig. 3).

Boundary	Pressure head	Concentration
ABCD	$h_{\text{ABCD}}(t) = H_{\text{ABCD}}(t) - y$	$\partial C/\partial y = 0.0$
EFGH	$h_{\text{EFGH}}(t) = H_{\text{EFGH}}(t) - y$	$\partial C/\partial y = 0.0$
ADEF	$h_{\text{ADEF}}(t) = H_{\text{ADEF}}(t) - y$	$C = 0.0\%$
BCHG	$h_p = (H_s - y) \cdot \frac{\rho}{\rho_f}$	$u > 0, \frac{\partial C}{\partial x} = 0$ $u < 0, C = 100\%$
ABGF	$-k_y \left[\frac{\partial h}{\partial y} + \frac{\rho}{\rho_f}\right] = 0$	$\partial C/\partial y = 0.0$
CDEH	$-k_y \left[\frac{\partial h}{\partial y} + \frac{\rho}{\rho_f}\right] = -Re(t)$	$\partial C/\partial y = 0.0$

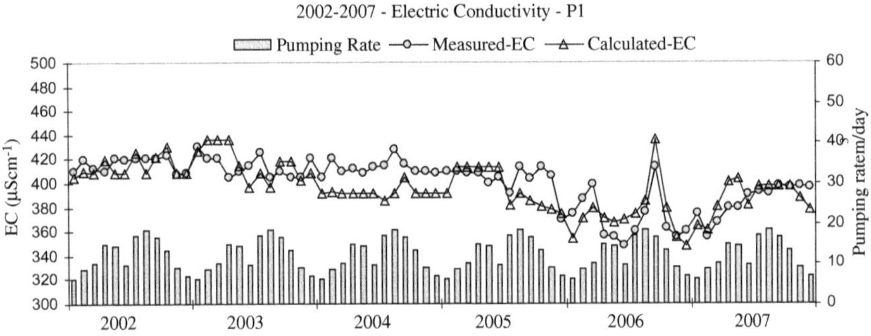

Fig. 6. Comparison of measured and calculated electric conductivity values of well P1.

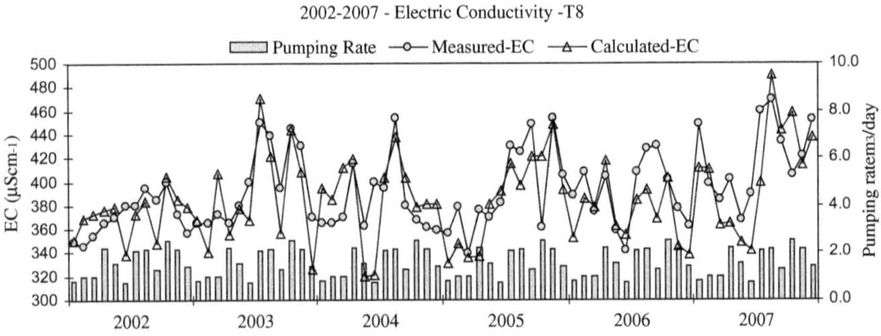

Fig. 7. Comparison of measured and calculated electric conductivity values of well T8.

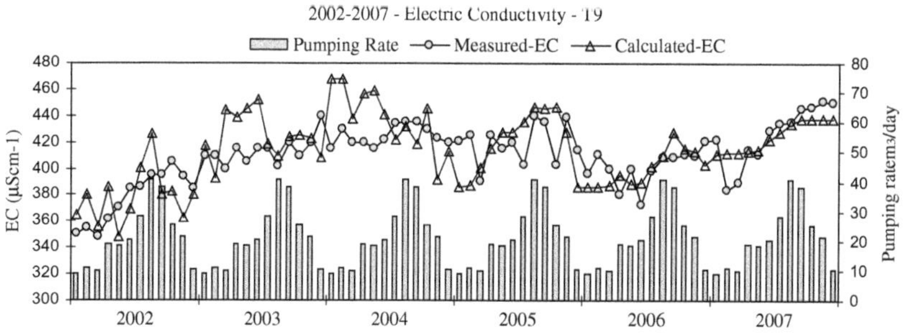

Fig. 8. Comparison of measured and calculated electric conductivity values of well T9.

of the well T8. Even if the pumping rate is low but the well screen is deep, the possibility of gaining relatively high salinity water is high. From year 2005 to 2007 reasonably good agreement of measured and calculated electric conductivities can be observed than from year 2002 to 2005.

4. Conclusion

Three dimensional numerical model sound more realistic than a two dimensional model since the geological and hydrological complexities of the coastal aquifers and the density dependent solute transport models are indispensable in modeling saltwater intrusion since the flow of two different densities are involved. Moreover, the sensitive issue of electric conductivity fluctuations observed in the pumped groundwater in the coastal aquifer of the Motooka has not been attracted much attention of saltwater intrusion modelers. The model developed in this study is capable of addressing the sensitive issue of electric conductivity fluctuation of pumped ground water for satisfactory extent. The numerical results and their comparison provide evidence for the credibility of the model for this particular simulation. The variation of electric conductivity of the pumped groundwater which is used for greenhouse farming and drinking is a critical issue in the Motooka area. In this numerical study it is able to address the problem and in future further more descriptive numerical studies are possible using the model developed in this study.

Acknowledgments

The authors are grateful to the Obayashi Foundation of Japan for their generosity to provide research funds to carry out this research successfully.

References

1. C. A. J. Appelo and D. Postma, Geochemistry, groundwater and pollution, 2nd Edition, A. A. Balkema publishers, Leiden (2007).
2. K. Jinno and T. Ueda, On the numerical solutions of convective dispersion equation by shifting particles, *Tranactions of JSCE.* **10** (1978) 126–129.
3. A. Tsutsumi, K. Jinno and R. Berndtsson, Surface and subsurface water balance estimation by the groundwater recharge model and a 3-D two-phase flow model, *Hydrological Sciences* **49**(2) (2004) 205–215.
4. C. Zheng and P. P. Wang, MT3DMS: A Modula three dimensional multi-species transport model for simulation of advection, dispersion, and chemical reactions of contaminants in groundwater systems; documentation and user's guide, U.S. Army crops of engineers. Engineer research and development center (1999), pp. 1–76.

Advances in Geosciences
Vol. 17: Hydrological Science (2008)
Eds. Namsik Park *et al.*

GEOCHEMISTRY OF ARSENIC IN THE HOLOCENE AQUIFER; SOUTH-WESTERN BANGLADESH*

MD. TAUHID-UR-RAHMAN†, AKIRA MANO‡ and KEIKO UDO§
Disaster Control Research Centre, School of Civil Engineering,
Tohoku University, Aoba 6-6-11-1110,
Sendai, 980-8579, Japan
†*liton2005@gmail.com*
‡*mano@civil.tohoku.ac.jp*
§*udo@potential1.civil.tohoku.ac.jp*

YOSHINOBU ISHIBASHI
Department of Civil and Environmental Engineering,
Tohoku Gakuin University, 1-13-1 Chuo,
Tagajo, Miyagi 985-8537, Japan
yishi@tjcc.tohoku-gakuin.ac.jp

This paper is targeting to explore the geochemical partitioning of Arsenic, As in the solid-water interface of the contaminated Holocene deltaic aquifer of the South-western part of Bangladesh. Observed relatively high HCO_3 and DOC in groundwater, GW may have significant role in leaching As from the brown clay and the grey sand in the sediment-water interface. Sequential extractions followed by spectroscopic investigation of the subsurface sediment reveal that Fe and Al minerals are the two principal adsorbent surfaces for As. Potential As source mineral iron-pyrite was detected with XRD and TEM image successfully captured the Fe_2O_3 contents in the clay sediments. Vivianite and siderite are the two minerals as suggested by the geochemical code PHREEQC that may control the GW Fe and As concentrations. Leaching of As from the upper clay layers and partial adsorption of that in the deeper sandy layers are the key transport mechanism. A parametric partition coefficient, K_d model ($R^2 = 0.69$) was derived from the solid-water interface parameters like GW-pH and the two other Gaussianised independent variables particularly logarithm of oxalic acid extractable Fe and Al contents of the contaminated aquifers. Such promising K_d model can easily be incorporated into the solute transport code for assessing the vulnerability of aquifers.

*This work is supported by the Grant in Aid for Scientific Research from the Japan Society for the Promotion of Science.

1. Introduction

High Arsenic contamination detected in shallow sandy Holocene aquifers in different parts of Bangladesh has been recognized as a severe environmental catastrophe, since the early 1990s.[1,2] Nearly about 45 million people[3–5] who are heavily dependent on in collecting drinking water from these groundwater sources, are increasingly exposed to this elevated As pollution. Sediment containing this As bearing minerals has been deposited at the lower part of the Bengal delta basin after being transported by the three mighty rivers, (Ganges, Brahmaputra and Meghna) from the eroded and weathered rock of the Himalaya mountain.

Subsurface As in groundwater in the South-Western Bangladesh has been identified to be trapped mainly within the shallow aquifer (depth $<$ 100 m) composed of finer to fine and fine to medium grained reduced grey sand, where most of the shallow hand tube wells have been installed.[1,4,5]

Depending on the pH, redox conditions and pattern of mineral dissolutions, distribution of As in sediment may be rearranged.[2] Arsenite (As III) and Arsenate (As IV) are the two principal As species that are mostly reported to dominate in the sediment as well as pore-water. In the oxidizing geo-environment, As (IV) ($H_2AsO_4^-$, $HAsO_4^{2-}$, AsO_4^{3-}) is the stable species whereas species like As (III) (H_3AsO_3) governs in the reduced environment over a wide pH range. Among all of the ever documented hypotheses for occurrence of high As in aquifer pore water, reductive dissolution of iron-oxy hydroxide as was postulated by Nickson *et al.*[3] is the well respected one. Leaching and desorption of As from its associated mineral surfaces such as iron, aluminum and manganese oxides under the influence of the aquifer complex geochemistry, largely take part in its transport from sediment to aquifer pore-water.[5] Adsorption has widely been considered as the retardation of As transport.[2]

The objective of this paper is to evaluate the dominant geochemical mechanisms of the Holocene sandy aquifers so as to focus particularly the partitioning of As in the sediment-water interface.

2. Materials and Methods

2.1. *Study site*

Samples of groundwater and subsurface sediments were collected from one of the Arsenic hot spots located at a village named as Koyla (22°51′ 03.1″ N, 89°03′ 50.9″ E, around 300 km from capital city Dhaka) under Kolaroa upazilla of Satkhira district of South-Western part of Bangladesh (Fig. 1).

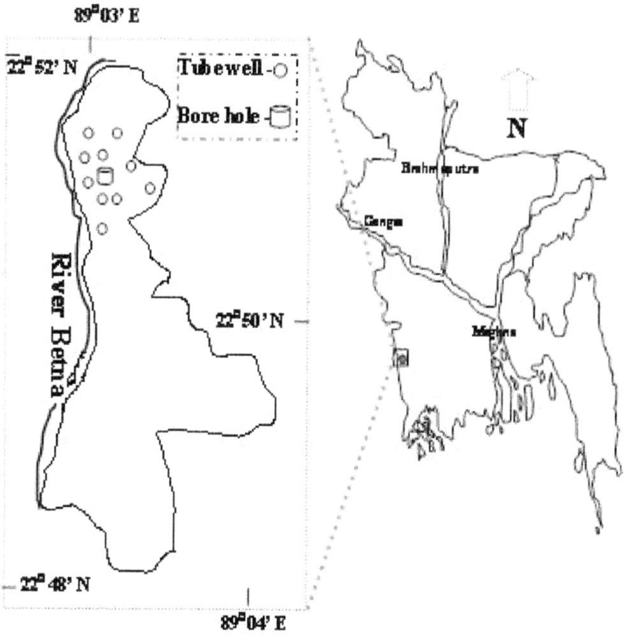

Fig. 1. Location of the study site Koyla, Kolarora, Satkhira, Bangladesh.

This site is situated on the river Betna and also not so far from a tributary, Kobodak of the Ganges river. The topographic surface of that area is very flat and the land elevation with respect to the mean sea level does not exceed even 5 m.[5]

According to the borehole investigations of the sub-surface geology, brown clay layers in the upper shallow part predominately were observed to have most of the detectable As. Just after this layer fine grey reduced sandy layers containing significant amount of As were found up to a depth of 40 m where most of the shallow hand tube wells have been screened.

2.2. *Sample collection*

Ten monitoring wells were installed using local drilling practice such as percussion method. This method includes a systematic and synchronized action of continuous raising and lowering of a steel pipe attached with a lever and having a diamond cutter at its mouth. Five sub-surface aquifer sediment samples at different depths (10, 50, 100, 150, 200 m) from each of the 10 bore holes were collected and then packed quickly in a sealed polythene bags to be transported to Japan. The sediment types (i.e. clay and sand) along with the associated As contents are illustrated in Fig. 2.

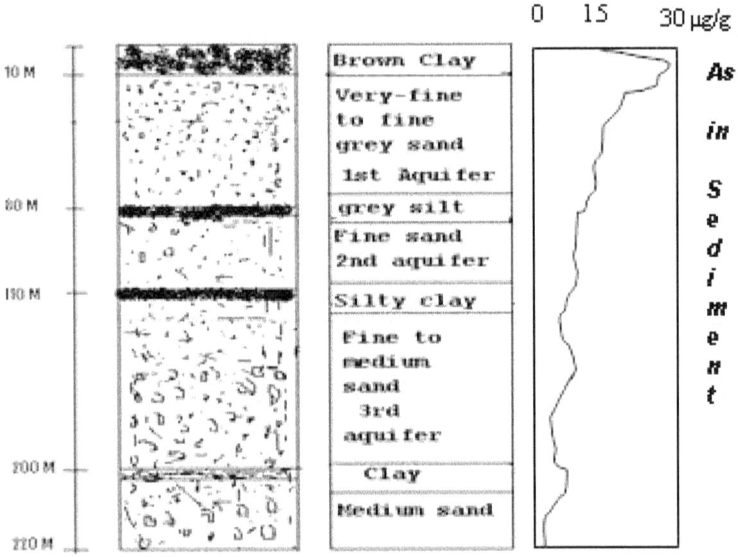

Fig. 2. Aquifer sediment and Arsenic distribution according to depth.

Groundwater (GW) was collected using peristaltic pumps that were maintained a nearly constant flow rate of 200 ml/min along with an attached multi probe flow channel accommodating pH, DO (dissolved oxygen), ORP (Eh, redox potential) and EC (electrical conductivity) meters' electrodes in the flow line. Water samples from each well were collected in three different ways for serving different purposes such as filtered (0.45 μm filter) for examining anions, unfiltered for dissolved organic carbon (DOC) analysis and filtered and then acidified samples (ultra pure HNO_3) for analyzing major cations. By using the anion exchange resin (Supelco), inorganic As species (III and V) were separated in the field.

During the field survey, each and every well was geo-referenced by utilizing the portable Global Positioning System (GPS) device. In addition to this, pH, DO, ORP, EC, temperature and approximate As concentration (with HACH field kit) of all of examined the water wells were noted down after measuring carefully with the portable instruments.

2.3. *Analysis methods*

2.3.1. *Groundwater analyses*

The groundwater was analyzed for identifying major anions (PO_4, SO_4, Cl, NH_4–N, NO_3–N and HCO_3), cations (Fe, Na, Ca, K, Al, Mn, Mg, Zn, Pb and Cr), inorganic As (total) and DOC. ICP-MS was utilized for identifying cations and inorganic As whereas anions like Cl, SO_4 were detected by using ion chromatography. HCO_3 was measured from alkalinity analysis after following the standard methods. PO_4, NH_4–N and NO_3–N were analyzed with TRAACS (BRAN-LUEBBE AACE 5.0). DOC was examined with TOC-5000 Shimadzu Total Organic Carbon analyzer.

2.3.2. *Sediment analysis*

2.3.2.1. Solid phase analysis

Sediment aliquots of subsurface clay and sand collected from different depths were analyzed for detecting As, Fe, Al and Ca by using energy dispersive X-ray fluorescence (EDXRF, JEOL-3200). Minerals of core samples were also examined by X-ray diffraction by using Cu Kα radiation coupled with a graphite monochrometer on a Philips vertical diffractometer ($0°$ to $100°$, 2Θ). Iterative mineral identification was accomplished by using JADE software. Elemental compositions of minerals were performed by employing scanning electron microscope (SEM) and transmission electron microscope (TEM).

2.3.2.2. Sequential extractions

Potential forms of As associated with minerals of the subsurface sediment grains (including both brown clay and grey sand) that were extracted from different depths, such as easily exchangeable As (by 1M $MgCl_2$), strongly adsorbed As (1M NaH_2PO_4), As coprecipitated with metal oxides (1M HCl followed by 0.2 M oxalic acid), coprecipitated As with crystalline Fe oxides (0.5 $TiCl_3$ with EDTA) and residue As (hot concentrated $HNO_3 + H_2O_2$) were determined after following Keon *et al.*[7] method that was also applied by other researcher such as Swartz *et al.*[6]

2.3.3. *Data analysis*

Speciation analysis and mineral saturation indices were performed by using the geochemical code PHREEQC where groundwater data were the main input. Sediment-water equilibrium distribution coefficient, K_d was computed from the ratio of adsorbed As that was found in sediment extraction with the As that was determined to be remaining as dissolved in the groundwater. Multiple regression analysis software SPSS-16.0 was employed to explore the parametric K_d model by selecting some dominating parameters of sediment and pore-water (Fe, Al and pH) as the Gaussianized independent variables.

3. Results and Discussions

3.1. *Geochemistry of sediment*

The brown clay (As, 28.4 μg/g) as well as the grey sand (As, 15.74 μg/g) of the upper shallow aquifer (Fig. 2) where most of the As was observed to be sorbed (either adsorbed or co-precipitated) might be considered as the potential source of As from where As can be leached gradually depending on the dominant geochemical reactions such as reductive dissolution of Fe-oxyhydroxides suggested by Nickson *et al.*[3] along with variations in pH and redox conditions. Some fractions of this As that may be leached by the likely interactions of the surface runoff infiltration through the fissures with the As rich clay layer, may possibly be re-adsorbed again onto the surfaces of the fine sand grains containing iron and other mineral hydroxides, before being mixed with the pore-water of sediment-water interface. Consequently, the fine and medium sand (As, 5.2 μg/g) found in successive layers may implicitly be considered as the sink of the leached As. Low As content (1–3 μg/g) was found in the sediments (i.e. black silty clay and medium sand) of the deeper part.

Solid phase such as bulk mineralogical analysis by XRD, elemental analysis by XRF, elemental mapping by SEM and TEM could further reflect the wide-spread presence of the adsorbent minerals like iron oxy-hydroxide minerals.[9] Roughly, 0.18% of the total mineral compositions containing Fe_2O_3in the brown clay was detected by XRD. Other minerals like quartz, K-feldspar, muscovite, microcline, mica and pyrite were also identified by XRD data (Table 1). Pyrite that was detected by XRD in both brown clay and grey sand is the most important minerals for being the source minerals of As.[5,8] However, As that may have relatively smallest share in the mineral compositions was not detected by XRD so clearly.

Noteworthy presences of Fe and Al in brown clay and in grey sand that may form compounds with other elements were evidently recognized by the XRF investigation (Table 2).

This major element like Fe and Al may form potential mineral surfaces that might be considered as the adsorbent of the As as well as other trace

Table 1. Quantitative mineral analysis by XRD of the sediment sample.

Wt (%)	Quartz	K-Feldspar	Micro-cline	Mus-covite	Mica	Chlorite	Fe_2O_3	Kaolinite	Pyrite
clay	53	20	11.8	7.3	4.29	2.78	0.18	0.53	0.12
sand	68	14	10.2	4.8	2.1	0.45	—	0.38	0.07

Table 2. Solid phase elemental analysis by XRF.

Wt (%)	Fe	Al	Si	Ca	Mn	K	Mg
Clay	34.21	16.20	37.2	7.12	1.2	2.5	1.57
Sand	18.2	14.3	63.42	2.12	0.45	0.89	0.62

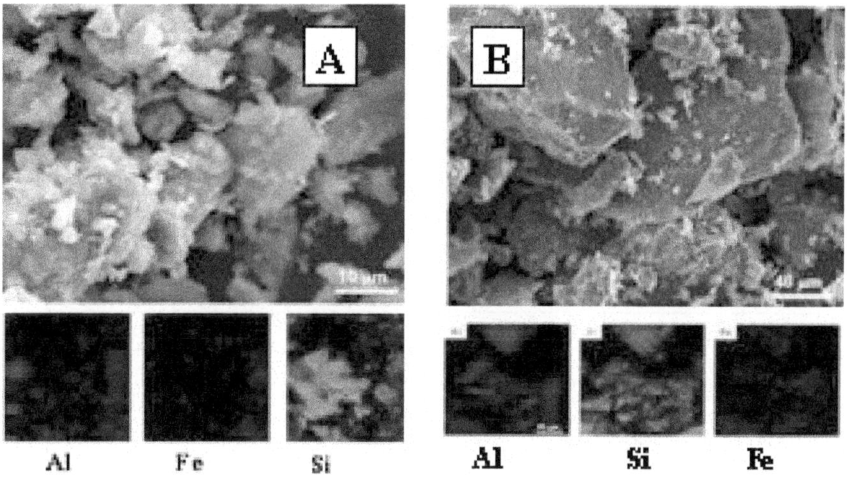

Fig. 3. Elemental map drawn by the SEM (A) brown clay (B) grey sand.

elements. Adsorbent surfaces can implicitly be visualized by the elemental composition map drawn by the SEM-EDS. The elemental map of a typical brown clay sample that was investigated with SEM shows the abundance of the Al, Fe and Si (Fig. 3). This investigation is quite consistent with the XRD, XRF and TEM (Fig. 4 right part) observations and is also nearly analogous with the findings of Swartz *et al.*[6]

The very irregular mineral surfaces of the sediment grains that may suitably offer other anions or cations to be adsorbed can clearly be recognized with this SEM image.

Observations of the same clay sample in a very finer scale (500 nm) by TEM, may help to identify the potential Fe-oxyhydroxide surfaces attached on a sediment grain (Fig. 4). Moreover, abundances of wide spread oxides that may have formed oxide minerals after binding with other elements such as Si, Fe and Al can easily be recognized from the right part of the TEM image (Fig. 4).

Data from chemical extractions (Table 3) show that relatively higher amount of elements like As (28.74 μg/g), Fe (59.5 mg/l), Ca (49.3 mg/g),

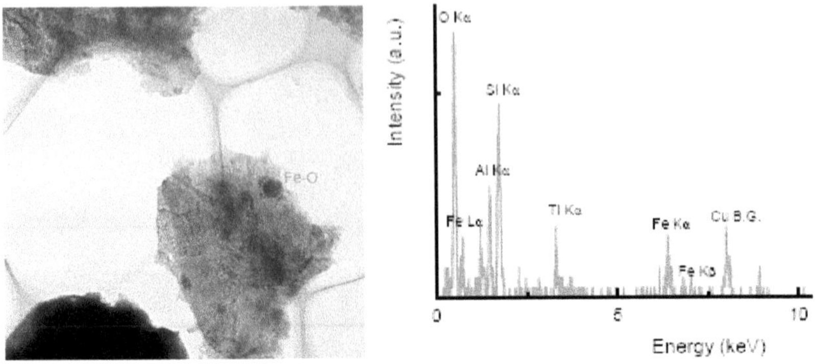

Fig. 4. Fe Oxide found in a brown clay sample by TEM observation.

Table 3. Average (avg) and Maximum (max) chemical composition of Clay (C) and Sand (S).

	As (μg/g)	Fe (mg/g)	Ca (mg/g)	Al (mg/g)	Mn (mg/g)	Na (mg/g)	K (mg/g)	Mg (mg/g)
C-avg	2.54	10.9	18.9	6.9	0.23	0.095	1.1	4.8
C-max	28.74	59.5	49.3	37.2	1.6	0.26	4.0	11.2
S-avg	1.02	3.47	6.14	1.23	0.11	0.13	0.41	1.03
S-max	15.74	26.1	30.12	14.25	0.87	0.31	1.12	2.71

Mn (1.6 mg/g) and Al (37.2 mg/l) may be observed in brown clay sampled from beneath the top soil surface, whereas lowest of them (As, 0.48 μg/g; Fe, 3.34; Ca, 1.12; Mn, 0.07; Al, 1.5 mg/g) may be observed in grey sand collected from the deeper part of the aquifers (depth, 207 m). This can be comparable with other As affected areas in Bangladesh.[1]

Selected sequential extractions of a clay samples presents again the various forms of dominant mineral phases (Table 4). Although, very insignificant amount of easily exchangeable As and Fe was found in this extraction procedure, but considerable As (13.7 μg/g) that was determined during the HCl and oxalic acid phase, may at least indicate its strongly adsorbed phase with the other dominant oxide minerals such as Fe and

Table 4. Sequential extraction result of a brown clay sample.

Target elements	$MgCl_2$	NaH_2PO_4	HCl	Oxalic	$TiCl_3$	HNO_3	$HNO_3+H_2O_2$
As (μg/g)	0.002	0.004	10.3	3.4	2.42	6.62	13.4
Fe (mg/g)	0.09	0.53	8.2	2.8	1.12	7.8	44.2
Al	5.13	0.35	5.72	3.4	0.64	10.2	20.41

Table 5. Value of major elements obtained during aqueous analysis.

	As (μg/l)	Fe (mg/l)	Ca (mg/l)	Al (mg/l)	Mn (μg/l)	Na (mg/l)	K (mg/l)	Mg (mg/l)
Minimum	1.05	0.85	19.21	0.04	4.23	3.82	2.11	2.31
Average	68.64	5.71	89.82	0.23	66.56	50.14	3.77	15.24
Maximum	211.14	11.24	122.3	0.56	81.23	145.21	6.51	44.92

Table 6. Value of major anions, DOC and ORP in GW obtained during aqueous analysis and field investigation respectively.

	HCO_3	NH_4–N	NO_3–N	PO_4	SO_4	DOC	pH	ORP
Minimum	187	0.31	0.12	0.01	0.11	1.09	6.15	−11
Average	345.2	2.54	1.28	0.05	1.27	5.14	7.02	25
Maximum	489.7	5.32	3.87	0.87	3.14	17.2	7.68	37

Al oxides. This finding is very much consistent with the solid phase spectroscopic analysis.

3.2. *Groundwater geochemistry*

As in a higher concentration (211.1 μg/l) may be found in the relatively shallow depth (> 80 m), whereas smaller amount of that (1.05 μg/l) may be observed in the deeper part (< 210 m) (Table 5). This unusual high As has also been found as coupled with relatively high Fe concentrations (10.2 mg/l) in ground water which has further been reported by other researchers such as Stollenwork *et al.*[2] The very low value of ORP (−11 to 37 mV) observed in the groundwater resembles with the characteristics of a reductive geo-environment. Other criteria like relatively higher concentration of PO_4, DOC, HCO_3 and NH_4–N and lower concentration of SO_4 are also harmonized consistently with that of reduced geo-environment (Table 6). More over, widespread presence of dissolved Fe and Ca may reasonably reflect the occurrence of the likely reductive dissolution of Fe and Ca minerals that would have taken place in the reductive aquifer that also was found in other As affected places.[7,9] Consequently, As that was adsorbed on that mineral surfaces may get released into the nearby pore-water. Degradation of organic matter may also contribute in this As releasing process. High HCO_3 and DOC may usually indicate the occurrence of the likely organic matter degradation.[5,9]

Saturation indices computed with PHREEQC[10] illustrate that siderite and vivianite which already have reached at the supersaturated phase may

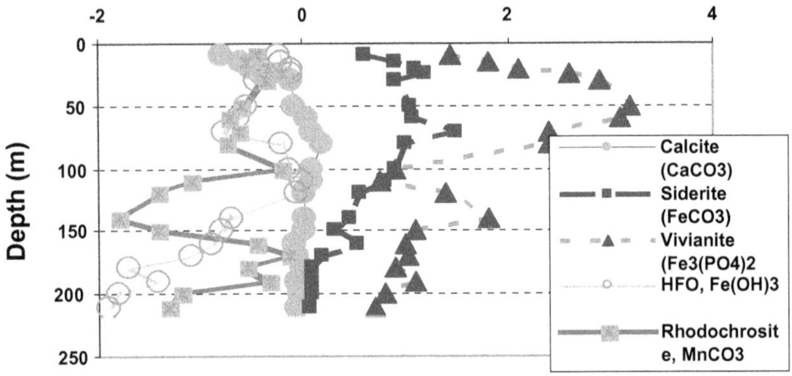

Fig. 5. Saturation Index computed with PHREEQC.

possibly control the dissolved Fe concentrations (Fig. 5). This also could control the As mobilization since it might have co-precipitated with the iron minerals.

Moreover, considerable adsorption of As onto Fe mineral surfaces should essentially be considered as its dominant natural attenuation mechanism in the groundwater aquifers.

3.3. *Parametric K_d model*

The partitioning of As in the aquifer solid-water interface can best be explained with the distribution coefficient, K_d (a ratio of solute adsorbed in sediment to that of dissolved in groundwater). Due to simplistic in nature, K_d has long been well appreciated as well as applied by the geochemical modelers. Apart from simple K_d model, parametric K_d model[11] offers a robust perception that may be predicted from interactively varying parameters taking part in solid-water interface such as groundwater pH and dominant adsorbent minerals like Fe and Al of the sediment matrix. In this study, K_d computed with the sediment and GW data was found to vary within 0.03 to 15.85 L/kg. Lower K_d was noticed in the upper layer where desorption is the dominant mechanism whereas high K_d was found in the following layers where adsorption is supposed to be the key to control the groundwater As transport.[12] However, pore water pH was found to have an adverse influence on K_d (Fig. 6(a)). Again, in high pH, As desorption is promoted whereas in low pH adsorption usually dominates. On the other

$LogK_d = 0.58LogFe + 0.19LogAl - 0.49pH + 3.66$, ($R^2 = 0.69$, $p < 0.05$ ...... (1))

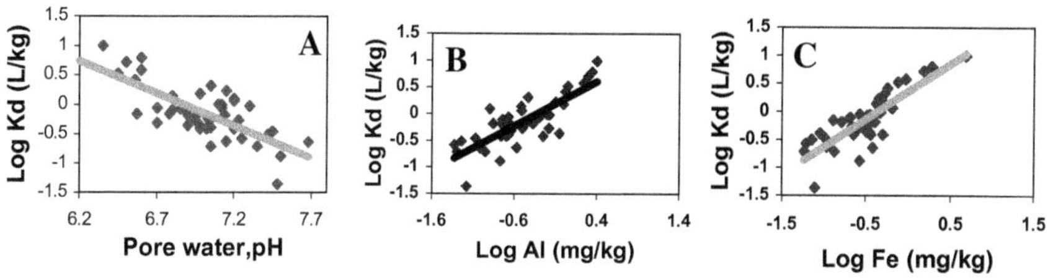

Fig. 6. Log K_d variation with (a) pore water pH, (b) oxalic acid extractable aluminum content in sediment, Log Al and (c) Iron content in sediment, Log Fe

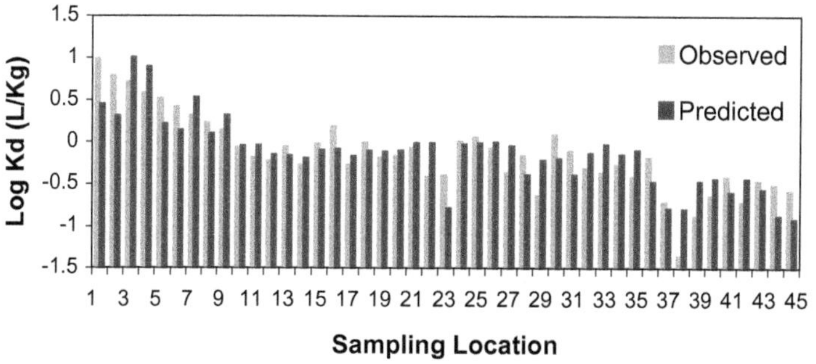

Fig. 7. Comparison of observ ed and predicted Log K_d.

hand K_d was observed to be influenced positively with Fe and Al, this is due to having the adsorptive capacity of these minerals (Figs. 6-b and 6-c). This finding resembles with that of spectroscopic evidence and chemical extractions of sediment matrix. Data of the oxalic acid extractable Fe and Al contents from different sediment samples were treated with Gaussian distribution first. Then, further treatment like Fe/(1-Fe) and Al/(1-Al) were done and logarithm of these two along with pH were taken as the Gaussianised independent variables in the multiple regression analysis. This resulted in a parametric K_d model that can explain 69% variation considering groundwater pH and logarithm of the other two sediment parameters particularly oxalic acid extractable Fe and Al. Such type of K_d model has also been derived by Brouwere *et al.*[13]

$$Log\,K_d = 0.58\,Log\,Fe + 0.19\,Log\,Al - 0.49\,pH + 3.66,$$
$$(\mathrm{R}^2 = 0.69, \mathrm{p} < 0.05 \ldots\ldots (1))$$

The equation (1) was applied for another data set that has recently been explored from another As contaminated site by the authors. The predicted K_d can nearly be comparable with that of the observed one (Fig. 7). This equation may also be incorporated with solute transport code for assessing As contamination in the successive aquifers.

4. Conclusion

High As in the reduced shallow sandy aquifer may primarily come from the leaching of the source layer (brown clay and reduced grey sand) due to the potential dominant geochemical reactions like reductive dissolution of the As bearing minerals like Iron pyrite. High HCO_3 and DOC observed in groundwater in the reduced zone may further promote As mobilization by inhibiting its adsorption in the successive layers. On the other hand, lower aquifers composing of Fe and Al minerals may take a role by offering themselves as sinks for the dissolved As, which is also supported by the spectroscopic study. This can further be supported strongly by the computed high K_d values for the lower layers. Parametric K_d model, that has promising predictive capability derived in this study can easily be applied for assessment of As transport from contaminated sediment matrix to the uncontaminated one when that will be incorporated with the solute transport code.

Acknowledgments

The financial support by the Grant in Aid for Scientific Research from the Japan Society for the Promotion of Science (B-18404009) is sincerely acknowledged. Priceless support from Mr. Takahashi regarding ICP-MS analysis and Mr. Miyazaki for SEM and TEM analysis are also greatly acknowledged. Three anonymous reviewers and the volume editor are thanked for their great suggestions in improving the quality of this manuscript.

References

1. BGS and DPHE, *BGS Technical Report.* **WC/00/19** (2001).
2. K. G. Stollenwork, G. N. Breit, A. H. Welch, J. C. Yount, J. W. Whitney, A. L. Foster, M. N. Uddin, R. K. Majumder and N. Ahmed, *Sci. Tot. Env.* **379** (2007) 133–150.

3. R. Nickson, J. McArthur, W. Burgess, K. M. Ahmed, P. Ravenscroft and M. Rahman, *Nature* **395** (1998) 338.
4. M. T. Rahman, A. Mano, K. Udo and Y. Ishibashi, *Int. Conf. Proc. 2nd Env. Management, Zagreb, Croatia.* (2007) 197–205.
5. M. T. Rahman, A. Mano, K. Udo and Y. Ishibashi, *Int. Conf. Proc, JSCE, Tokyo, Japan*, 2008, pp. 229–232.
6. C. H. Swartz, N. K. Blute, B. Badruzzaman, A. Ali, D. Brabander, J. Jay, J. Besancon, S. Islam, Niedan, H. F. Hemond and C. F. Harvey, *Geochemica et Cosmochemica Acta* **68**(22) (2004) pp. 4539–4557.
7. N. E. Keon, C. H. Swartz, D. J. Brabander, C. Harvey and H. F. Hemond, *Environ. Sci. Technol.* **35** (2001) 2778–2784.
8. A. H. Welch, D. B. Westjohn, D. R. Helsel and R. B. Wanty, *Ground Water* **38** (2000) 589–604.
9. A. H. Welch and K. G. Stollenwerk, *Arsenic in Groundwater: Geochemistry and Occurrence* (Kluwer Academic Publishers, Boston, 2003).
10. D. L. Parkhurst and C. A. J. Appelo, *User's Guide to PHREEQC (Version 2), USGS*, 1999.
11. United States Environmental Protection Agency, *EPA-**402-R-04-002C***, 2004.
12. M. T. Rahman, A. Mano, K. Udo and Y. Ishibashi, Book of Abstract, Annual GSA meeting, Houston, Texas, 2008.
13. K. D. Brouwere, E. Smolders and R. Merckx, *European J. of Soil Sci.* **55** (2004) 165–173.

Advances in Geosciences
Vol. 17: Hydrological Science (2008)
Eds. Namsik Park *et al.*

AN EXTENDED DYNAMICAL MODEL OF A GEYSER INDUCED BY INFLOW OF GAS: CONSIDERING EFFECTS OF A COMPLICATED UNDERGROUND WATERCOURSE

HIROYUKI KAGAMI
Department of Preschool Education,
Nagoya College, 48 Takeji, Sakae-cho,
Toyoake, Aichi, 470-1193, Japan
kagami@nagoyacollege.ac.jp

We have proposed a mathematical model (a static model), a dynamical model and a modified dynamical model of a geyser induced by inflow of gas (a periodic bubbling spring) based on observation and model experiments of the geyser and have also proposed a combined model combining above 2 models and improved them. But effects of a complicated underground watercourse during spouting have not added to above models yet. In this paper, we derive an extended dynamical model to which effects of a complicated underground watercourse during spouting are added. Concretely, effects of underground watercourse's sudden expansion and sudden contraction are added. Then we discuss the effects of underground watercourse's sudden expansion and sudden contraction through numerical simulation of the extended dynamical model.

1. Introduction

Geysers are classified into two types dependent on inducer. One is a geyser induced by boiling and the other is a geyser induced by inflow of gas (a periodic bubbling spring). The latter is a target of a series of our studies. There are many geysers induced by boiling and some theories about its mechanism have been proposed.[1] On the other hand, there are a few geysers induced by inflow of gas (a periodic bubbling spring) and a few theories about its mechanism have been proposed.[2] Iwasaki (1962)[2] did model experiments of a geyser induced by inflow of gas and showed that injection of higher pressure gas spouted water from a spouting spout intermittently. Then he calculated spouting time and pause time using the gas supply rate as a parameter based on a mathematical model of gas balance. But this model had not discussed spouting dynamics of a geyser induced by inflow of gas.

We have proposed a mathematical model (a static model),[3,4] a dynamical model[4,5] and a modified dynamical model of a geyser (a periodic bubbling spring)[4,6,7] based on observation of Hirogawara Geyser (Yamagata, Japan)[8] and model experiments of the geyser[9] and have also proposed a combined model combining above 2 models.[4] And numerical simulations of the modified dynamical model or the combined model has reappeared dynamics of spouting of geysers (periodic bubbling springs) and it becomes possible that parameters (volume of the underground space, depth of spouting hole and so on) under a geyser are estimated due to comparison between results of simulation and those of observation.[4,7] Moreover we have verified above models through geological exploration, analysis of hot spring water and radioactive prospecting.[10] And last year we reported the modified combined model considering evaporation effect of gas dissolved in hot spring water.[11]

In this study, we add effects of a complicated underground watercourse during spouting to the combined model so as to re-create more practical spouting of a geyser induced by inflow of gas (a periodic bubbling spring) and estimate parameters as stated above more accurately. In the case of the former model, it was assumed that the underground watercourse was vertically straight for simplicity. Practically the measurable underground watercourses of geysers induced by inflow of gas are almost straight because of past remains of boring and so on. But it is expected that the underground watercourse has a complicated shape at deep underground region. And for example a curved shape, a taper one or other one give a kind of resistance. In this study, we add resistances of sudden expansion and sudden contraction to the former model as an example and discuss effects of them quantitatively through numerical simulation of the extended model.

2. Model

In this chapter, at first we introduce main fruits of the former dynamical model, the mathematical model (a static model) and the combined model of a geyser induced by inflow of gas (a periodic bubbling spring). Then we introduce derivation of an extended dynamical model to which above-mentioned effects are added. Then we introduce some results of numerical simulation of the modified combined model and minutely discuss the effects of above-mentioned resistances of sudden expansion and sudden contraction.

2.1. *The former model*

Detailed derivation of the former dynamical model, the mathematical model (a static model) and the combined model of a geyser induced by inflow of gas (a periodic bubbling spring) was already stated in Refs. 4 and 7. So we show only main fruits of above models here.

At first, in case friction between walls of a spouting pipe and water packed in it is not taken into account, an evolution equation of temporal variations of height of top of a water pole packed in the spouting pipe of a periodic bubbling spring is written as;

$$(n_0 + \beta t)(V_0 + Sx)\rho H \frac{d^3x}{dt^3} + (n_0 + \beta t)pS\frac{dx}{dt} = (V_0 + Sx)p\beta \qquad (1)$$

where n_0 represents molar number of gas in a underground space just before the water pole's beginning to move up to the upper entrance of the spouting pipe, β is constant concerning gas supply rate, V_0 represents volume of gas packed in a underground cave, S represents an area of a cross section of the spouting pipe, H represents length (height) of a lump of water packed in the pipe, p represents pressure of gas packed in the underground cave, x is regarded as a position of the lower interface between water and gas of the water pole and an upper direction of a vertical line is regarded as a plus direction of x-axis.

In case friction between walls of a spouting pipe and water packed in it is taken into account, Eq. (1) is changed to following Eq. (2).

$$\begin{aligned}(n_0 + \beta t)(V_0 + Sx)\rho H \frac{d^3x}{dt^3} \\ + \frac{8\pi\eta H}{S}(n_0 + \beta t)(V_0 + Sx)\frac{d^2x}{dt^2} + (n_0 + \beta t)pS\frac{dx}{dt} \\ = (V_0 + Sx)p\beta \qquad (2)\end{aligned}$$

where η represents viscosity coefficient. As a result, second term of Eq. (2) is added and it represents effects of viscosity.

A sample of numerical simulation of Eq. (2) is shown in Fig. 1. A graph of numerical simulation of Eq. (2) is drawn in comparison with graphs of observational results in Fig. 1. We are able to guess underground parameters which we cannot measure easily because of geological difficulties through

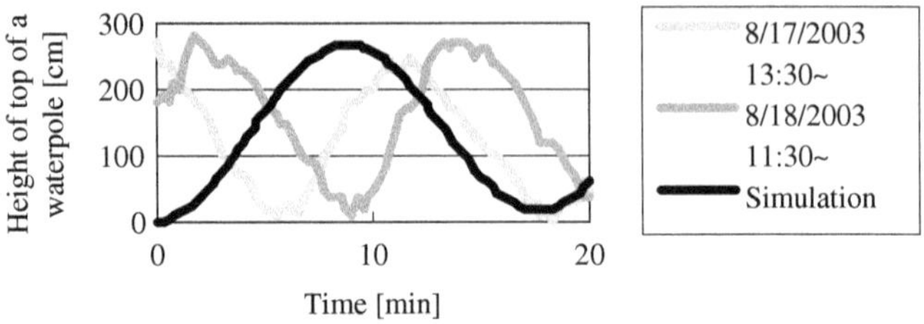

Fig. 1. A graph of numerical simulation of Eq. (2) comparison with graphs of observation of a geyser induced by inflow of gas (Kibedani geyser (Yamagata, Japan)).

comparing results of numerical simulation with those of observation. Concretely we try to fit data of numerical simulation to those of observation changing values of parameters. When both data fit most, we can guess chosen values of parameters are ones of real underground parameters.

Using the mathematical model (a static model), a spouting period τ and so on are represented as functions of various parameters. For example, a spouting period τ can be written as;

$$\tau = \frac{V_0}{\alpha\beta} + \frac{f_k S}{\rho g \alpha\beta}(f_k + P_0 + \rho g H) \tag{3}$$

where $\alpha = RT$, R represents gas constant, T represents temperature, f_k represents pressure due to a surface tension on an interface between water packed in the spouting pipe and gas in the underground cave, P_0 represents pressure of the atmosphere and g represents gravity acceleration. For example, when we observe a spouting period of a geyser induced by inflow of gas (a periodic bubbling spring), we can evaluate underground parameters (volume of the underground space (V_0), depth of spouting hole(H) and so on) through Eq. (3).

And the combined model is one combining above the former dynamical model and the mathematical model (a static model). The combined model enables more reliable estimation of underground parameters of a geyser induced by inflow of gas. Concretely, the dynamical model or the mathematical model (a static model) can determine (underground) parameters of a geyser induced by inflow of gas independently as shown above. On the other hand, some sets of parameters enable resemble spouting dynamics or period in each model. That is, we cannot decide most correct parameters' set using only one model. But using independent 2 models, that is, the dynamical model and the mathematical model (a static model), we can decide more correct parameters' set. These are essential qualities of the combined model.

2.2. *An extended dynamical model*

In this section, we explain introduction of a complicated underground watercourse during spouting to the combined model. Here we introduce effects of resistances of sudden expansion and sudden contraction as an example of a complicated underground watercourse. For it is thought that sudden expansion and sudden contraction can be frequently seen in underground watercourses. By the way, effects of other resistances can also be dealt with similarly.

Sudden expansion and sudden contraction are shown in Figs. 2 and 3, respectively. D_1 or D_2 represents a diameter of a wide part of a pipe or one of a narrow part of it respectively in each figure. And V_1 or V_2 represents

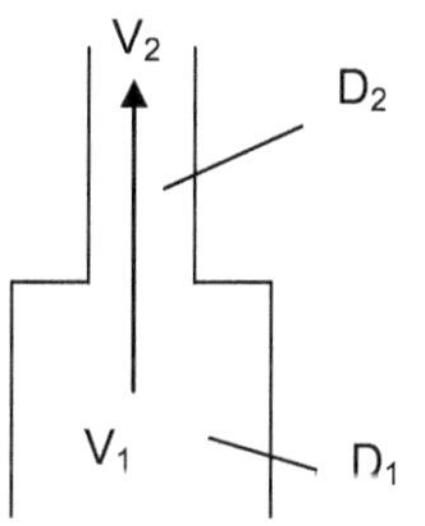

Fig. 2. Illustration of sudden contraction.

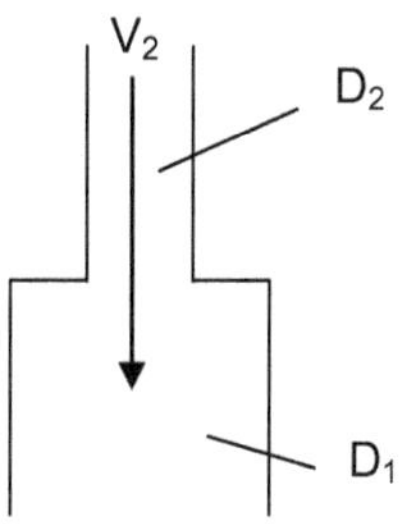

Fig. 3. Illustration of sudden expansion.

velocity of a flow in a wide part of the pipe or one in a narrow part of it then respectively in each figure. A shape of sudden expansion is the same as one of sudden contraction as shown in both figures. But though in the case of the former water flows from a narrow part of the pipe to a wide one, in the case of the latter water flows from a wide part of the pipe to a narrow one, reversely. For example, if hot spring water passes a sudden contraction region when it flows upward, the region replaces a sudden expansion region when it flows downward.

Next we discuss loss water head of these singular shapes. Loss water head of sudden expansion h_{se} and that of sudden contraction h_{sc} are defined using loss coefficient of sudden expansion f_{se} and that of sudden contraction f_{sc} respectively as;

$$h_{se} = f_{se}\frac{V_1^2}{2g} \tag{4}$$

$$h_{sc} = f_{sc}\frac{V_2^2}{2g} = f_{sc}\left(\frac{D_1}{D_2}\right)^4\frac{V_1^2}{2g} \tag{5}$$

where g represents gravity acceleration.

Now we assume the direction from a wide part of the pipe to a narrow one, that is, the direction of an up-pointing arrow in Fig. 2 coincides with the vertically upward direction. A spouting mode begins when a surface tension on an interface between water packed in the spouting pipe and gas in the underground cave (and weight of a lump of water packed in the spouting pipe and pressure of the atmosphere) becomes smaller than pressure of gas in the underground cave.[7] Then an equation of motion of the lump of water is written as[7];

$$\rho SH\frac{d^2x}{dt^2} = pS - \rho gSH - p_0S \tag{6}$$

where x is regarded as a position of the lower interface between water and gas of the water pole and an upper direction of a vertical line is regarded as a plus direction of x-axis. Here adding effects of sudden contraction and sudden expansion to the former model, that is, Eq. (6), we get the following equations.

(1) In case of $\frac{dx}{dt} \geq 0$

$$\rho SH\frac{d^2x}{dt^2} = pS - \rho gSH - p_0S - h_{sc}. \tag{7}$$

(2) In case of $\frac{dx}{dt} \le 0$

$$\rho SH\frac{d^2x}{dt^2} = pS - \rho gSH - p_0S - h_{se}. \tag{8}$$

Finally, we can get the following equations instead of Eq. (1).
(1) In case of $\frac{dx}{dt} \ge 0$

$$\begin{aligned}(n_0 + \beta t)(V_0 + Sx)\rho H\frac{d^3x}{dt^3}\\ + (n_0 + \beta t)(V_0 + Sx)f_{sc}\frac{S}{g}\frac{dx}{dt}\frac{d^2x}{dt^2} + (n_0 + \beta t)pS\frac{dx}{dt}\\ = (V_0 + Sx)p\beta.\end{aligned} \tag{9}$$

(2) In case of $\frac{dx}{dt} \le 0$

$$\begin{aligned}(n_0 + \beta t)(V_0 + Sx)\rho H\frac{d^3x}{dt^3}\\ + (n_0 + \beta t)(V_0 + Sx)f_{se}\left(\frac{D_2}{D_1}\right)^4\frac{S}{g}\frac{dx}{dt}\frac{d^2x}{dt^2} + (n_0 + \beta t)pS\frac{dx}{dt}\\ = (V_0 + Sx)p\beta.\end{aligned} \tag{10}$$

Above Eqs. (9) and (10) are basic equations adding effects of sudden contraction and sudden expansion.

3. Results and Discussion

In this chapter, we show some results of numerical simulation of the above extended dynamical model and discuss them.

Here temporal variation of a top of a water pole depended on the ratio of $D_2 : D_1$, that is, D_2/D_1 is shown in Fig. 4. And corresponding values of f_{sc} and f_{se} in case of each $D_2 : D_1$ are shown in Table 1. At first, we can see that the larger the ratio of $D_2 : D_1$ is, the shorter period of height's oscillation is. And we can also see that the larger the ratio of $D_2 : D_1$ is, the larger amplitude of height's oscillation is. The reason why above 2 tendencies occur is common. That is, because the smaller the ratio of $D_2 : D_1$ is, the larger values of f_{sc} and f_{se} are, consequently effects of resistances are larger and hinder water flow more in the case of larger D_2/D_1.

But, we can also see that though we can see dependence of change of height's oscillation on the ratio of $D_2 : D_1$, it is not large. From these results,

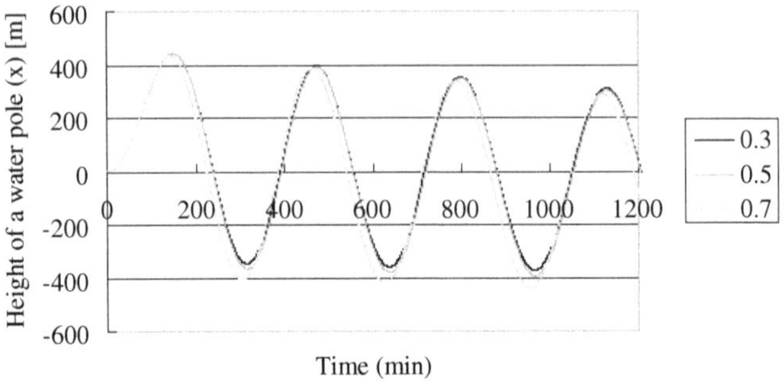

Fig. 4. Temporal variation of a top of a water pole depended on the ratio of $D_2 : D_1$.

Table 1. Corresponding values of f_x and f_{se} in case of each $D_2 : D_1$.

$D_2 : D_1$	f_{sc}	f_{se}
0.7	0.29	0.26
0.5	0.43	0.56
0.3	0.49	0.82

it may be thought that effects of complicated shape of the underground watercourse are not very large.

4. Conclusions

We derived extended dynamical model equations considering effects of underground watercourse's sudden expansion and sudden contraction. Through numerical simulation of the extended dynamical model equations, we discussed effects of complicated shape of the underground watercourse.

As a result, we see that the smaller the ratio of a diameter of a wide part of a pipe to one of a narrow part of it is, the longer a spouting period is and the smaller amplitude of height's oscillation is. For the smaller the ratio of a diameter of a wide part of a pipe to one of a narrow part of it, the larger head loss is.

After this, we will study effects of underground watercourse's shapes except sudden expansion and sudden contraction and confirm the tendency of this study's results.

References

1. K. Honda and T. Terada, *Publ. Earthq. Inv. Com.* **22B** (1906).
2. I. Iwasaki, Bull. Tokyo Institute of Technology, No. 46, 1962.
3. H. Kagami, Abstracts of the 53rd Annual Meeting of the Balneological Society of Japan, **27** (2000).
4. H. Kagami, Advances in Geosciences Vol. 4 : Hydrological Science (HS) **191** (2006).
5. H. Kagami, Abstracts of The 55th Annual Meeting of the Balneological Society of Japan **33** (2002).
6. H. Kagami, The 2003 IUGG General Assembly, HW04/09P/C31-004, 2003.
7. H. Kagami, *Proceedings of The 38th Conference of Sciete Internationale des Techniques Hydrothermales and The 56th Annual Meeting of the Balneological Society of Japan*, 2003, pp. 55–60.
8. E. Ishii, *et al.*, Abstracts of The 52th Annual Meeting of the Balneological Society of Japan **28** (1999).
9. M. Katase, *et al.*, Abstracts of a meeting for presenting research papers of Kanto Gakuin University College of Engineering 1999, pp. 99–100.
10. H. Kagami, Advances in Geosciences Vol. 6 : Hydrological Science (HS) **203** (2007).
11. H. Kagami, Advances in Geosciences 2007 : Hydrological Sciences (in press).

Advances in Geosciences
Vol. 17: Hydrological Science (2008)
Eds. Namsik Park *et al.*

AN EXTENDED DYNAMICAL MODEL OF A GEYSER INDUCED BY INFLOW OF GAS IN CASE OF PLURAL UNDERGROUND GAS SUPPLY SOURCES

HIROYUKI KAGAMI
Department of Preschool Education,
Nagoya College, 48 Takeji, Sakae-cho,
Toyoake, Aichi, 470-1193, Japan
kagami@nagoyacollege.ac.jp

We have proposed a mathematical model (a static model), a dynamical model and a modified dynamical model of a geyser induced by inflow of gas (a periodic bubbling spring) based on observation and model experiments of the geyser and have also proposed a combined model combining above 2 models and improved them. As a result, spouting mechanism of a geyser induced by inflow of gas spouting regularly has been clarified. But, concerning a geyser induced by inflow of gas, there are not only one spouting regularly but also one spouting irregularly. In this paper, we propose an extended dynamical model which assumes plural underground gas supply sources so as to clarify mechanism of irregular spouting geysers induced by inflow of gas. Then we discuss the effects of plural underground gas supply sources through numerical simulation of the extended dynamical model.

1. Introduction

Geysers are classified into two types dependent on inducer. One is a geyser induced by boiling and the other is a geyser induced by inflow of gas (a periodic bubbling spring). The latter is a target of a series of our studies. There are many geysers induced by boiling and some theories about its mechanism have been proposed.[1] On the other hand, there are a few geysers induced by inflow of gas (a periodic bubbling spring) and a few theories about its mechanism have been proposed.[2] Iwasaki (1962)[2] did model experiments of a geyser induced by inflow of gas and showed that injection of higher pressure gas spouted water from a spouting spout intermittently. Then he calculated spouting time and pause time using the gas supply rate as a parameter based on a mathematical model of gas balance. But this model had not discussed spouting dynamics of a geyser induced by inflow of gas.

We have proposed a mathematical model (a static model),[3,4] a dynamical model[4,5] and a modified dynamical model of a geyser induced by inflow of gas (a periodic bubbling spring)[4,6,7] based on observation[8] and model experiments of a geyser induced by inflow of gas[9] and have also proposed a combined model combining above 2 models.[4] And numerical simulations of the modified dynamical model or the combined model reappear dynamics of spouting of geysers induced by inflow of gas (periodic bubbling springs) and it becomes possible that parameters (volume of the underground space, depth of spouting hole and so on) under a geyser are estimated due to comparison between results of simulation and those of observation.[4,7] Moreover we have verified above models through geological exploration, analysis of hot spring water and radioactive prospecting.[10] And we have modified above model as occasion demands.[11,12] As a result, spouting mechanism of a geyser induced by inflow of gas spouting regularly has been clarified.

But, concerning a geyser induced by inflow of gas, there are not only one spouting regularly but also one spouting irregularly. For example, Hirogawara Geyser (Yamagata, Japan) is an instance of a geyser induced by inflow of gas spouting irregularly. In case of a geyser induced by inflow of gas spouting irregularly we cannot explain its spouting mechanism based on above-mentioned usual dynamical model which assumes single underground gas supply source. In such a case it is natural that we think there are plural underground gas supply sources and interaction of them produces irregular spouting period.

In this study, we propose a dynamical model which assumes plural underground gas supply sources by extension of above-mentioned usual dynamical model. Then we show complicated spouting period occurs as a result of interaction of plural underground gas supply sources through numerical simulation of this extended dynamical model. We aim at clarification of spouting mechanism of a geyser induced by inflow of gas spouting irregularly through this study.

2. Model

In this chapter, at first we introduce main fruits of the former dynamical model, the mathematical model (a static model) and the combined model of a geyser induced by inflow of gas (a periodic bubbling spring). Then we introduce derivation of an extended dynamical model which assumes plural (two) underground gas supply sources by extension of above-mentioned former dynamical model. Then we introduce some results of numerical simulation of the extended dynamical model and discuss the effects of plural (two) underground gas supply sources.

2.1. *The former model*

Detailed derivation of the former dynamical model, the mathematical model (a static model) and the combined model of a geyser induced by inflow of gas (a periodic bubbling spring) was already stated in references.[4,7] So we show only main fruits of above models here.

At first, in case friction between walls of a spouting pipe and water packed in it is not taken into account, an evolution equation of temporal variations of height of top of a water pole packed in the spouting pipe of a periodic bubbling spring is written as;

$$(n_0 + \beta t)(V_0 + Sx)\rho H \frac{d^3x}{dt^3} + (n_0 + \beta t)pS\frac{dx}{dt} = (V_0 + Sx)p\beta \tag{1}$$

where n_0 represents molar number of gas in a underground space just before the water pole's beginning to move up to the upper entrance of the spouting pipe, β is constant concerning gas supply rate, V_0 represents volume of gas packed in a underground cave, S represents an area of a cross section of the spouting pipe, H represents length (height) of a lump of water packed in the pipe, p represents pressure of gas packed in the underground cave, x is regarded as a position of the lower interface between water and gas of the water pole and an upper direction of a vertical line is regarded as a plus direction of x-axis.

In case friction between walls of a spouting pipe and water packed in it is taken into account, Eq. (1) is changed to following Eq. (2).

$$\begin{aligned}(n_0 + \beta t)&(V_0 + Sx)\rho H \frac{d^3x}{dt^3}\\ &+ \frac{8\pi\eta H}{S}(n_0 + \beta t)(V_0 + Sx)\frac{d^2x}{dt^2} + (n_0 + \beta t)pS\frac{dx}{dt}\\ &= (V_0 + Sx)p\beta\end{aligned} \tag{2}$$

where η represents viscosity coefficient. As a result, second term of Eq. (2) is added and it represents effects of viscosity.

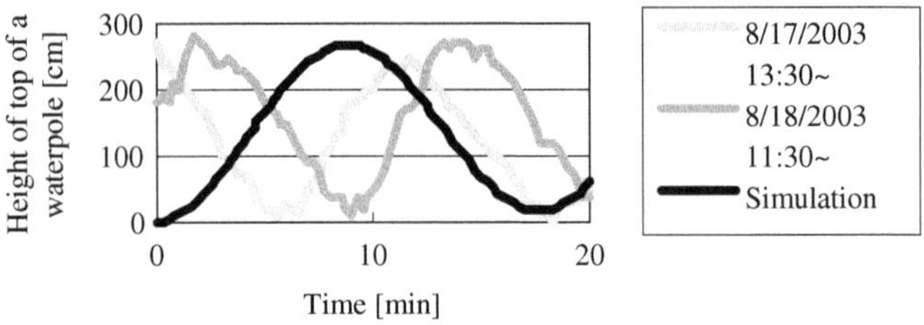

Fig. 1. A graph of numerical simulation of Eq. (2) comparison with graphs of observation of a geyser induced by inflow of gas (Kibedani geyser (Yamagata, Japan)).

A sample of numerical simulation of Eq. (2) is shown in Fig. 1. A graph of numerical simulation of Eq. (2) is drawn in comparison with graphs of observational results in Fig. 1. We are able to guess underground parameters which we cannot measure easily because of geological difficulties through comparing results of numerical simulation with those of observation. Concretely we try to fit data of numerical simulation to those of observation changing values of parameters. When both data fit most, we can guess chosen values of parameters are ones of real underground parameters.

Using the mathematical model (a static model), a spouting period τ and so on are represented as functions of various parameters. For example, a spouting period τ can be written as;

$$\tau = \frac{V_0}{\alpha\beta} + \frac{f_k S}{\rho g \alpha\beta}(f_k + P_0 + \rho g H) \tag{3}$$

where $\alpha = RT$, R represents gas constant, T represents temperature, f_k represents pressure due to a surface tension on an interface between water packed in the spouting pipe and gas in the underground cave, P_0 represents pressure of the atmosphere and g represents gravity acceleration. For example, when we observe a spouting period of a geyser induced by inflow of gas (a periodic bubbling spring), we can evaluate underground parameters (volume of the underground space (V_0), depth of spouting hole(H) and so on) through Eq. (3).

And the combined model is one combining above the former dynamical model and the mathematical model (a static model). The combined model enables more reliable estimation of underground parameters of a geyser induced by inflow of gas. Concretely, the dynamical model or the mathematical model (a static model) can determine (underground) parameters of a geyser induced by inflow of gas independently as shown above. On the other hand, some sets of parameters enable resemble spouting dynamics or period in each model. That is, we cannot decide most correct parameters' set using only one model. But using independent 2 models, that is, the dynamical model and the mathematical model (a static model), we can decide more correct parameters' set. These are essential qualities of the combined model.

2.2. *An extended dynamical model which assumes plural (two) underground gas supply sources*

In this section, we simply introduce derivation of an extended dynamical model which assumes plural (two) underground gas supply sources by extension of above-mentioned former dynamical model.

In the beginning, we suppose simple underground system having two underground gas supply sources as shown in Fig. 2. There is a water pole on each gas supply source. The upper surface of the water is common to both water poles. A central dotted line in Fig. 2 shows that 2 water poles are separated under the ground. Each water pole moves almost independently except for interaction at the upper surface of the water. Now we define x-, y- and z-axis as upward vertical direction and set the origin of x-, y- and

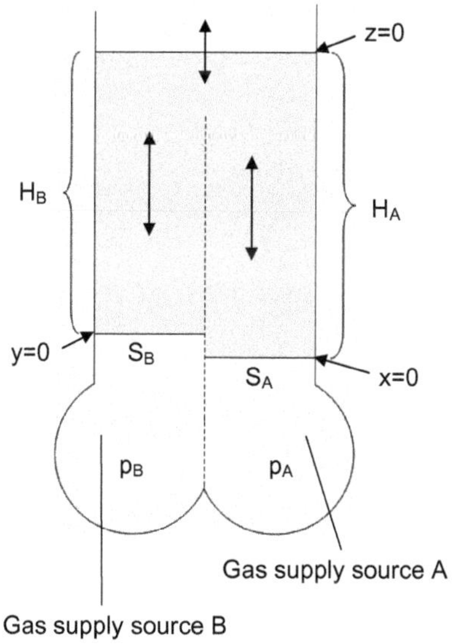

Fig. 2. Assumed system of an extended dynamical model which assumes plural underground gas supply sources.

z-axis at the lower surface of the right water pole, one of the left water pole and the upper surface of the water, respectively. And we name the right watercourse and the left one for watercourse A and watercourse B, respectively.

Then a basic dynamical equation of each watercourse (A, B) based on the former model is written as;

$$(n_{0A} + \beta_A t)(V_{0A} + S_A x)\rho(z - x)\frac{d^3x}{dt^3} + (n_{0A} + \beta_A t)p_A S_A \frac{dx}{dt} = (V_{0A} + S_A x)p_A \beta_A \tag{4}$$

$$(n_{0B} + \beta_B t)(V_{0B} + S_B y)\rho(z - y)\frac{d^3y}{dt^3} + (n_{0B} + \beta_B t)p_B S_B \frac{dy}{dt} = (V_{0B} + S_B y)p_B \beta_B \tag{5}$$

where n_{0A} or n_{0B} represents molar number of gas in a underground space just before the water pole's beginning to move up, β_A or β_B is constant concerning gas supply rate, V_{0A} or V_{0B} represents volume of gas packed in a underground cave, S_A or S_B represents an area of a cross section of the spouting pipe and p_A or p_B represents pressure of gas packed in the underground cave in watercourse A or B, respectively.

And an equation concerning conservation of water volume is written as;

$$S_A(z - x) + S_B(z - y) = S_A H_A + S_B H_B \tag{6}$$

where H_A or H_B represents initial height of a water pole of watercourse A or one of watercourse B, respectively.

3. Results of Numerical Simulation of Above Model and Discussion

In this chapter, we show some results of numerical simulation of the above extended dynamical model and discuss them.

In the beginning, we show dependence of variation of height of a water pole on cross section of one side's watercourse (S_A) in case of one of the other side's watercourse (S_B) $= 1.1\,$E-02[m^2] in Fig. 3. We may see that irregular spouting dynamics is realized. And we see that the larger difference between cross section of one side's watercourse (S_B) and one of the other side's watercourse (S_A) is, the smaller an amplitude of temporal variation of water pole's height is and the shorter an irregular period of it is.

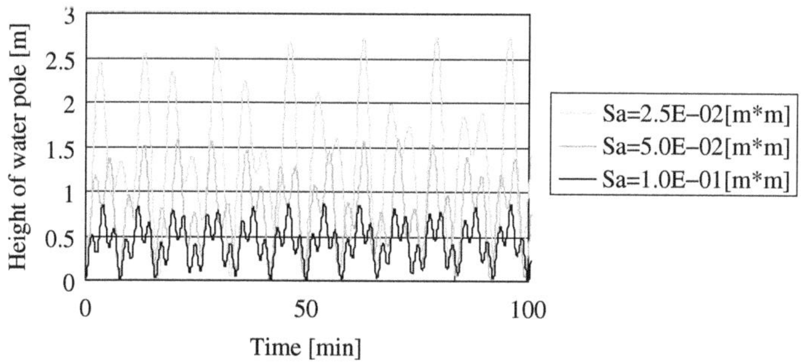

Fig. 3. Dependence of variation of height of a water pole on cross section of one side's watercourse (S_A) in case of one of the other side's watercourse (S_B) = 1.1 E-02[m^2].

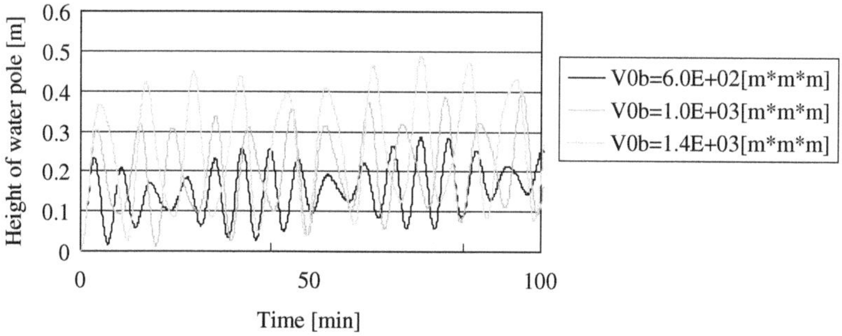

Fig. 4. Dependence of variation of height of a water pole on volume of one side's underground cave (V_{0B}) in case of the other side's underground cave (V_{0A}) = 4.0E + 02[m^3].

Then we show dependence of variation of height of a water pole on volume of one side's underground cave (V_{0B}) in case of the other side's underground cave (V_{0A}) = 4.0E+02[m^3] in Fig. 4. We also see that irregular spouting dynamics is realized. And we see that the larger difference between volume of one side's underground cave (V_{0B}) and one of the other side's underground cave (V_{0A}) is, the smaller degree of irregularity of temporal variation of water pole's height is because the cave whose volume is larger than the other controls whole temporal variation of water pole's height more. And the larger volume of one side's underground cave (V_{0B}) is, the larger an amplitude of temporal variation of water pole's height is and the longer an irregular period of it is. The reason of this is the same one seen in case of the former dynamical model in which a single underground gas supply source is assumed. That is, in case of the former dynamical model in

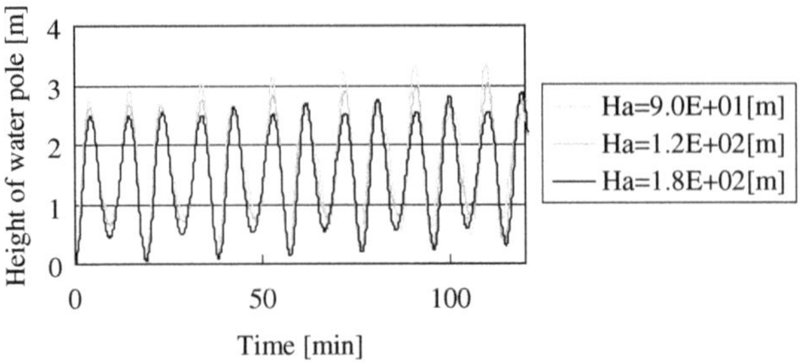

Fig. 5. Dependence of variation of height of a water pole on depth of one side's gas supply source (H_A) in case of one of the other side's gas supply source (H_B) = 9.1E + 01 [m].

which a single underground gas supply source is assumed, the larger volume of an underground cave is, the larger an amplitude of temporal variation of water pole's height is and the longer a period of it is.

Then we show dependence of variation of height of a water pole on depth of one side's gas supply source (H_A) in case of one of the other side's gas supply source (H_B) = 9.1E + 01 [m] in Fig. 5. We see that irregular spouting dynamics is realized except for spouting period. And we see that in obedience to interaction between depth of one side's gas supply source (H_A) and one of the other side's gas supply source (H_B), an amplitude of temporal variation of water pole's height varies with time complicatedly.

4. Conclusions and Future View

We derived a dynamical model of a geyser induced by inflow of gas in case of having two underground gas supply sources based on the former dynamical model in case of having one underground gas supply source.

Irregular spouting dynamics was realized through numerical simulation of the dynamical model. And dependence of spouting dynamics of the model on various parameters is clarified.

Then real spouting dynamics of a geyser induced by inflow of gas spouting irregularly should be reproduced through numerical simulation of this model. Then as a result, various underground parameters should be estimated.

And we should investigate a progressed model in case of having over two underground gas supply sources.

References

1. K. Honda and T. Terada, *Publ. Earthq. Inv. Com.* **22B** (1906).
2. I. Iwasaki, Bull. Tokyo Institute of Technology, No. 46, 1962.
3. H. Kagami, Abstracts of The 53rd Annual Meeting of the Balneological Society of Japan **27** (2000).
4. H. Kagami, Advances in Geosciences Vol. 4 : Hydrological Science (HS) **191** (2006).
5. H. Kagami, Abstracts of The 55th Annual Meeting of the Balneological Society of Japan **33** (2002).
6. H. Kagami, The 2003 IUGG General Assembly, HW04/09P/C31-004, 2003.
7. H. Kagami, *Proceedings of The 38th Conference of Sciete Internationale des Techniques Hydrothermales and The 56th Annual Meeting of the Balneological Society of Japan*, 2003, pp. 55–60.
8. E. Ishii, *et al.*, Abstracts of The 52nd Annual Meeting of the Balneological Society of Japan **28** (1999).
9. M. Katase, *et al.*, Abstracts of a meeting for presenting research papers of Kanto Gakuin University College of Engineering 99–100, 1999.
10. H. Kagami, Advances in Geosciences Vol. 6: Hydrological Science (HS) **203** (2007).
11. H. Kagami, Advances in Geosciences Vol. 11: Hydrological Sciences, **37** (2009).
12. H. Kagami, Advances in Geosciences 2010 : Hydrological Sciences (in press).

Advances in Geosciences
Vol. 17: Hydrological Science (2008)
Eds. Namsik Park *et al.*

SEDIMENT DELIVERY AND ENVIRONMENTAL CHANGE IN ASIAN RIVERS: LINKING ARCHIVES WITH FIELD INVESTIGATIONS

DAVID L. HIGGITT*
Department of Geography,
National University of Singapore,
1 Arts Link, Singapore 117570
geodlh@nus.edu.sg

The sensitivity of sediment yields to changes in catchment land use has been demonstrated widely at restricted spatial scales, particularly in headwater catchments, but in larger river basins the link between land use change and sediment yield is less distinct. There are multiple sediment delivery processes operating which moderate overall catchment response. Seemingly simple questions about the downstream impact of environmental change or management interventions can be notoriously difficult to predict. In large river basins the spatial scale is challenging for field-based approaches. The paper considers examples from Asian monsoon-dominated river basins to examine the extent to which long term trends in sediment flux at the basin scale can be linked to information about changing erosion and sediment transport dynamics from constituent parts of the river basin. In part this reflects the counter-balancing impacts of increasing land degradation and sediment mobility versus the rapid development of water resources and reduction in sediment conveyance. In turn, changes in fluvial systems have implications for the functioning of the carbon cycle. We have approached the problem from three main angles. Firstly, attention to database assembly and multivariate analysis of within-catchment sediment yields, from which basin response to environmental change can be identified. Secondly, responsive field investigation can be planned to validate desk-top analysis, which in turn requires the development of rapid appraisal techniques and geoindicators. Thirdly, the empirical approach can be extended by constructing geomorphologically-informed predictive models of catchment disturbance on sediment yields. The prediction of longer-term sediment yield in large catchments, which may be critical to water resource management and continent-ocean nutrient fluxes, is sensitive to many aspects of sediment delivery.

*Work partially supported by Royal Geographical Society (with Institute of British Geographers) Ralph Brown Expedition Award.

1. Introduction: The Sediment Delivery Concept

A sediment budget is an accounting procedure to identify and quantify the production, transfer and export of sediments through a river catchment. The sum of the transfer processes that result the export of sediment beyond the defined catchment outlet (usually represented by the location of a gauging station) is termed sediment delivery. It has been well established that the proportion of sediments 'delivered' to the catchment outlet is only a small fraction of those mobilized on catchment slopes or within channels. It is now 25 years since the crux of the issue was elucidated by Walling[1] as the "Sediment Delivery Problem". Conceptually, untangling the sediment delivery problem is necessary for identifying the relative importance of particular sediment transfer mechanisms under different hydrological or climatological conditions, for articulating the linkages and pathways of sediment transfer and describing its temporal and spatial variability. Practically, the sediment delivery concept provides a means of evaluating the response of erosion and sedimentation processes to changes occurring within the river basin, including the prospect of predicting sensitivity to future climatic or land use change. In Asian rivers, the challenge of the sediment delivery problem is particularly significant given the combined impacts of climate change, widespread and dramatic land use change and direct intervention in river systems (e.g. dam building, water resource distribution and aggregate mining).

Furthermore, the Asia-Pacific Rim represents one of the most significant interchanges of sediments and associated nutrients between terrestrial and ocean stores on the planet. The high suspended sediment loads of Asian rivers are driven by the tectonic setting, rapid weathering rates and monsoon rainfall regimes, aggravated by the continued reduction of forest cover and the expansion of agricultural land onto steep slopes. Conversely, the construction of water impoundment structures, ranging from mega dams of large rivers to many thousands of smaller structures in the upper reaches of catchments disrupts the downstream transfer of sediments. Emerging evidence now suggests that the combined impact of water resource development and impoundment structures is reducing water and sediment delivery from Asian rivers. Concern about the impact of reduced freshwater and nutrient input in coastal shelf seas on the productivity of coastal shelf seas has to some extent replaced the previous alarm about accelerated soil erosion and land degradation within Asian river basins.

Such a switch in sentiment, however, belies the need to understand the sediment delivery mechanisms with large Asian river catchments in some detail. This paper reports on some of the methodological considerations and constraints that have influenced the understanding of large scale sediment budgets.

2. The Challenge of Large Rivers: Methodological Considerations

The majority of geomorphological studies concerned with sediment delivery have tended to focus on relatively small-scale catchments. Typically, field investigations use a combination of river gauging, field mapping, interpretation of sediment archives (e.g. lake, reservoir or floodplain sediments) and sediment geochemistry to infer contemporary or reconstruct historical sediment delivery dynamics and identify key source areas. Field data may drive the interpretation or may provide calibration data for distributed hydrological models. When such approaches are scaled up to large river basins data acquisition becomes much more challenging and field-based approaches are difficult to manage logistically or to justify in terms of representativeness. By contrast early studies of land to ocean sediment transfer in large rivers focused on explaining global scale patterns by considering only the sediment load measured at the catchment outlet (Milliman and Syvitski[2]; Summerfield and Hulton[3]). From the early 1990s onwards the first generation of geodatabases and the increasing availability of satellite imagery offered new opportunities to examine sediment delivery patterns within large rivers. The developing interest in understanding sediment budgets of large rivers has placed emphasis on the construction and analysis of suspended sediment data archives but the role of fieldwork, so crucial in deciphering sediment delivery dynamics at smaller scales is a major challenge. In the paper an example of linking the macroscale analysis of secondary data with rapid appraisal style fieldwork is discussed for an example tributary catchment of the Upper Yangtze. The three elements of the methodological challenge are presented as establishment and analysis of suspended sediment data archives; integrating fieldwork within large river research; and linking sediment budgets to other aspects of biogeochemical cycling.

2.1. *Establishing suspended sediment archives*

Our work on sediment budgets in large rivers was initiated by the discovery that large volumes of secondary data on hydrological and water quality can be obtained from selected Asian rivers but that much effort is required to put the records into a usable form. The initial focus was the Upper Yangtze River, defined as the catchment area above the Three Gorges Dam. Data was published in Hydrological Year Books and released into the public domain until 1987. After 1987, data were not disclosed but have subsequently been made available for purchase. Information from printed yearbooks was transcribed into a database over a period of several months. The database contained annual and monthly sediment discharge data (kg s^{-1}) from which a sediment flux could be calculated. A total of 255 gauging stations operated upstream of Yichang at various stages during the period 1956–1987. The database contains 3820 station years within which 56 stations have 25 or more years of data. These long-term stations, together with another 6 stations of slightly shorter time duration in the Wu tributary formed the main subset of data to explore spatial and temporal patterns of sediment delivery. There are several issues that impact on the analysis of suspended sediment records. First, scale is an important factor since sediment delivery ratios have been shown to be strongly scale dependent. As basins increase in size, opportunities for sediment storage increase disproportionally as mean gradient reduces. Thus, as a general rule there is an inverse relationship between specific sediment yield (t km^{-2} a^{-1}) and catchment area. Second, sediment concentrations are strongly influenced by weather variations with the majority of the material transported during storm events in the summer monsoon period (June to September). Identifying underlying trends in sediment yield require removing the influence of hydrological variation. Popular methods include using double mass plots which represent cumulative sediment discharge against cumulative runoff (e.g. Walling and Webb[4]). However, we preferred to employ the more robust method of Helsel and Hirsch[5]) by examining temporal trends in the residuals of the sediment yield — discharge relationship. Third, the monthly sediment discharge data may not be representative of the mean value during the month in question. To some extent, the response of large monsoon driven rivers results in sediment concentration rising steadily for periods of days, such that spot sampling may be less of a problem than systems where suspended sediment is strongly controlled by individual storm events. Fourth, there is an obvious issue of data quality control which is inevitable when using archive data.

The spatial patterns of sediment yield in the Upper Yangtze were explained using multivariate techniques to predict sediment yield in terms of catchment characteristics. Extracting data from geodatabases for a variety of climatic, topographic and land use data, mean subcatchment characteristics can be calculated in a GIS framework. In earlier work (Lu and Higgitt[6]) it was demonstrated that more than 95% of the variation in sediment yields in the western part of the Upper Yangtze could be explained by the topographic variables mean slope and man elevation. However, in the eastern part of the basin, where there is greater human impact through agriculture the multiple regression technique was less successful in explaining the spatial variability of sediment yields, even when land use data and a sloping agricultural land index was introduced (Higgitt and Lu[7]). To some extent this reflects the limited (1 km) resolution of geodatabases available at the time and there is increasing potential to undertake future analyses. While China has the most comprehensive hydrological records for large Asian Rivers, there is also potential to develop databases for the Mekong basin and for a number of Thai rivers. There are, however, very significant rivers for which there are very limited data. The two large rivers of Myanmar, the Irrawaddy and Salween are examples of data-scarce rivers.

When the Three Gorges Dam was being planned, the impact of future sedimentation was a major concern. Analysis of the sediment yield at Yichang, the hydrometric station immediately downstream of the dam site, displayed no significant trend despite widespread evidence of an increase in extent and magnitude of soil erosion across much of southern China. This apparent paradox between evidence for increasing sediment mobilization without a corresponding increase in sediment export from the catchment outlet can be investigated by looking at the spatial variability of temporal trends within the basin itself. Dividing the gauging record into three time slices, Lu and Higgitt[8] identified heterogeneity in sediment yield time series with 7.8% of the basin area comprising subcatchments with statistically significant upward trends of sediment yield and 2.8% with declining sediment yield. The paradox is partly explained by the presence of existing dams on major tributaries as many of the stations exhibited declines in sediment yield lie downstream of major HEP schemes. Comparison of sediment yields above and below dams in the Wu, Jialing and Dadu tributaries show departure in sediment yield trends following dam closure. The case for the Dadu is shown in Fig. 1. The parts of the basin where sediment yields have increased in the decade to 1987 have been mapped. At a gross scale such a map is useful in demonstrating the role of

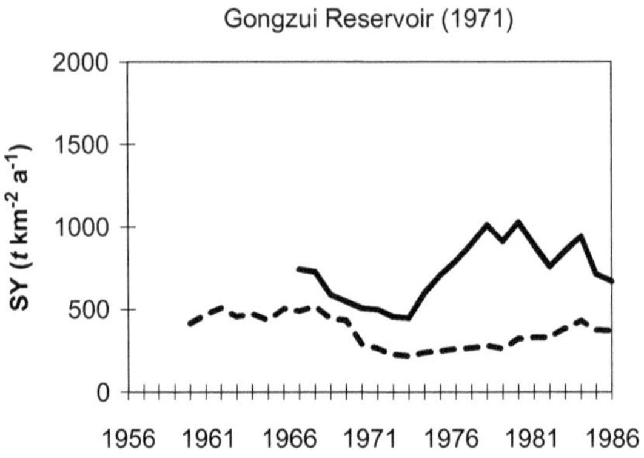

Fig. 1. Trends in annual specific sediment yield (t km^{-2} a^{-1}) at gauging stations above (solid line) and below (dashed line) the site of the Gongzui Reservoir on the Dadu River. The dam was closed in 1971.

specific tributaries in supplying sediment to the main river, but it provides limited scope for assessing the reasons for the changing sediment yields. The map of subcatchments with upward trending sediment yields (Fig. 2) identifies the Three Gorges Area itself and the Dadu tributary as two sections of the Upper Yangtze where sediment yields had been increasing during the 1980s. It was suspected that deforestation and conversion to agricultural land was the principal reason for the observed trends in the Dadu (based on the gauging record at Shaping where the catchment area is 75,016 km^2), but the evidence is circumstantial and required verification from fieldwork.

2.2. *The role of fieldwork: Rapid appraisal and monitoring*

The computer-based analysis of the suspended sediment data archive identified the Dadu as an increasing significant source in the Upper Yangtze. In order to test the notion that the increase was due to the expansion of agricultural land, a short field expedition was conducted. Mapping and quantifying sediment sources are standard techniques in field-based sediment budget studies but are logistically challenging in large river basins. Furthermore, as Reid and Dunne[9] note, many land managers seeking information about sediment-related problems do not require a detailed and meticulous scientific investigation of erosion and sediment transport processes but rather a concise and quickly produced guidance on the key

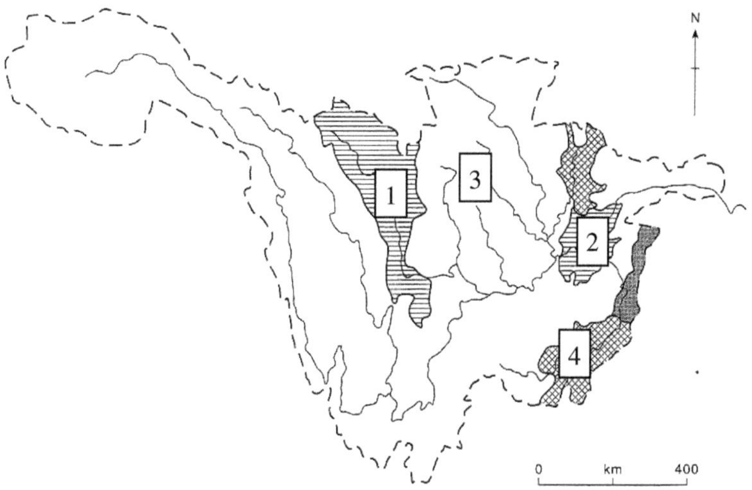

Fig. 2. Map of the Upper Yangtze catchment indicating areas which experienced significant increases (horizontal = $0 - 250\,t\ km^{-2}\ a^{-1}$; cross hatch = $> 250\,t\ km^{-2}a^{-1}$) and significant decreases (solid shading = $0 - 250\,t\ km^{-2}\ a^{-1}$ decrease) during the 1980s. Locations: 1 = Dadu catchment (increase); 2 = Upper part of Three Gorges (increase); 3 = Jialing and Sichuan Basin — the major source of sediment in the Yangtze but not indicating a significant trend at this time; 4 = Upper Wu (major increase).

processes and source areas and the likely scenarios if a particular land use change is applied.

In the case of the Dadu River, it was hypothesized that the increase in sediment yield determined by the time series analysis had resulted from the removal of forest cover and expansion of agricultural land in the 1980s. An erosion audit was conducted by traversing a distance of just under 200 km along the Dadu valley, mapping the location of erosion features. The mapping commenced 25 km north of Luding, the major town in the Dadu valley where the main road leaves the valley and heads west towards Kangding. The Dadu flows almost due south before turning abruptly east at Shimian. From there the audit continued to Fulin. An additional 20 km of erosion features was also audited in the Lu Sa He tributary. The erosion features comprised mainly debris flows feeding into talus cones, rotational slides and gully complexes. The rapid appraisal does not provide an opportunity to fully record the dimensions of every erosion feature but an arbitrary designation to major feature was given to large and active erosion features (>500 m width). The density of erosion features per 5 km reach is plotted in Fig. 3. In addition the extent of agricultural land on

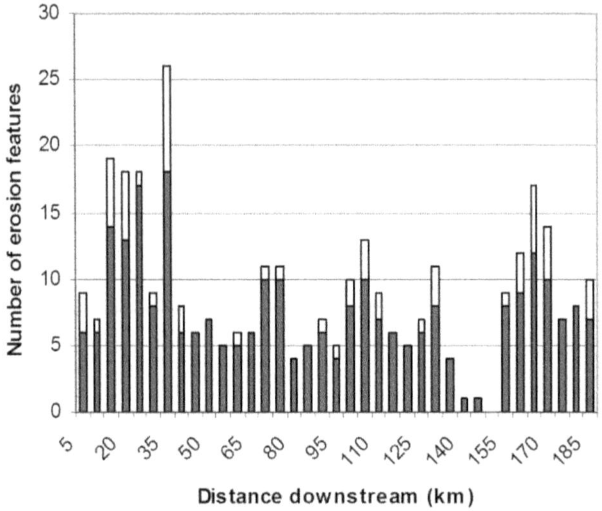

Fig. 3. Distribution of erosion features in the Dadu River Valley over a 195 km reach from north of Luding to Fulin. The total number of features is indicated by the full bar of which the upper part indicates "major features".

the valley slopes was noted. The density of landslide scars reaches close to 5 km^{-1} (i.e. 25 features per 5 km reach) in the vicinity of Luding, where the valley sides had been extensively cleared for terrace field agriculture. South of Luding, the extent of agricultural conversion is slightly less and the density of erosion features decreases. At Shimian where the river turns east, the valley is very narrow and the forest cover remains intact. Where there has been limited disturbance of the land cover, very few erosion features are observed. The rapid appraisal therefore confirms that the conversion to agricultural land is the main driver of increased sediment yields.

Many of the fields on the steep valley slopes have been abandoned as a major conservation initiative has sought to reafforest the valley, yet these abandoned fields represent a major sediment source (Fig. 4). The field observations have been backed up by analyzing Landsat imagery. False colour images from 1975, 1986 and 2001 show the progressive destruction of the vegetation cover along the river corridor. Figure 5 reproduces black and white images of the 1975 and 2001 images which depict the opening up of the river which was not visible in the earlier scene. This also confirms that almost all of the land cover change taking place in the catchment is along the axis of the valley and some tributaries.

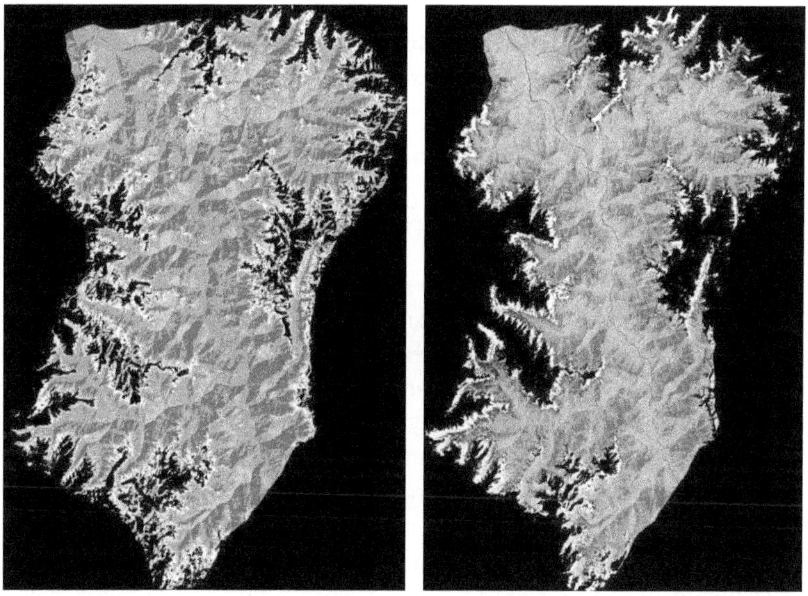

Fig. 4. Land Cover Changes in the study reach of the Dadu River, 1975–2001. The course of the river is not clearly visible in 1975 as forest cover extends along most valley slopes. By 2001, conversion of valley slopes to agriculture has revealed the river. Away from the main valley relatively little of the forest cover has been disturbed.

Fig. 5. Two views of erosion features along the Dadu Valley. (a) talus fan at km 30 (b) abundant sediment load in a right bank tributary of the Dadu at km 28. Gully systems crossing abandoned agricultural terraces are conspicuous in background.

The land cover characteristics of each 5 km reach was classified into three categories: agricultural land extending more than half way up the valley side; agricultural land extending 20% up the valley side with abandoned land at higher elevations; and agricultural land less than 20% up the slope (i.e. mainly forested). The mainly forested areas are found from 60–95 km and 110–145 km. The average density of erosion features per 5 km reach section is 10.9, 7.6 and 5.1 for the three landcover categories, respectively.

In other Asian rivers with limited gauging records, the key question is whether field-based research can make a valid contribution to understanding sediment delivery dynamics. For example, a project has recently started to install instrumentation at fixed stations on the Irrawaddy and Salween Rivers of Myanmar. The Irrawaddy was subject to a decade of detailed discharge measurements in the late nineteenth century, including one year of suspended sediment measurements (in 1876) and these data have subsequently been used in global analyses of land to ocean sediment transfer. Consequently, starting a monitoring programme to measure water and sediment discharge from two of the world's largest rivers can help to refine information on the denudation of the Himalayas. Though the field data thus collected has enabled reappraisal of the nineteenth century data and a preliminary estimate of sediment and carbon flux (Robinson *et al.*,[10] Bird *et al.*[11]), it will be many years before the representativeness of the contemporary measurements can be assessed.

2.3. *Predictive models and biogeochemical cycles*

The above analysis demonstrates that a largely desk-based study of the spatial and temporal variability of sediment yields within a large river basin can be supplemented with fieldwork and satellite image interpretation to explore the processes that resulted in a change in the sediment dynamics in part of the basin. Though there are many caveats relating to the reliability of rapid appraisal technique it is clearly not feasible to conduct thorough field investigations in a short time span across a major tributary. As an alternative to detailed investigations at a type site, where there are well known difficulties of extrapolating the results to a larger scale, the rapid appraisal provides confirmation that the active erosion in the valley is strongly associated with recent land cover change.

However, both aspects of the analysis are highly empirical in nature and by their nature are focused on the discovering past erosion-sediment delivery dynamics. A major challenge for geomorphologists working with catchment managers is to provide predictions of likely future scenarios. Whereas sediment delivery investigations in smaller basins often reveal a strong signal of disturbance, the signal from large Asian rivers is obscured by a number of competing influences taking place in different parts of the basin. A simple simulation model (Higgitt and Lu[12]), for example demonstrated that the direction of landcover change (whether deforestation proceeds from the lowest point in a tributary and proceeds upstream towards the headwaters or vice versa) can have a profound impact on the sediment yield time series at a catchment outlet. This is explained by the increased likelihood of sediment going into storage as the distance between source and catchment outlet increases.

While sediment dynamics are of intrinsic interest to geomorphologists, the growing interest in global environmental change has diverted attention away from rivers as a terrestrial to ocean conveyor of sediment towards consideration of the role of large rivers as a lateral flux in global biogeochemical cycles. In particular the continued dam construction on Asian rivers and increased use of water resources has resulted in declining freshwater and nutrient fluxes to the ocean, which in turn has impacts on primary productivity in coastal shelf seas. There are limited historical data available on water chemistry and nutrients, so it is difficult to conduct empirical investigations of long term time series. However, some recent work in the Pearl (Xijiang River) has enabled some of the trends in water chemistry to be determined and there is a strong correspondence between suspended sediment and particulate organic carbon fluxes, suggesting that the methodological challenges to understanding sediment delivery in large rivers will also aid understanding of biogeochemical cycles.

3. Summary

This paper reflects on the methodological challenges of investigating sediment delivery in large rivers and the need to reconcile empirical investigation of databases with field-based evaluation of erosion and sediment dynamics. The scale is a major challenge to conventional fieldwork initiatives and the paper reports a combined approach from empirical, remote sensing and rapid appraisal methods to examine erosion sensitivity in a large tributary of the Yangtze River. Since the initial work on developing suspended sediment archives for the Yangtze was undertaken, there has been a significant improvement in the resolution of data on catchment characteristics that can be derived from geodatabases though not all relevant datasets are yet in the public domain. The opportunity to combine archived data with representations of catchments characteristics in a GIS framework, supported, where appropriate, with field investigations therefore provides an opportunity to elucidate the impacts on environmental change on sediment delivery.

References

1. D. E. Walling, The sediment delivery problem, *Journal of Hydrology* **65** (1983) 209.
2. J. D. Milliman and J. P. M. Syvitski, *Journal of Geology* **100** (1992) 525.
3. M. A. Summerfield and N. J. Hulton, *Journal of Geophysical Research* **99** (1994) 13871.
4. D. E. Walling and B. W. Webb, in: eds. D. E. Walling and B. W. Webb *Erosion and Sediment Yield: Global and Regional Perspectives* **3** (IAHS Publ 236) (1996).
5. D. R. Helsel and R. M. Hirsch, *Statistical Methods in Water Resources* (Elsevier, Amsterdam, 1992).
6. X. X. Lu and D. L. Higgitt, *Earth Surface Processes and Landforms* **24** (1999) 1077.
7. D. L. Higgitt and X. X. Lu, *Hydrobiologia* **410** (1999) 269.
8. X. X. Lu and D. L. Higgitt, *Environmental Management* **22** (1998) 697.
9. L. M. Reid and T. Dunne, *Rapid Evaluation of Sediment Budgets.* Catena Verlag, Reiskirchen (1996).
10. R. A. J., Robinson, M. I. Bird, N. Win Oo, T. B. Hoey, M. Maung Aye, D. L. Higgitt, X. X. Lu, A. Swe, T. Tun and S. H. Win, *Journal of Geology* **115** (2007) 629.
11. M. I. Bird, R. A. J. Robinson, N. Win Oo, M. Maung Aye, X. X. Lu, D. L. Higgitt, A. Swe, T. Tun, S. H. Win, K. Sandar Aye, K. Mi Mi Win and T. B. Hoey, *Quaternary International* **186** (2008) 113.
12. D. L. Higgitt and X. X. Lu, *Geomorphology* **41** (2001) 143.

Advances in Geosciences
Vol. 17: Hydrological Science (2008)
Eds. Namsik Park et al.

CHARACTERISTICS OF BACKFLOW IN IWAKI RIVER MOUTH*

MIKIO SASAKI†,‡
Department of Civil and Environment Engineering,
Hachinohe Institute of Technology, 88-1 Ohbiraki Myo
Hachinohe, Aomori Prefecture, 031–8501, Japan
‡*msasaki@hi-tech.ac.jp*

HITOSHI TANAKA and MAKOTO UMEDA
Department of Civil Engineering, Tohoku University, 6-6-06 Aoba,
Sendai 980-8579, Japan

Saltwater flow characteristics in the mouth of the Iwaki River were investigated. The flow of water at the river mouth was highly dependent upon the water level of the coastal lake located immediately upstream, then the speed of the flow was calculated in highly accuracy by using the water level changes of the lake. Conditions in the river mouth are also markedly affected by tidal fluctuations, and the discharge of the fluctuation flow reaches to $200\,m^3/s$. Flow characteristics in the reaches river mouth were modeled and comparisons of calculations with measurement in the field revealed good agreement for periodicity and magnitude of flow rates. Incoming and outgoing water velocities reached 50 cm/s, which was sufficiently strong to transport the coastal drift that has entered in the river channel from near the shoreline back into the ocean. Backflow was lowest in April due to snowmelt, increasing in July due to the relative decrease in precipitation.

1. Introduction

The Iwaki River in Aomori Prefecture, Japan, flows across the Tsugaru Plains into Lake Jyusan, which empties into the Japan Sea. Iwaki River is a class 1 river managed by the Ministry of Land, Infrastructure and Transport. The mouth of the Iwaki River, which is formed by Lake Jyusan,

*This research was partially supported by a Grant from Ministry of Education, Culture, Sport, Science, and Technology (Head investigator: Professor Akira Mano, No. 17360230).
†A part of this study also was conducted by the Iwaki River Working Group (Head investigator; Mikio Sasaki) of the River Ecology Research Group. We are grateful to the Foundation for Riverfront Improvement and Restoration.

is locally referred to as Mitoguchi. Previously, the engineering works at the river mouth Mitoguchi were destroyed during rough seas caused by strong westerly winds in winter, causing the river mouth to become blocked four to five times a year as well as extensive flooding of the area as the lake overflowed. These floods caused considerable damage to the lower reaches of the Iwaki River and areas adjacent to Lake Jyusan. This flood damage could generally be classified as being one of two types of natural disaster, flooding due to increased rainfall and flooding due to blockage of the river mouth. Local inhabitants therefore petitioned the national government to implement flood control measures in the Iwaki River.

The government established a construction office in Goshogawara, and the flood control measures by the nation were initiated in December 1918. The construction of the Mitoguchi jetty for preventing blockage of the river mouth was started as a pier from land to sea in 1926 after eight-year study of the geographical characteristics, waves and current in the estuary had been completed. While initial efforts were focused on the construction of a pier on the northern side of the river mouth, in 1928 it was decided to construct jetties instead given the effectiveness of the pier as a jetty. Construction was initiated on the south pier in 1930, and the entire Mitoguchi pier construction project was completed 16 years later in 1946.

As a result, the project provided environments of safe sailing of boat in the river channel, and of preventing blockage of the river mouth and of seawater entering the lake (Sasaki *et al.*, 1997; Takahashi *et al.*, 1994). However, given the marked impact these changes have had on the flow characteristics of the estuary system, we clarified the flow characteristics of the river channel at the river mouth in this study.

2. Iwaki River Mouth

The Iwaki River has it origins near the world heritage site of the Shirakami Mountains. The river catchment is 2540 km^2 in size and the length of the main river channel is 102 km, making it the 24th largest river catchment and 24th longest river in Japan. Figure 1 shows the geographical location of the Iwaki River.

The construction of the jetties on both sides of the river mouth successfully stabilized the area; interestingly, the jetties had been referred to locally as piers because they were initially constructed as piers.

The present Mitoguchi channel is shown in Fig. 2 neither the basic shape nor the position of the jetties have changed since the time they were

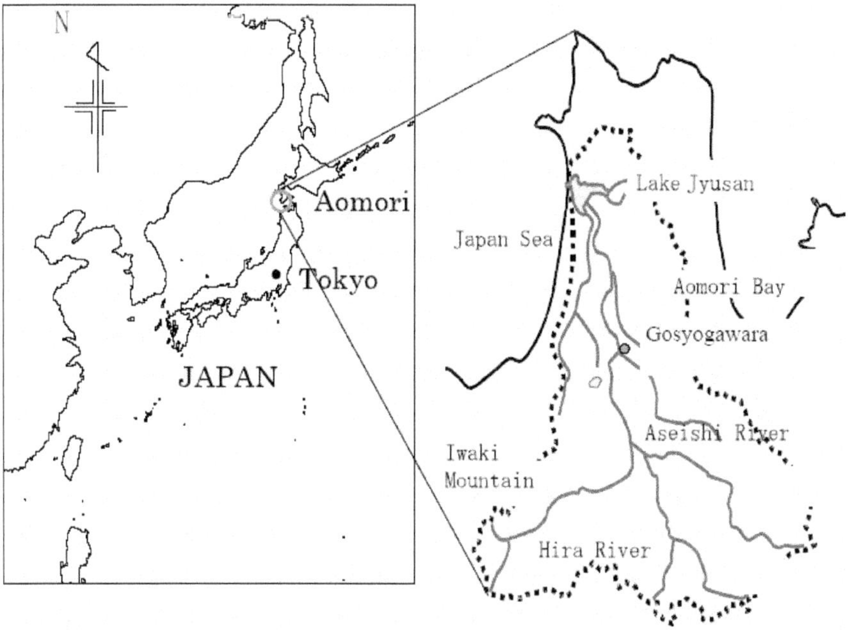

Fig. 1. Geographical location of the Iwaki river catchment. The dotted line means the catchment boundary.

Fig. 2. Present river mouth taken by author in 2004. View is from the Japan Sea to the river mouth in an easterly direction.

constructed (Sasaki *et al.* 1990). The northern jetty measures 389 m, the southern jetty measures 337 m, and the distance between the two jetties is 180 m. About 60 m of the head of the northern jetty inclines at the south and it is curved.

The mouth of the Iwaki River had historically always been susceptible to becoming blocked. The first attempt to control the flow of water through the river mouth was by the feudal lord, Nobuyoshi, who engineered a waterway at the river mouth using 3000 laborers from July to September in 1649. However, the new waterway became blocked, and as with other attempts at designing a river channel in the river mouth, the new other waterway had become blocked.

Most of the previous attempts at designing river channels in the area of the river mouth were excavated in the south, primarily because waves come predominantly from the north in rough seas. The area where the old river channels have formed marshes and a lagoon can currently be found to the south of the present river channel. In fact, the scars left by the old river channels can still be seen today as shown in Fig. 3. The figure is showing the present river mouth on the center and the previous river channel on right.

The last time the river mouth was blocked was in January 1930, when construction of the 175 m jetty on the southern shore was completed. This

Fig. 3. Photograph of the previous river channel and present river mouth of the Iwaki River taken by the author. View is in a northerly direction.

suggests that the effectiveness of the jetty was finally demonstrated when length of the jetty on both shores were sufficiently long. In so far as the river channel is concerned, the jetties on both shores function well for the control of coastal sediment. It can be easily understood by knowing the characteristics of the shoreline change in the river mouth when there was on pier.

Records at the Construction Office in Goshogawara show that the river mouth was blocked 39 times from 1918 to 1930 (Sasaki *et al.*, 2003). Since the construction of the jetties, there have been no blockages of the river mouth. The jetty at the river mouth of the Iwaki River still works well today and it is one of the very few river structures to have remained effective in Japan. Specifically, construction of the jetties has prevented silting up of the river mouth, and, the flooding due to the lake overflowing.

3. Flow in the River Mouth

After the jetties are completed, incoming and outgoing water velocities are alternately generated in the mouth of the Iwaki River. Seawater comes with incoming flow to the mouth of the river and the lake. Seawater go back in the upstream, and water environment in the estuary is changed though the blockage of the river mouth disappeared due to the jetties. Because seawater movement is affected by the flow, we investigate the flow of the mouth of the Iwaki River.

Now, in order to calculate the rate of flow from the lake to the sea Q_2, we use the equation

$$Q_2 = Q_1 - A_b \frac{\partial \eta_b}{\partial t}, \qquad (1)$$

where Q_1 is the river discharge into the lake, and A_{bs}, η_b and t are the lake area, the water level of the lake, and time respectively.

Figure 4 shows the flow rate in the river mouth calculated by Eq. (1) for October 2003. The effect of the tides is clearly visible in the figure with seaward and backflow currents reaching approximately $200\,\text{m}^3/\text{s}$.

Figure 5 shows a comparison between the field observation data and the data derived using Eq. (1). As can be seen in the figure, there is good agreement between the calculation and the observation in the periodicity and magnitude of flow velocity. Both show current velocities as reaching $50\,\text{cm/s}$. The magnitude of the flows has the ability to sweep out coastal drift that has moved to the river channel from the shore.

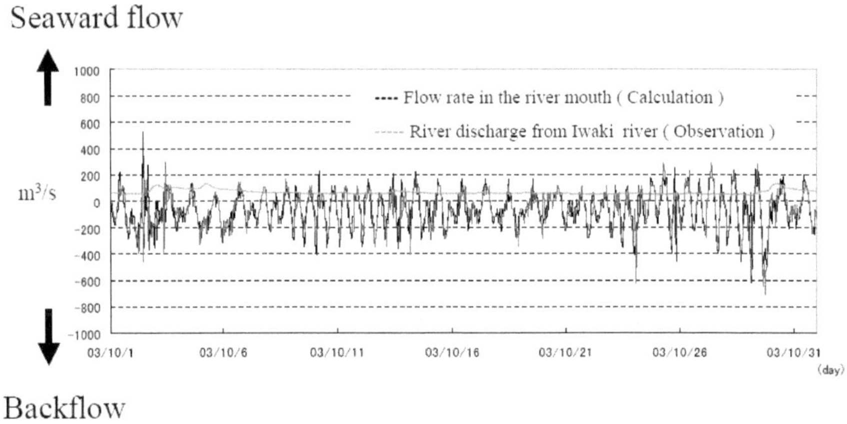

Fig. 4. Flow rate in the river mouth in October 2003.

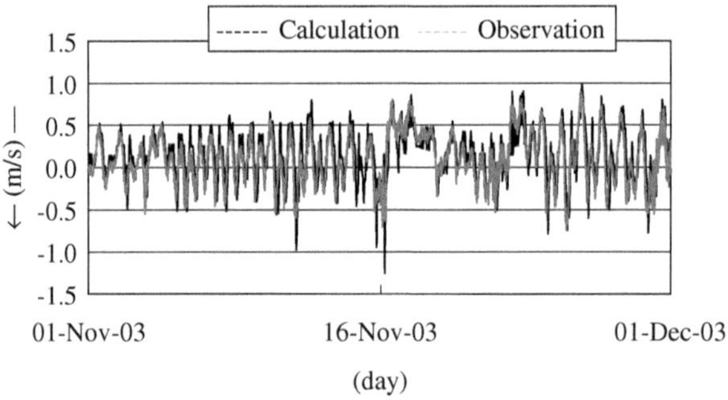

Fig. 5. Comparison between the modeled and observed channel flow.

4. Backflow in the River Mouth

Figure 6 shows backflow flow rates at monthly intervals form April to October. Flow rates are observed to increase from April to July and decrease in September before increasing again in October. In April, snowmelt from the mountains causes an increase in the river discharge. The increase raises the water level of the lake, then the backflow doesn't occur easily so that the water level of the lake becomes higher consequentially than the water level of the sea. Namely, the water level of the lake increases as the runoff to the lake increases. Then, the water level of the lake lowering more than surface of the sea decreases. Consequentially, the backflow decreases. However, as can be seen in the figure, backflow was more pronounced in July 2007 than

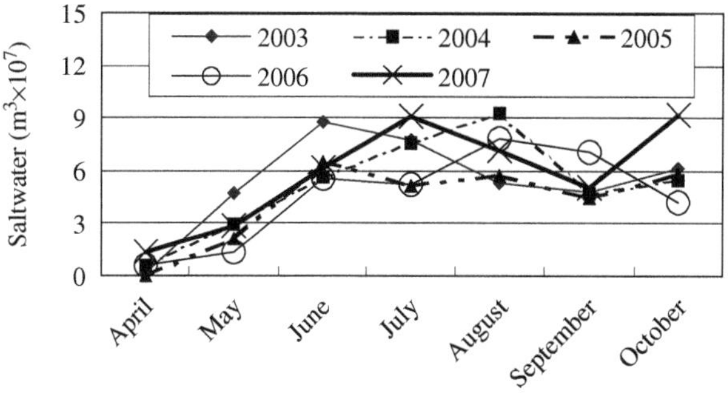

Fig. 6. Flow rate of the backflow current over a six-month period between 2003–2007.

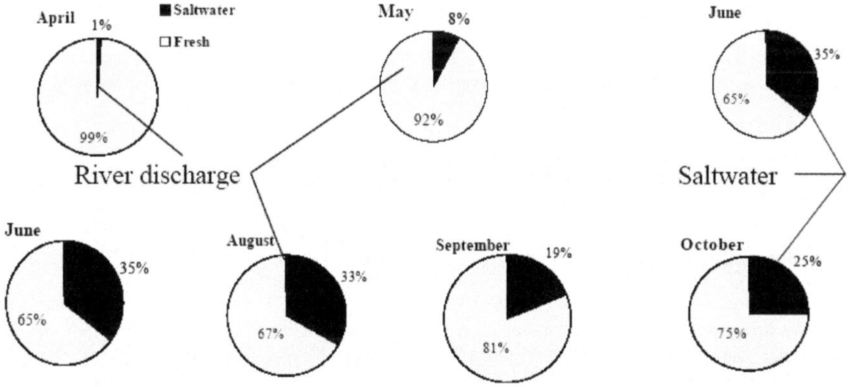

Fig. 7. Relationship between river runoff and backflow rate for five years from 2003.

in other years. The reason for this increase in backflow is because the water level of the lake decreased as runoff into the lake decreased due to small rain. In fact, the magnitude of backflow is affected by river discharge.

Figure 7 shows the relationship between river runoff into the lake and backflow rate for the months April to October. The values shown are five-year means for the period 2003 to 2007. The percentage of backflow to runoff into the lake was 1% in April, 8% in May, 35% in June, 36% in July, 33% in August, 19% in September, and 25% in October. Backflow rates are lowest in September when there is an increase in precipitation due to the numerous typhoons. Similarly, backflow rates are also low in April because of snowmelt. Backflow increases in July when the amount of precipitation is low.

5. Conclusions

The present study investigated the characteristics of water flow in the mouth of the Iwaki River. The flow discharge of the waterway in the estuary is calculated from the water level in the lake. The following were obtained by the present study.

(1) Low and high tides were clearly observed in the river mouth, with seaward and backflow currents measuring approximately 200 m^3/s.
(2) Data obtained from the calculations described in the present study to estimate the flow rates in the river mouth corroborated those obtained from direct observations. Specifically, the periodicity and magnitude of the flows in the river channel obtained from the calculation were in good agreement with the observation.
(3) Velocities of the seaward current and backflow reached 50 cm/s, which is sufficiently strong to keep out coastal sand from the mouth of the river.
(4) The backflow rate is low in April due to the increase in runoff caused by snowmelt, increasing in July when precipitation decreases.

References

1. M. Sasaki, K. Muraoka, M. Atsumi and Y. Komatsu, History of river channel blockage and stability of river channel, *Proceedings of Coastal Engineering in Japan*, 1990, pp. 344–348. (in Japanese)
2. M. Takahashi, Y. Fujita, M. Sasaki, M. Ikeda and K. Kawana, Field observations of spatial-temporal changes in water quality after flooding in Lake Jyusan, *Proceedings of Hydraulics Engineering in Japan*, 1994, pp. 183–191.
3. M. Sasaki, Y. Tamura and Y. Fujita, Characteristics of salt water movement in Lake Jyusan, *Proceedings of Hydraulics Engineering in Japan*, 1997, pp. 501–507. (in Japanese)
4. M. Sasaki and T. Takeuchi, History of river blockage and stability of the Iwaki River mouth, *Proceedings of 15th IAHR Congress*, Thessaloniki, Greece, CD-ROM, A3, 24–29, August 2003.

Advances in Geosciences
Vol. 17: Hydrological Science (2008)
Eds. Namsik Park *et al.*

HOW REMOTE SENSING TECHNOLOGY CONTRIBUTES TO WATER STUDIES

MOHAMAD RUKIEH
Counselor at Minister of Communications and Technology-Syria
Syria, Damascus, Mazeh, PO Box 9916
mrukieh@gmail.com

This research deals with the manner of using remote sensing data in water studies. We refer in this study to the unique properties of space imagery and the spatial survey systems that are useful for this kind of studies (spectral, thermal, radar, T.V surveys), also we indicate the knowledge levels of hydrological maps resulted from using remote sensing data. This research refers to the work steps in water studies and to the remote sensing data analyses and interpretation. Then we refer to the surface water study and the hydrological mapping by using remote sensing techniques with explanation for all characteristics and conditions related to surface water. After that we refer to the ground water study and the hydrological mapping by taking benefit from remote sensing techniques, as we refer to several important factors such as rainfall quantity, thermal system, relief, watercourses system, feed and surface water basins, vegetation cover, filtering specifications of soil and rocks, tectonic characteristics, and the degree of faulting, fracturing and cracking, and the structural position of the study area.

All those factors find direct or indirect reflection in the physiographical structures of the land surface which can be determined with the help of remote sensing techniques and the thematic mapping with scales range from 1/15,000,000 to 1/3,000.We obtained important results through this research, which can be depended on for water studies that save time, efforts and costs.

1. Introduction

The remote sensing techniques and space images can contribute — through their unique specifications, (Synoptic Coverage, Repetitive Coverage, Different Scales, Spatial Resolution, the Wide Spectral Characteristics of Space Images, The High Informatics of Space Images, process these images by different methods of processing and new software's, and using them within geographic information systems.)

Various kinds (spectral, thermal, radar) and different scales — in studying the geological mediums, and water resources (surface and ground water), actually water studies depend mainly on studying and recognizing the tectonic and geologic situation of the studied areas.

2. Surface Water Studies and the Hydrological Mapping

Surface water is one of the most important phenomenons which can be recognized by space images and remote sensing techniques, it is typical and natural sign, because it forms an absorbent material to the sun radiation, in both visible and infrared range.

Therefore, it appears as dark color in the space images. And the absorption increases as more as the water is turbid and polluted.

The soil contents of water is also a good material to be determined by remote sensing techniques, where is the increase moisture in the soil leads to the increase of the radiation absorption in the visible and infrared range, because of that, the wet soils seem darker than the dry ones, where the darkness increases according to the increase of the moisture ratio in the soil.

In spite of the possibility of studying the surface water simply and directly through remote sensing techniques, the indirect methods are used for studying and determining the ground water. The spectral coefficient differs according to the wave length and the radial characteristics related to the nature of water and the other surrounding conditions; however we can obtain the information related to the conditions and kinds of the water bodies through the reflective measurements in the visible field.

The study of the spectral properties of the water bodies showed that they take a wide range of the conditions related to the absorption characteristics and the conductivity, whereas the clean water absorb little quantity of the radiation coming in the visible field, less than 0.6 micrometer. On the contrary side, the transmissibility power is high in the short part of the visible field, and it decreases according to the increase of the turbid degree and the quantity of the materials which are existed in the water.

Using remote sensing techniques for studying the conditions of forming surface and ground water is very important, because these techniques help in studying all the factors which are connected to water formation, as geomorphologic, climatic, lithological, botanical, structural, tectonic position and other factors. Since we can define a lot of characteristics and conditions related to surface water through these techniques such as:

1. Determining of the area, dimensions and the shapes of the water bodies, and the way they change seasonally and yearly. As well as remote

sensing techniques help in the floods prediction and determine the consequences resulting from it and the areas they covered, and follow up the changes directly. We can also follow-up the water abatement in the rivers and water bodies in the dry seasons and its effect on environment.

2. Determining the boundaries of the water shed basins and defining the supplying areas. (Fig. 1)
3. Preparing Drainage system schemes accurateily. (Fig. 2)
4. Determining the fresh springs in the seas. (Fig. 3)
5. Determining the sites of the hot springs and studying the underground thermal energy. (Fig. 4)

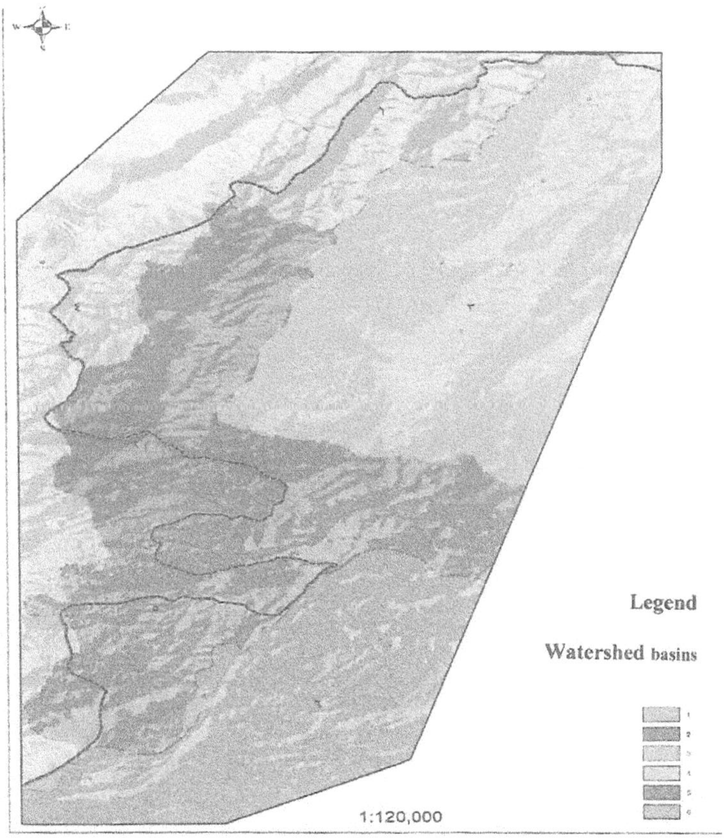

Fig. 1. Scheme watershed basins west of Damascus — SYRIA By interpretation of space images — GORS, 2005.

Fig. 2. Drainage system in wadi almiah Syria by interpretation of space images — Rukieh, 2004.

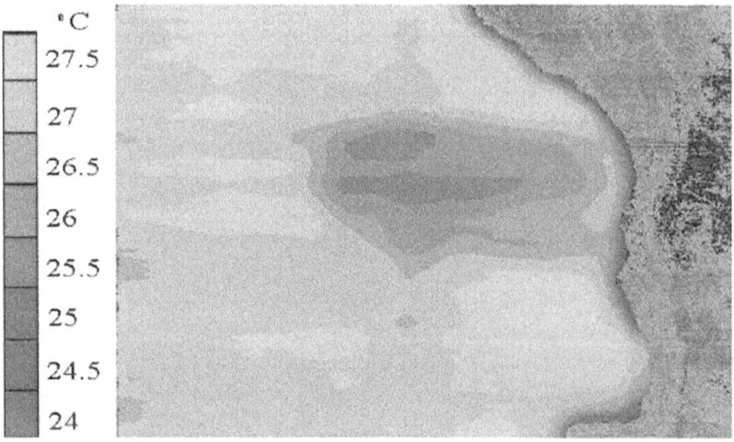

Fig. 3. Thermal survey image show fresh sprigs in the med. Sea south of Banyass — Syria, 1998 (dark tone).

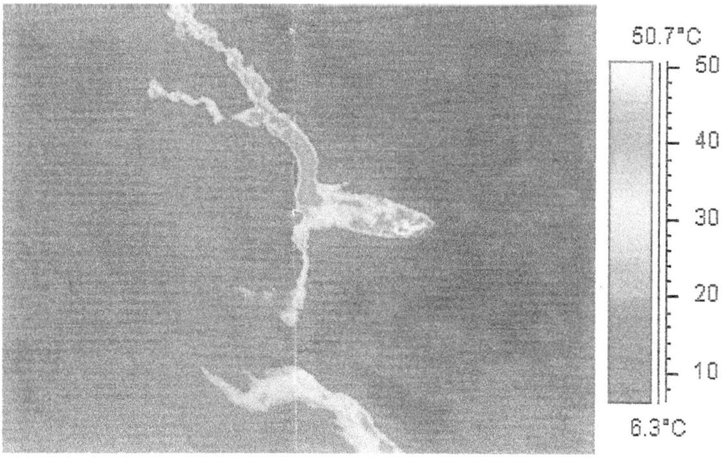

Fig. 4. Bear valley creek, hot spring is 52° C.

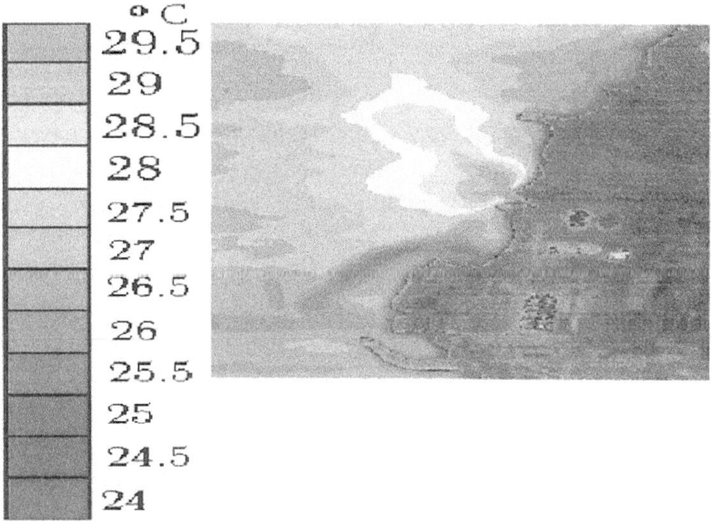

Fig. 5. Thermal survey image show hot pollution in sea water Syria, 1998.

6. observating volcanic irruption.
7. Determining thermal pollution in water (Fig. 5)
8. Determining soil moisture degree.
9. Determining the pollution types in the water bodies, and their seasonal and yearly changes, including the pollution by petroleum derivatives,

Fig. 6. Spot image show petroleum pollution in Arabic Golf (Red color) result war 1990.

especially in the coastal areas. This can be done through wide spectrum range starting from ultraviolet then blue, green and red till the thermal infrared, depending on the kinds and thickness of the pollutants. (Fig. 6)

10. Determining the chlorophyll ration in water, the chlorophyll concentration in water affect the spectrum range, where the increase of the concentration leads to decrease of reflective energy ration in the blue range, but it increases in the green one. This sign is an indicator of the productivity of the green material and growing of the mass cover, and its rottenness in the water bodies.
11. Choosing more perspective areas and sites for detailed studies in the future.

3. Ground Water Study and Hydrogeological Mapping

The ground water plays an increasing role in the development of national economy, especially in the dry and semidry areas, which characterize a lot of areas in the Arab world. And for that we must achieving and designing the best suitable methods for ground water investigation; and the most important achievement in this field is the use of remote sensing and space images in studying ground water.

The ground water doesn't find direct reflection in the space surveying data, however the analysis depend on the interpretation, and determining of the physiography of the landscape , which reflect the rules of formation and distribution of the groundwater.

The ground water study and its relation with the surface water can be done through the analysis of several important factors such as: the yearly rainfall, thermal system, relief, Drainage system, surface water basins, vegetation cover,filtring characteristics of soil and Rocks, Tectonic characteristics, structural position of the studied areas etc... these factors find direct or indirect reflection in the physiographic structures of the landscape, which can be determined by remote sensing techniques.

It is very important to study the connection zones between surface and groundwater, in order to evaluate the exchangeable relation between the surface and ground water, which can be studied only by using remote sensing data, the study and analysis of the water exchange systems by these data allow determining the hydrological conditions which affect the formation of the fresh and less salty groundwater positions.

The space images are widely used for studying the geological position in the visible and infrared ranges, especially 0.5–0.86 micrometer where this ranges gives good information about the geological status and its hydrological and tectonic signs. But for studying the thermal characteristics of the geological medium, especially studying the hydrogeological conditions, we use the thermal band, which has great importance, particularly in the range 2.5–14 micrometer. The thermal micrometer survey of the microwaves in the range 0.3–30 cm used for studying soil moisture and determining the depths of the groundwater, also the radar discover the undersurface structure and the depth in the point radar to 100 m in the dry and sandy lands.

The investigation for groundwater by spatial remote sensing methods can be defined through showing the following factors and indicators:

- The general formation rules of groundwater positions, which means local, temporal and fixed original relation of the groundwater with land natural compositions.
- The determination factors of water which include some of the composition of hydrogeological conditions which affect mainly the water formation in the studied areas among these factors: the geostructural,

Geomorphological and lithological; they are divided into positive factors which ensure the position of the groundwater, and negative ones.
- The investigation signs which describe the possibility of the ground water availability inside the perspective areas according to the determine factors.

The remote sensing methods are mainly used in the regional research, since they are the most affective in directing the prospective works on the ground water, besides the shortness of time and costs.

Using remote sensing techniques in the regional research depend on preparing group of maps of small and medium scales (1/5 million, 1/100,000). On the base of these maps, we can choose the perspective areas for the detailed exploration.

In the detailed works, we prepare maps of 1/50,000, 1/25,000 1/10,000 and 1/5,000, then we can choose the suggested wells sites for drilling. It is known that water is found in the affected tectonic areas and in the high penetrative rocks, for this reason, remote sensing methods are basic for the detection of these areas. Preparing lithological and faulting maps and their density (Rukieh 1991, 1993, 2005).

In the volcanic areas, remote sensing techniques help in preparing better maps for the various lava flow materials and sedimentation between them. The field checks help in recognition the porous rocks which can be considered water.

At the end, a description of the general rules of the different kinds of the ground water positions and their determining factors is possible, as well as the exploration signs which can be seen clearly on the space images, for example, the ground water in the river valleys, in the sandy Rocks, faulting zones karsts areas *et al.* ...

3.1. *The vegetation cover role*

The vegetation cover and its density, especially in the semidry areas, reflect the moisture rate in the atmosphere and the water kind, however, it is an important indicator on the water availability, although some researchers think that the vegetation cover can restrict the interpretation of space and air images, but at the same time it represents a recognized application to the geophytes analysis, whereas it gives fantastic results.

3.2. *Results*

1. *Remote sensing survey methods, spectral, thermal, and radar form quick and very active facility in water resources studying, preparing regional maps and choosing perspective Areas for details studies.*
2. *The Remote Sensing methods study wide range matters relation with fresh surface water, or hot water, mineral water, and its pollution.*
3. *These methods with field check work cave various thematic maps (lithological, tectonic, Geomorphological, water drainage, water shed basins, splay Zones, which help direct or indirect in determination of ground water basins and sites suitable for wells drilling and its exploiting.*

References

1. Rukieh M, Water and Remote Sensing Techniques, *Remote Sensing Journal* **14** (December 2002), Damascus GORS.
2. M. Rukieh, A. A. Alkafri and A. A. Khalaf, New data about the tectonic activity in Wadi Almiah in Syria by space images interpretation, *Remote Sensing Journal* **16** (September 2004), Damascus GORS.
3. M. Rukieh, Physiographic analyze and letological mapping from space images (case study from Syria), *Remote Sensing Journal* **18** (September 2005), Damascus.
4. H. Ibrahim and A. Ammar, Report: Use R.S. and GIS in studying fresh Springs in sea water, in Syrian coastal area, IRZ report Company, 2002, Damascus GORS. Use near and thermal infrared survey for some Rivers in Idaho Region, USA.
5. M. Rukieh and H. Mnini, Report about using R.S. to study water Resources in West area at Damascus. Syria, GORS, 2005. Rukieh M. 1993 Report — New data about Volcanic Rocks in south Syria by interpretation of space images, Damascus-GORS.
6. M. Rukieh, *Re port on the tectonics of the Western part of Syria with Lebanon and the Structure of the Arabian Rift by interpretation of space images.* GORS-Damascus, 1991.

Advances in Geosciences
Vol. 17: Hydrological Science (2008)
Eds. Namsik Park *et al.*

WATER BALANCE USING MODIS SATELLITE DATA*

PREEYAPHORN KOSA
Department of Science and Engineering,
Kasetsart University Chalermphrakiat
Sakonnakhon Province Campus,
59 Moo1 Sakonnakhon-Nakhonphranom Rd.
Moung District Sakonnakhon Province 47000,
Thailand
preeyaphorn.k@ku.ac.th

KOBKIAT PONGPUT
Department of Water Resources Engineering,
Kasetsart University, 50 Phahon Yothin Rd,
Chatuchak Bangkok 10900, Thailand

In the large area, there is a tendency towards higher water demands than in the past, so significant water storage, such as reservoirs and canals, has been added in the watershed area. On the one hand floods often occur during the rainy season, because there are rain storms and excessive water in the reservoirs. On the other hand, during the dry season there is insufficient water, because water in the reservoirs is released without accurate planning during the rainy season. Thus, improving water resources management is becoming more and more necessary. The purpose of this session is to calculate water budget for the accurate planning of water management. Then, there are three main parts of this research: actual evapotranspiration; rainfall; and water budget. First, actual evapotranspiration was calculated by SEBAL and the FAO Penman-Montieth equation. The time period for actual evapotranspiration calculations from SEBAL and from the FAO Penman-Monteith computations is one year. In the SEBAL method, MODIS images were used for daily actual evapotranspiration estimation, while weather data from weather stations was used to produce daily actual evapotranspiration using the FAO Penman-Monteith equation. Thereafter, actual evapotranspiration values calculated from the FAO Penman-Monteith method were compared with actual evapotranspiration values calculated from SEBAL to determine the suitable actual evapotranspiration for the calculation of water budgets. Second, rainfall was obtained from TRMM images and rain gauge stations. The time period for rainfall data from TRMM images and from rain gauge stations is one year. Thereafter, monthly rainfalls from TRMM image were compared with monthly rainfall from rain gauge station to

*The Royal Golden Jubilee Ph.D. Program and International Water Management Institute

determine the suitable monthly rainfall for the calculation of water budgets. Finally, monthly water budgets were calculated. The main inflow for basin is rainfall, while the main outflow for each basin is actual evapotranspiration. Then, monthly actual evapotranspiration and monthly rainfall were utilized for the monthly water budget calculations. Regarding, the contribution of this session, the combination of evapotranspiration and rainfall from satellite imagery to compute water budget in large scale is new. Accordingly, after the results that can be applied to river basins and analyzing water management, the methodology can also be applied in other countries around the world.

1. Introduction

Water availability is a key factor for constraining future agricultural development in the upper Chao Phraya river basin, because it is the core of multi-cropping and for crop diversification. Water management for agricultural areas must aim to achieve greater equity and efficiency in water deliveries at the basin, irrigation project, and farm level. For example the achievement of high water use efficiency through crop diversification depends on much greater local level control on water deliveries.

The availability of water can be evaluated from evapotranspiration, rainfall, and water budget or water balance studies. Evapotranspiration is an outflow but rainfall is an inflow in water budgets. This study evaluated water budgets using evapotranspiration and rainfall, which were calculated from satellite images and also measured data. Recent remote sensing technological developments in evaluating evapotranspiration and rainfall over large areas have not been applied in Thailand. This technology is interesting to apply in the field of water resources management regarding planning, development and operation.

The Chao Phraya river basin is a large basin in Thailand. During the wet season the outflow from this basin is the main factor of flooding in the lower Chao Phraya river basin contained many economic areas. On the other hand, during the dry season there is insufficient water because water in the reservoirs is released without accurate planning during the rainy season. Thus, improving water resources management is becoming more and more necessary. Then, the purpose is to calculate water budget for the accurate planning of water management.

2. Methodology

2.1. *The chao phraya river basi*

In Thailand, the Chao Phraya river basin is the largest and most important geographical unit in the terms of land and water resources development.

It is located in the north and central regions of the country. The Chao Phraya river basin includes the upper and lower Chao Phraya river basin. The upper Chao Phraya river basin is consisted of Ping, Wang, Yom and Nan sub-basins while the lower Chao Phraya river basin is contained Sakae Krang, Pasak, Tha Chin and Chao Phraya sub-basins.[1]

In the upper Chao Phraya river basin, there are four large rivers that are the Ping, the Wang, the Yom, and the Nan Rivers. These four upstream rivers flow southward to meet at Nakhon Sawan and form the Chao Phraya River. The Chao Phraya river flows southward through a large alluvial plain or the lower Chao Phraya river basin, splitting into four channels that are the Tha Chin, the Noi, the Lop Buri and the Chao Phraya Rivers. The average yearly rainfall is about 1,200 mm, where the peak is in September and the dry months are November through April. The rainy season starts from the middle of April and ends in late October.[2]

2.2. *Actual evapotranspiration*

Actual evapotranspiration was calculated by SEBAL and the FAO Penman-Montieth equation. The time period for actual evapotranspiration calculations from SEBAL and from the FAO Penman-Monteith computations is one year (January–December 2002). In the SEBAL method, 27 MODIS images were used for daily actual evapotranspiration estimation, while weather data from 33 weather stations was used to produce daily actual evapotranspiration using the FAO Penman-Monteith equation.

The SEBAL algorithm is a tool to compute actual evapotranspiration for flat areas with the most accuracy and confidence. Satellite image and weather data are used in the SEBAL model to calculate actual evapotranspiration by using a surface energy balance. SEBAL evaluates an instantaneous actual evapotranspiration flux for the image time, because the satellite image provides information for the overpass time only. The actual evapotranspiration flux can be calculated for each pixel of the image as a residual of the surface energy budget equation[3]:

$$LE = (R_n - G) - H \tag{1}$$

where LE is the latent energy of evaporation (W/m^2), R_n is the net radiation flux at the soil surface (W/m^2), G is the soil heat flux (W/m^2), and H is the sensible heat flux to the air (W/m^2).

After LE was computed, the Evaporative Fraction (Λ) is the next value that is obtained using Eq. (2). The Evaporative Fraction at each pixel of a satellite image can be estimated using the 24-hour evapotranspiration for the day of the image. The Evaporative Fraction is assumed to be a constant value over the full 24-hour period.

$$\Lambda = \frac{LE}{R_n - G} = \frac{LE}{LE + H} \tag{2}$$

To estimate 24-hour actual evapotranspiration, the following equation is utilized.

$$ET_{24} = \frac{86400\Lambda(R_{n24} - G_{24})}{\lambda} \tag{3}$$

where R_{n24} is daily net radiation, G_{24} is daily soil heat flux, 86,400 is the number of seconds in a 24-hour period, and λ is the latent heat of vaporization (J/kg). The 24-hour actual evapotranspiration, ET_{24}, can be expressed in mm/day.

Basically, SEBAL is suitable for flat areas but the upper Chao Phraya river basin, on which this study is based, includes both flat and mountainous terrain. The Digital Elevation Model or DEM was used to adjust the value in mountain area. DEM is a representation of the topography for the Earth's surface by an array of numbers representing heights above a reference datum. In this research, both contour map elevation and recorded elevation data were used as input data for the GIS software to calculate the spatial DEM in the upper Chao Phraya River Basin.

Landsat 7 images were also used to estimate 24-hour actual evapotranspiration by using the concept of SEBAL. Six Landsat 7 images were used for actual evapotranspiration calculation. These six Landsat 7 images were selected because they are also based on less cloud condition.

To calibrate actual evapotranspiration of 27 MODIS images, the relationship between 24-hour actual evapotranspiration from six MODIS images and from six Landsat 7 images in the same date was obtained. Landsat 7 (30 m and 60 m) image was selected for this calibration because it is included a higher resolution than MODIS image (250 m and 1000 m).

The FAO Penman-Monteith method includes the calculation of actual evapotranspiration (ET_c) that can be estimated from the multiplication of crop coefficient (K_c), and reference crop evapotranspiration (ET_o) as following equation.

$$ET_c = K_c \times ET_o \tag{4}$$

The reference crop evapotranspiration (ET_o) can be calculated on a daily basis using the FAO Penman-Monteith equation as following[4]:

$$ET_o = \frac{0.408\Delta(R_n - G) + \gamma\frac{900}{T_M+273.2}u_2(e_s - e_a)}{\Delta + \gamma(1 + 0.34u_2)} \quad (5)$$

where ET_o is reference crop evapotranspiration (mm d^{-1}), R_n is net radiation (MJ $m^{-2}d^{-1}$), G is soil heat flux (MJ $m^{-2}d^{-1}$), T is air temperature (C), e_s is saturation vapor pressure at air temperature (kPa), e_a is vapor pressure of air (kPa), u_2 is wind speed at 2 m (m s^{-1}), Δ is slope of saturation vapor pressure curve at air temperature (kPa C^{-1}), and γ is psychrometer constant (kPa C^{-1}).

Thereafter, reference crop evapotraspiration was interpolated from point value to be spatial value using the concept of Co-Kriging. Co-kriging is a geostatiatical technique developed to improve the estimation of a variable using the information on other spatially correlated variables which are generally better sampled.[5,6] Consider a set of experimental values of reference evapotranspiration or ET_o: $[ET_{o,i}, i = 1, 2, \ldots, 12]$ and a set of experimental values of elevation or Y: $[Y_k, k = 1, 2, \ldots, 6252]$. 12 is number of month in one year and 6252 is number of elevation point that is converted from an elevation contour map by a grid of 5 km × 5 km. This elevation contour map was supported by Royal Irrigation Department. In this study, the value of ET_o at space location x_o from ET_o and Y was estimated. A following equation presents the estimation of ET_o (ET_o^*) at any point x_o.

$$ET_o^*(x_o) = \sum_{i=1}^{12} \lambda_i^1 ET_o(x_i) + \sum_{k=1}^{6252} \lambda_k^2 Y(x_k) \quad (6)$$

where λ_i^1 and λ_k^2 are the weighing factors for first and second known variables.

The crop coefficients (K_c) depend on the crop type and the development of crop. There may be several reference crop values for a single crop depending on the crop's stage of development. For this study, three values of the crop coefficient for description the crop coefficient value were used; those during the initial stage ($K_{c,ini}$), the mid-season stage ($K_{c,mid}$) and the end of the late season stage ($K_{c,late}$).[7] Since the crop coefficients represent the crop type, the land use map and the cropping pattern were input data to indicate the crop type for each area. The main crop types cultivated in the study area consist of rice, sugarcane, maize, cassava, fruit tree and small vegetation. Also the crop coefficients depend on the stage of crop development so cropping calendar was used to denote the stage of crop development in one year.

To determine the linear relationship between the variables of actual evapotranspiration, actual evapotranspiration from SEBAL were compared with that from the FAO Penman-Monteith method using the concept of the correlation coefficients. The result is that the averaged correlation coefficient is 0.79. This result presents that there is a high linear relationship between actual evapotranspiration from SEBAL and the FAO Penman-Monteith method.

The advantage of actual evpaotranspiration computed by SEBAL is spatial resolution while the advantage of actual evpaotranspiration computed by the FAO Penman-Monteith method is temporal resolution. Since actual evapotranspiration computed by SEBAL was calculated only for 27 days, actual evapotranspiration the FAO Penman-Monteith method on other days was used. Thereafter daily actual evpaotranspiration was summarized to be monthly actual evpaotranspiration.

2.3. *Rainfall*

Monthly rainfall was obtained from TRMM images and the time period is one year (January–December 2002). Monthly rainfall from TRMM images were provided by NASA. For the rainfall calculation using TRMM image, Algorithm 3B-42 is used. The Algorithm 3B-42 will use Tropical Rainfall Measuring Mission (TRMM) merged high quality (HQ)/infrared (IR) precipitation and root-mean-square (RMS) precipitation-error estimates to calculate rainfall. These girded estimates are on a 3-hour temporal resolution and a 0.25-degree by 0.25-degree spatial resolution in a global belt extending from 50 degrees south to 50 degrees North latitude.

Before monthly rainfall from TRMM images was used, it was compared with monthly rainfall from recorded monthly rainfall using the concept of the correlation coefficients to determine the linear relationship between the variables of monthly rainfall. There are 230 rain gauge stations measured by RID. For each month, there are 100 coordinates used for comparison and these 100 coordinates are the same location. The result is that the correlation coefficients from January to December consist of 0.87, 0.63, 0.73, 0.78, 0.49, 0.81, 0.48, 0.53, 0.84, 0.73, 0.87, and 0.85, respectively. In May, July and August the correlation coefficients are low because there are storm rainfall in many place. This result presents that there is a high linear relationship between monthly rainfall from TRMM image and from rain gauge station.

2.4. *Water balances*

In hydrological cycle, water is always in movement and the hydrological cycle is described by the continuous movement of water on, above and below the Earth's surface. There is no beginning or end in the hydrological cycle because it is truly a cycle. Water is evaporated from the land and ocean by the latent heat and the water vapor rises. Thereafter, it condenses into clouds and they will be rain. The rain hits the ground and part runs off, part infiltrates into the ground, and part evaporates back up into the sky. The water from soil will be taken by the plants and vegetation and the water will be released into the air through transpiration. Other parts of the ground water enter streams and flows onto the surface. Eventually, the surface water that runs off in streams, towards the ocean, will be evaporated and continued the cycle. To concern the amount of water in the hydrological cycle, the concept of the water budget is considered.[8,9]

The water balance concerns the amount of water entering a system (through rainfall and river and groundwater flows) and the amount leaving a system (through evaporation, plant transpiration and river and groundwater flows) as presented in Fig. 1.

The water balance may be expressed as following equation

$$(P + Q_{si} + Q_{ssi}) - (ET + Q_{so} + Q_{sso}) = \Delta S \tag{7}$$

where P is rainfall (MCM), Q_{si} is surface inflow (MCM), Q_{ssi} is subsurface inflow (MCM), ET is evapotranspiration (MCM), Q_{so} is surface outflow (MCM), Q_{sso} is subsurface outflow (MCM), and ΔS is change in storage within the domain that consists of change in surface water, subsurface water and percolation.

For this study, the domain of interest at each site represents a sub-basin that consists of agricultural, municipal, industrial and forest uses of water. The size of each sub-basin is sufficient to analyze separately from the entire basin in which it is located. The time period of interest is from January to December in 2002.

To compute water balance in eight sub-basins, Ping, Wang, Yom, Nan, Sakae Krang, and Pasak sub-basin were thought as the upper sub-basin because they are included no surface inflow at the apex of their

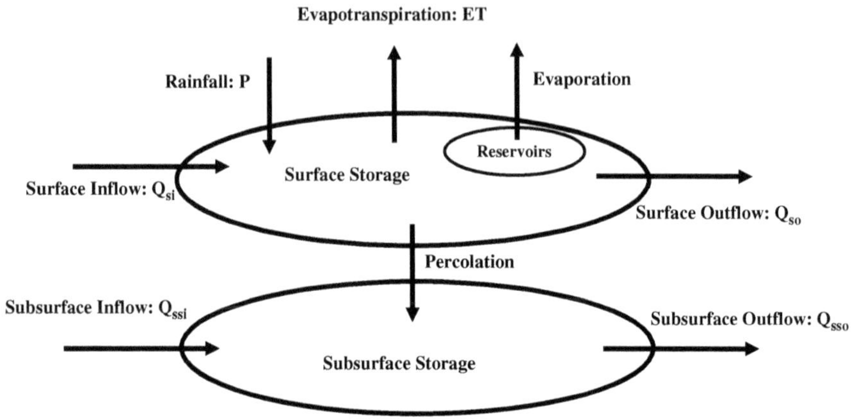

Fig. 1. The components of water balance concept.

sub-basins. Water balance in Wang sub-basin was firstly computed to determine subsurface outflows and the change in storage for each month. The subsurface and surface outflows of this sub-basin were side flow entering Ping sub-basin. Water balance in Ping sub-basin was then calculated to concern subsurface outflows and the change in storage. Thereafter, the water balance in Yom, Nan, Sakae Krang, and Pasak sub-basin were also calculated to obtain subsurface outflows and the change in storage for each sub-basin. Finally, subsurface and surface outflows from Ping, Yom, Nan, Sakae Krang, and Pasak sub-basin were inflow entering Chao Phraya sub-basin. For surface inflow entering Tha Chin sub-basin, water was diverted from Chao Phraya sub-basin. Since it's difficult to separate subsurface inflow between Tha Chin and Chao Phraya sub-basin, water balance of these two sub-basins was calculated together. Then, surface outflow and the change in storage of two sub-basins were results.

In the upper watersheds, the surface inflow (Q_{si}) and the subsurface inflow (Q_{ssi}) equals to zero. The rainfall was derived from meteorological data and TRMM (3B42 V6) image while the actual evapotranspiration was derived from MODIS satellite imagery, as well as from meteorological data. This actual evapotranspiration was for non-irrigated, irrigated and forest area. However, it can be presented the evaporation from reservoirs and bare soil. The surface outflow (Q_{so}) was obtained from existing stream gauging records that are consisted of P17, W4A, Y17, N10A, CT8, and S2 station for Ping, Wang, Yom, Nan, Sakae Krang, and Pasak sub-basin, respectively. For surface outflow of Chao Phraya and Tha Chin sub-basin, Molle (2001) said that, to control the intrusion of saline water, minimum flows for the Chao Phraya and Tha Chin Rivers are respectively 50 and 35 m^3/s. Also, the main difficulty is that the discharge of the Chao Phraya and Tha Chin Rivers at its mouth cannot be easily measured because of the effect of tidal. Then, minimum flows (50 and 35 m^3/s) are used in this study. The subsurface outflows (Q_{sso}) and the change in storage (ΔS) were derive by the concept of trial and error under the condition assumed that annual change in storage for each sub-basin equal to zero because storage value recovers close to zero at the beginning of each year. Since the subsurface water and the change in storage are results of this computation, it's important to validate the significance level of these results using the statistical significance.

2.5. *Validation*

The subsurface water and the change in storage were validated by the significance level using the statistical significance. In normal English, "significant" means important, while in Statistics "significant" means probably true (not due to chance). A research finding may be true without being important. When statisticians say a result is "highly significant" they mean it is very probably true. They do not (necessarily) mean it is highly important. Then, significance is a statistical term that tells how sure you are that a difference or relationship exists.

To determine the significant level, the concept of t-test was concerned. The t-test assesses whether the means of two groups are statistically different from each other. This analysis is suitable whenever the means of two groups is compared. The equation for the t-test is a ratio between the difference of two means or average and a measure of the variability or dispersion of the scores. The t-test equation can be presented by following equation

$$t = \frac{\overline{X}_1 - \overline{X}_2}{\sqrt{\left(\frac{(N_1-1)S_1^2+(N_2-1)S_2^2}{N_1+N_2-2}\right)\left(\frac{N_1+N_2}{N_1 N_2}\right)}} \tag{8}$$

and

$$df = N_1 + N_2 - 2 \tag{9}$$

where df is degrees of freedom, $\overline{X}_1$ and $\overline{X}_2$ are the mean of 1st and 2nd group data, respectively, N_1 and N_2 are the size of 1st and 2nd group data, respectively, and S_1^2 and S_2^2 are the variance of 1st and 2nd group data, respectively, that can be computed by following equation

$$S^2 = \frac{N\sum X^2 - (\sum X)^2}{N(N-1)} \tag{10}$$

3. Result and Discussion

In Ping sub-basin, there is water shortage in January, February, March, April, and December that are −2110.5, −3238.0, −3741.3, −1672.0, and −1704.0 million m^3 (MCM), respectively. During May to November, there is excess water that are 4096.0, 606.2, 1160.8, 4240.0, 1548.8, 563.8, and 250.3 MCM, respectively. An annual subsurface outflow is 9716.4 MCM. The highest values of subsurface outflow can be found in May that is 3162.5 MCM. The significance level of the result is 81.99 %.

In Wang sub-basin, there is water shortage in January, February, March, April, November, and December that are −481.2, −702.8, −692.0, −145.9, −160.3, and −409.7 MCM, respectively. During May to October, there is excess water that are 828.2, 255.2, 479.6, 727.0, 157.4, and 144.5 MCM, respectively. An annual subsurface outflow is 3265.1 MCM. The highest values of subsurface outflow can be found in May that is 1487.4 MCM. The significance level of the result is 88.76 %.

In Yom sub-basin, there is water shortage in January, February, March, April, September, October, November, and December that are −842.1, −1381.3, −1309.0, −1201.3, −30.9, −79.6, −372.8, and −591.7 MCM, respectively. During May to August, there is excess water that are 2503.6, 152.4, 707.3, and 2445.4 MCM, respectively. An annual subsurface outflow is 6531.8 MCM. The highest values of subsurface outflow can be found in August that is 2684.7 MCM. The significance level of this result is 86.89%.

In Nan sub-basin, there is water shortage in January, February, March, April, September, October, November, and December that are −1102.1, −1961.5, −2093.5, −1223.17, −1073.3, −1065.2, −1857.9, and −456.9 MCM, respectively. During May to August, there is excess water that are 3220.5, 478.4, 2168.4, and 4965.2 MCM, respectively. An annual subsurface outflow is 6089.7 MCM. The highest values of subsurface outflow can be found in August that is 2089.4 MCM. The significance level of this result is 76.36%

In Sakae Krang sub-basin, there is water shortage in January, February, March, April, November, and December that are −185.5, −200.3, −275.3, −95.3, −61.6, and −207.0 MCM, respectively. During May to October, there is excess water that are 124.5, 49.7, 10.1, 382.8, 230.9, and 227.1 MCM, respectively. An annual subsurface outflow is 1312.7 MCM. The highest values of subsurface outflow can be found in September that is 362.6 MCM. The significance level of this result is 93.71%.

In Pasak sub-basin, there is water shortage in January, February, March, April, September, October, November, and December that are −683.2, −836.2, −239.6, 394.1, −144.0, −483.6, −629.1, and −421.7 MCM, respectively. During May to August, there is excess water that are 636.9, 217.2, 241.3, and 2736.1 MCM, respectively. An annual subsurface outflow is 3092.9 MCM. The highest values of subsurface outflow can be found in May that is 802.1 MCM. The significance level of this result is 79.99%.

In Chao Phraya and Tha Chin sub-basin, there is water shortage in January, February, March, April, and December that are −3179.1, −3400.3, −3796.2, −2726.5, and −568.4 MCM, respectively. During May to November, there is excess water that are 4396.0, 484.8, 755.7, 5036.4, 1238.1, 636.5, and 1123.0 MCM, respectively. An annual subsurface outflow is 32651.4 MCM. The significance level of this result is 98.05 %.

4. Conclusion

As monthly rainfall and monthly actual evapotranspiration were utilized to concerned water balance calculation, the result of this calculation can be concluded that there is water shortage during dry season (January, February, March, April, November, and December) while there is excess water during wet season (May to October). In addition, in Ping, Wang, Yom, Nan, Sakae Krang, and Pasak sub-basin, the summation of annual rainfall is more than the summation of annual actual evapotranspiration for all land use while, in Chao Phraya and Tha Chin sub-basin, the summation of annual rainfall is less than the summation of annual actual evapotranspiration for all land use. It means that, in Ping, Wang, Yom, Nan, Sakae Krang, and Pasak sub-basin, rainfall can support actual evapotranspiration if the excess water during wet season is storage to release during dry season. However, it is important to consider water loss from runoff and infiltration. For Chao Phraya and Tha Chin sub-basin, rainfall cannot support actual evapotranspiration so these sub-basins should be supported water from other sub-basin.

Acknowledgments

This study was supported by The Royal Golden Jubilee Ph.D. Program (RGJ), International Water Management Institute (IWMI), and Geo-Informatics and Space Technology Department Agency (Public Organization) or GISTDA.

References

1. A. Buddhapalit, *Proceeding of IDNDR Symposium*, 1999.
2. P. Sathianpantarit, A. Mujalinvimuti, P. Klinkhachorn, A. Punyachom and S. Jirasirisopon, *International Chao Phraya Delta Conferen*, 2002.
3. W. G. M. Bastiaanssen, M. Meneti, R. A. Feddes and A. A. M. Holtslag, *Journal of Hydrology* 212–213: 213–229 (1998).
4. R. G. Allen, L. S. Pereira, D. Raes and M. Smith, FAO irrigation and drainage paper No 56. FAO, Rome (1998).
5. A. G. Journel and C. J. Huijbregts (Academic Press, 1978).
6. E. H. Isaaks and R. M. Srivastasva (Oxford Univ., 1989).
7. R. G. Allen, L. S. Pereira, D. Raes and M. Smith, *FAO irrigation and drainage paper No 56. FAO, Rome*, 1998.
8. D. Molden, *SWIM Paper 1.*, 1997.
9. P. Natalia, R. Salthivadivel, N. R. Scott, E. Kendy, and T. S. Steenhuis, *Journal of Hydrology.* **292** (2003) 1–22.

Advances in Geosciences
Vol. 17: Hydrological Science (2008)
Eds. Namsik Park *et al.*

ANALYSIS ON THE RELATIONSHIP BETWEEN VEGETATION COVERAGE AND SOIL MOISTURE IN THE LOESS PLATEAU REGION OF CHINA

WEIWEI SHAO*
Department of Hydraulic Engineering,
Tsinghua Universtiy, Beijing 100084, China
sww04@mails.tsinghua.edu.cn

DAWEN YANG
Department of Hydraulic Engineering,
Tsinghua Universtiy, Beijing 100084, China

The growth of vegetation is affected by intermittent water availability, while it also feeds back to influence regional water balance. Analysis on the relationships between vegetation and water balance is significant for the research on the catchment's ecohydrological response to the climate change. A better understanding of the relationships between vegetation coverage and soil moisture variation will be useful for the explanation of the complicated interactions among climate changes, vegetation dynamics, and water cycles. In the present study, 51 catchments in the Loess Plateau region of the Yellow River basin in China were selected as the study area. Through the simplified representation of the average canopy water balance model developed by Eagleson, the mean vegetation coverage in study area is shown well correlated with the meteorological factors. The function of vegetation cover, potential canopy conductance and soil moisture is proposed for the Loess Plateau and the spatial distribution relationship between the average vegetation coverage and soil moisture is analyzed. The result indicates that there is an approximate positive relation between the vegetation coverage and soil moisture in space for the Loess Plateau region. It implies that the alternation of vegetation coverage due to climate change and human activity can be potentially predicted through the soil moisture change, and *vice versa*. This result is useful for assessing the long-term average effect of vegetation changes on catchment hydrology, and hence it has implications to the water resources management in the non-humid regions of China.

*Corresponding author: Weiwei SHAO, Tsinghua University, sww04@mails.tsinghua.edu.cn.

1. Introduction

1.1. *Ecosystem degradation problem in the yellow river basin*

In recent decades, water shortage and soil erosion have been the main ecosystem degradation problems in the non-humid regions of China due to deforestation, over-cultivation and intensive water resources development.[1] The Yellow River dried up along the lower reaches during the irrigation seasons since 1972, and the dry-up situation became worse rapidly in the 1990s. Decrease in river discharge caused rapid rise of riverbed along the lower reaches of the Yellow River due to the heavy erosion in the Loess Plateau and the sedimentation along the lower reaches.[2]

In 1998 the Chinese government initiated the "Nature Forest Protection Project (NFPP)" to prevent severe soil erosion and to ensure sustainable development.[3] The Loess Plateau, which is located in the middle reach of the Yellow River basin, is one of the "hot-spots" of environmental degradation with erosion rates ranging from 20,000 to 30,000 ton km^{-2} per year.[4] The government mandated in 1999 that forest cover should be significantly increased in the Loess Plateau. Increased afforestation should consider the complex impacts of vegetation coverage on the regional water balance in the water-limited areas, otherwise additional environmental problems may emerge. If afforestation is excessive, soil moisture will be exhausted and a dry soil layer will develop. There will not be enough available water to maintain normal growth rates in the re-vegetation area.[3] Understanding the interaction among the atmosphere, vegetation and the water cycle in the Loess Plateau is required to derive a guideline for effective afforestation of the soil water conservation practice in the north China.

1.2. *The Eagleson's water balance model*

Eagleson[5] proposed a statistical-dynamic model of average annual water balance during vegetation growing season. The model was constructed in terms of three main state variables: average root-zone soil moisture, canopy cover, and canopy conductance. Several approximate analytical solutions illustrating the dynamic interactions between the long-term seasonal mean vegetation and soil moisture were given by Eagleson.[5]

The Eagleson's average annual water balance model expressed the equilibrium relationship among climate, soil, and vegetation [5, Eq. 6.65]:

$$\underbrace{Mk_v^*\beta_v}_{1} \approx \frac{m_h}{m_{tb'}E_{ps}}\left\{1 - \underbrace{\frac{\overline{h_0}}{m_h}}_{2} + \underbrace{\frac{\Delta S}{m_v m_h}}_{3} - \underbrace{\frac{m_{tb''}E_{ps}}{m_h}(1-M)\beta_s}_{4} - \underbrace{e^{-G-2\sigma^{3/2}}}_{5}\right.$$
$$\left. - \underbrace{\frac{m_\tau K(1)}{P_\tau}s_o^c}_{6} + \underbrace{\frac{m_\tau K(1)}{P_\tau}\left[1+\frac{3/2}{mc-1}\right]\left[\frac{\psi(1)}{z_w}\right]^{mc}}_{7}\right\} \tag{1}$$

where the numbered dimensionless terms in Eq. (1) are:

(1) the canopy moisture flux;
(2) the space-time average surface-retained precipitation;
(3) the seasonal climate carryover soil moisture storage;
(4) the bare soil evaporation;
(5) the surface runoff;
(6) the deep percolation;
(7) the capillary rise.

The physical meanings for the parameters in Eq. (1) are listed in Appendix.

Eagleson[5] assumed transpiration occurs at the potential rate, $\beta_v = 1$, and noted that for a semi-arid region, the sum of terms (4), (6), and (7) in Eq. (1) is less than 3% of m_h (storm depth), which can be neglected [5, Eq. 6.66]. Eagleson[5] further proposed an approximation strictly applicable to the very dry, non-seasonal climate [5, Eq. 6.70]:

$$Mk_v^* \approx \frac{m_h}{m_{tb}E_{ps}} \tag{2}$$

where m_{tb} is the mean duration between storms. In the following, it is assumed that $m_{tb} = m_{tb'} = m_{tb''}$, since the difference among them can be neglected in arid or semi-arid regions, although they have different meanings.

2. Study Area and Data

The 51 catchments in the Loess Plateau of China are selected as the study areas, see Fig. 1. All the 51 catchments have relatively minor human interference such as dam operation and irrigation diversion. Table 1 summarizes the basic basin characteristics of these 51 catchments.

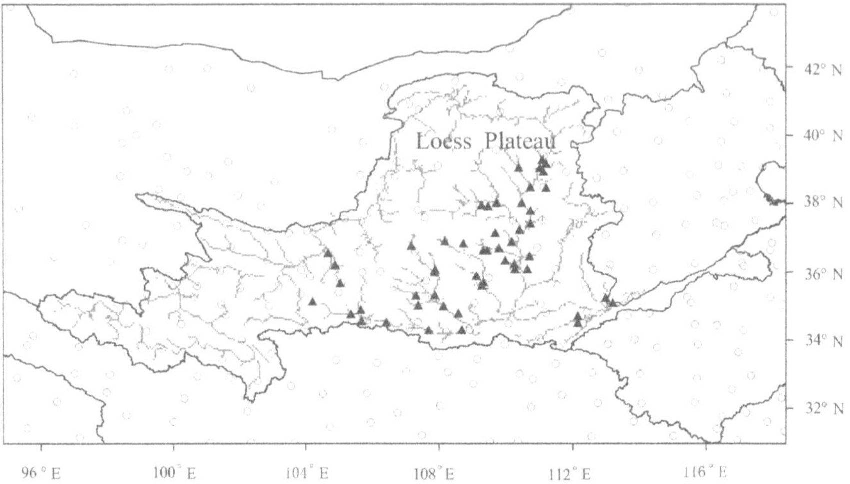

Fig. 1. The study area in the Loess Plateau of China. The triangles represent the 51 hydrological stations in the Loess Plateau, and the circles represent the 238 meteorological stations non-humid regions of China.

The long-term meteorological data at totally 238 meteorological stations over the study area were provided by the China Administration of Meteorology, see Fig. 1. The data consists of daily precipitation, air temperature, sunshine duration, wind speed, and relative humidity from 1956 to 2005. Additionally, daily solar radiation data at 47 of these 238 meteorological stations are available for the calculation of potential evaporation in this study. In addition, the 1951∼2000 monthly stream discharge data at the outlets of the 51 catchments were provided by the Hydrological Bureau of the Ministry of Water Resources of China.[6] By assuming the annual change of water storage to be zero, annual evapotranspiration from 1956 to 2000 is calculated for each catchment by subtracting annual discharge from annual precipitation.[7]

The catchment boundary is extracted using the 1-km DEM for the estimation of the area-average hydroclimatic variables. The procedures for calculating catchment — average precipitation and potential evaporation include the following:

- A 10-km gridded dataset covering the study area was interpolated from the station data by the distance-direction weighted average method (except the interpolation of temperature is modified with the altitude).

Table 1. The basin characteristics and the long-term (1982–2000) annual precipitation, actual evapotranspiration, potential evaporation, and vegetation coverage percentage of the 51 study catchments in the Loess Plateau.

Code	Area (km^2)	Mean values (mm/yr)			$\overline{M}$	Code	Area (km^2)	Mean values (mm/yr)			$\overline{M}$
		$\overline{P}$	$\overline{E}$	$\overline{E_0}$				$\overline{P}$	$\overline{E}$	$\overline{E_0}$	
42753	990	404	394	886	0.14	44261	436	468	447	887	0.41
42760	4853	378	368	867	0.12	45502	3440	395	364	858	0.12
42766	10647	375	365	837	0.13	45504	774	421	382	840	0.15
44003	2831	382	349	929	0.11	45511	17180	555	528	879	0.35
44018	1562	410	400	892	0.21	45546	4715	495	473	829	0.47
44037	1263	381	333	931	0.12	45562	2266	549	509	841	0.50
44039	2939	433	425	886	0.28	46005	600	465	426	751	0.32
44063	650	445	395	899	0.42	46024	4788	487	467	936	0.34
44071	3829	340	299	958	0.12	46037	37006	625	559	899	0.48
44077	8645	408	352	967	0.12	46043	46827	675	601	810	0.45
44091	1121	391	350	969	0.11	46069	2484	448	432	929	0.19
44099	283	426	386	982	0.24	46085	9805	410	394	914	0.25
44115	4102	428	388	934	0.39	46125	1019	470	426	922	0.33
44143	15325	331	306	988	0.10	46288	282	558	516	812	0.41
44153	29662	366	332	963	0.12	46517	14124	491	452	841	0.37
44161	2415	371	349	886	0.11	46520	40281	509	477	859	0.26
44167	327	386	348	918	0.08	46571	4640	372	353	967	0.12
44201	913	420	378	873	0.12	46591	10603	446	421	908	0.16
44203	3468	438	395	897	0.14	46593	2988	416	386	862	0.21
44227	3992	449	426	875	0.35	46597	19019	494	468	850	0.33
44235	5891	457	422	870	0.22	46623	928	488	449	949	0.38
44243	3208	420	378	867	0.15	47209	9713	824	674	871	0.37
44244	719	446	409	856	0.29	47275	829	585	500	1219	0.37
44253	1121	487	471	871	0.41	47320	12880	594	555	971	0.43
44254	1662	474	455	888	0.26	47373	3149	552	512	849	0.35
44259	2169	511	489	877	0.47						

- Potential evaporation at the daily time scale is estimated for each grid using the Penman equation following Shuttleworth.[8] The catchment average values are then calculated for each variable.[2]
- In the estimation of net radiation, solar radiation was calculated with an empirical equation involving the sunshine duration.[6] The first-order estimate of net long-wave radiation was derived from the relative sunshine duration, surface minimum and maximum air temperature, and vapor pressure, as per the method recommended by FAO.[9]

A global dataset of monthly NDVI from 1982 to 2000 with 1-km resolution was obtained from the NOAA/AVHRR. Monthly vegetation coverage was calculated from the NDVI data using the method proposed by Gutman and Ignatov.[10]

The observed soil moisture data of 420 stations over China are obtained from the Hydrological Bureau of the Ministry of Water Resources of China. The data show the volumetric soil moisture content at the 10 cm, 20 cm, 50 cm, 100 cm depth at the 10-day scale respectively from the year 1991 to 2007. The kriging interpolation method was used to get the average value of the top soil moisture of 10 cm, 20 cm and 50 cm depth during the growing season (April to September) from the year 1982 to 2000 in the validation the water balance model.

The long-term (1982–2000) average meteorology and vegetation data for the 51 catchments are summarized in Table 1.

3. Spatial Relationship between Vegetation Coverage and Soil Moisture Referring to the Eagleson's Water Balance Model

Following Eagleson's[5] water balance model (Eqs. (1) and (2)), the vegetation state variable Mk_v^* is considered to be related to the meteorological factor $(m_h/(m_{tb}E_{ps})$. In this study, the growing season is defined from the first storm in April to the last storm in September. During this period, one single rainfall event (storm) is identified whenever the accumulated precipitation amount is more than 1 mm. The statistics of the total rain depth and storm duration during the growing season for each catchment of the Loess Plateau are calculated. The mean seasonal potential evapotransporation E_0 and M are calculated for the same time period (April to September) for each year, where E_0 is the potential evaporation during the growing season calculated by the Penman equation.[8] When E_0 is averaged at the daily scale during the growing season, it is equivalent to E_{ps} [5, Eq. 5.20].

As can be observed from Fig. 2, a clear linear relationship is found between M and $m_h/(m_{tb}E_{ps})$ for the 51 catchments with correlation coefficients of 0.66, for the Loess Plateau. However, notice that the regression line in Fig. 2 has a slope less than 1; as $k_v^* \leq 1$, therefore, $Mk_v^* < m_h/(m_{tb}E_{ps})$, which suggests that in this region other water balance components such as base soil evaporation, surface-retained precipitation and surface runoff should also be considered when analyzing the canopy moisture flux using Eq. (2).

Here we assume that in the Loess Plateau with non-seasonal climate and deep groundwater conditions, the deep percolation, capillary rise,

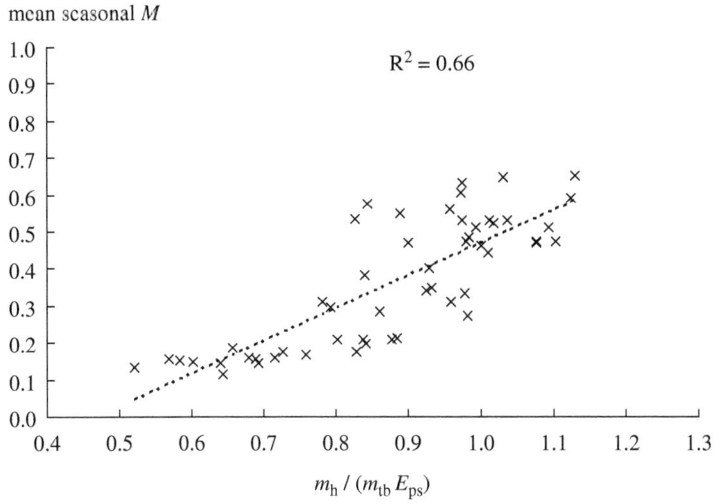

Fig. 2. Relationship between mean seasonal vegetation coverage (M) and the climate factor ($m_h/(m_{tb}E_{ps})$) in 51 catchments of the Loess Plateau.

and seasonal climate carry-over soil moisture storage can be neglected. However, the bare soil evaporation must be considered due to the non-dense vegetation condition in the Loess Plateau. Choudhury and Monteith[11] indicated that in arid or semi-arid regions, since the total energy available is limited by absorption of radiation, the increase of vegetation transpiration as its coverage expands decreases evaporation from the soil. Increasing vegetation cover shades the soil and reduces the relative contribution of soil evaporation to total evapotranspiration. This interaction of vapor fluxes is governed by the behavior of the saturation vapor pressure deficit within the vegetation.

In the present study, we hypothesize that the competition for water between plant transpiration and bare soil evaporation is the main driving mechanism that affects long-term vegetation dynamics through the soil water limitation. This interaction is also implicitly represented in the Eagleson's model (Eq. (1)). Based on the above arguments, Eq. (1) can be simplified as follows:

$$\underbrace{Mk_v^*}_{1} \approx \frac{m_h}{m_{tb}E_{ps}} \left\{ 1 - \underbrace{\frac{\overline{h_0}}{m_h}}_{2} - \underbrace{\frac{(1-M)E_{ps}f(\theta)m_{tb}}{m_h}}_{3} - \underbrace{\frac{s_d}{m_h m_v}}_{4} \right\} \quad (3)$$

where the numbered dimensionless terms represent:

(1) the canopy moisture flux;
(2) the space-time average surface-retained precipitation;
(3) the bare soil evaporation; note that the original term in Eq. (1), $m_{tb''}E_{ps}\beta_s$ is replaced by $m_{tb}E_{ps}f(\theta)$. We neglect the difference between m_{tb} (the mean duration between storms) and $m_{tb''}$ (the mean inter-storm duration available for bare soil evaporation), and assume $m_{tb} = m_{tb''}$. The β_s (bare soil evaporation efficiency) is mainly controlled by the soil varying between $0 \sim 1$ [5, Fig. 6.8]; here we use a simple linear function $f(\theta)$ to represent the effect of soil moisture on evaporation, $f(\theta) = (\theta - \theta_w)/(\theta_f - \theta_w)$,[12,13] where θ is the soil moisture content in the root-zone; θ_w is the moisture content at the wilting point and θ_f is the soil moisture content at the field moisture capacity; considering the major soil type (silt loam) in the Loess Plateau, and referring to the Chinese Natural Resources Database from Chinese Academy of Sciences, the volumetric soil moisture contents at the field capacity and at the wilting point are 0.28 and 0.06 respectively, and the saturated soil moisture content is about 0.4;
(4) the surface runoff; the original term in Eq. (1), $e^{-G-2\sigma^{3/2}}$ is replaced by s_d/m_v, where s_d [LT^{-1}] is the average surface runoff depth which will be estimated by using the hydrograph separation technique as will be detailed in the following.

The surface-retained precipitation during the growing season ($\overline{h_0}$), is mainly due to the canopy interception loss. To estimate, the following monthly interception loss equation based on the statistical characteristics of daily rainfall proposed by De Groen and Savenije is used[14]:

$$\overline{h_0} = \frac{\sum_{n=4}^{9} I_n}{m_v} = \frac{\sum_{n=4}^{9} P_n \left(1 - \exp\left(-\frac{Dn_r}{P_n}\right)\right)}{m_v} \tag{4}$$

where n is the index of the month; I_m is the monthly interception; P_m is the monthly rainfall; n_r is the number of rain days per month; D is the daily interception threshold parameter. The value of D represents either the interception capacity of the entire moisture surface or the potential evaporation from a wet surface in a rain day, whichever is largest.[10]

To estimate surface runoff depth from the observed discharge data, the digital filter technique proposed by Nathan and McMahon[15] and subsequently improved by Chapman[16] is used. Following the reasonable range (0.925∼0.995) suggested by Chapman,[16] the filter parameter is chosen to be 0.95. The separated surface runoff from the total daily discharge is found to be not sensitive to this parameter; a sensitivity test indicates that the difference is less than 10% of the estimated surface runoff when different filter parameter values used.

Through estimating of the long-term seasonal average water balance components in Eq. (3), the relationship between M and θ can be analyzed. Since the digital filter technique is operated at daily time scale, only 29 catchments, all located in the Loess Plateau with daily discharge data are used in this analysis. Moreover, since these 29 catchments are considered to have similar vegetation structure, k_v^* can be reasonably assumed to be a constant. According to the main tree species in the Loess Plateau (the Chinese pine and acacia), k_v^* is taken as 0.70 following Eagleson [5, Fig. 7.9]. The estimation of water balance components in Eq. (3) is summarized in Table 2. Figure 3 plots the relationship between M and θ derived from Eq. (3) with a clear positive correlation ($R^2 = 0.73$). This indicates that sites with higher vegetation coverage also have higher soil moisture. A regression equation of soil moisture (θ) with respect to vegetation coverage (M) is given as follows:

$$\theta = 0.23\,M + 0.14 \tag{5}$$

which shows the approximate positive relation between the average vegetation coverage and soil moisture in space for the Loess Plateau region. It also indicates that full vegetation coverage would yield a volumetric soil moisture content of 0.37, which exceeds the field capacity (0.28) and the moisture may not be available for plant growth. If the soil moisture drops below 0.14, it would be difficult for any vegetation to survive although still above the wilting point (0.06). The vegetation coverage corresponds to 0.61 as soil moisture reaches the field moisture capacity.

Based on Eq. (5), the seasonal average soil moisture data in the other 22 catchments of the Loess Plateau can be estimated from their seasonal average vegetation coverage. The observed soil moisture data of these 51 catchments are used to validate the model. The results approve the simulation with the positive correlation between the simulated volumetric soil moisture (θ) and the observed data, see Fig. 4. The correlation coefficient reaches 0.55, and the Root Mean Squared Error (RMSE) is 0.04.

The water balance model (i.e. Eq. (3)) can be potentially useful for estimating the catchment's ecohydrological response to the climate change

Table 2. The estimation of water balance terms in Eq. (4) for the 29 catchments during the growing season in the Loess Plateau region.

Code	M	m_h (mm)	m_{tb} (d)	$\overline{h_0}$ (mm)	s_d (mm/season)	m_v (season^{-1})	$f(\theta)$	θ
44003	0.14	15.07	5.65	3.12	26.35	37.13	0.44	0.16
44037	0.16	15.43	5.55	3.12	34.45	36.56	0.46	0.16
44071	0.15	14.54	6.06	3.23	19.08	36.00	0.37	0.14
44077	0.16	15.94	5.40	3.15	27.19	35.75	0.51	0.17
44091	0.14	16.10	5.71	3.13	18.54	36.50	0.51	0.17
44099	0.31	16.54	5.39	3.06	26.06	36.19	0.56	0.18
44115	0.53	16.79	5.33	2.94	14.45	37.94	0.62	0.20
44153	0.15	14.99	6.16	3.24	9.30	33.75	0.40	0.15
44161	0.16	15.80	5.94	3.00	5.85	33.75	0.52	0.19
44167	0.11	15.59	6.02	3.13	14.94	35.25	0.47	0.16
44201	0.17	17.10	5.57	2.95	24.54	39.19	0.64	0.20
44203	0.20	17.45	5.48	3.01	23.42	36.63	0.66	0.20
44227	0.47	18.07	5.41	2.93	11.95	40.25	0.77	0.23
44235	0.31	18.42	5.25	3.00	18.95	37.38	0.81	0.24
44243	0.21	17.57	5.44	2.98	25.58	41.38	0.70	0.21
44253	0.56	18.64	5.21	2.96	4.55	41.56	0.93	0.26
44254	0.35	18.43	5.31	2.88	7.57	39.31	0.82	0.24
44259	0.60	19.08	5.18	2.93	6.35	39.31	0.99	0.28
44261	0.55	18.22	5.53	2.95	7.80	39.56	0.78	0.23
46043	0.59	20.96	5.31	3.14	29.58	40.75	1.00	0.28
46069	0.27	14.94	4.43	2.64	8.03	44.81	0.83	0.24
46125	0.48	15.84	4.59	2.75	18.13	45.69	0.87	0.25
46571	0.15	14.63	6.08	3.22	16.51	35.81	0.41	0.15
46591	0.21	16.24	5.44	3.07	16.83	39.50	0.61	0.19
46597	0.44	18.52	5.09	2.79	14.87	43.19	0.95	0.27
47209	0.51	21.58	5.36	3.19	29.91	39.38	1.00	0.28
47275	0.46	21.00	5.61	2.95	27.30	39.56	0.94	0.27
47320	0.53	22.42	5.82	3.04	15.48	33.75	1.00	0.28
47373	0.47	20.59	5.42	3.13	11.52	34.56	1.00	0.28

or human activity. When the catchment soil moisture state changes, for example, the water and soil conservation projects like the silt retention dams increasing the soil moisture content, the model can be used for estimating the variation of the vegetation coverage accompanying with the variation of the soil moisture.

4. Discussion

In this study, the simplified equation (Eq. (3)) derived from the Eagleson's water balance model (Eq. (1)) develops the relationship between the

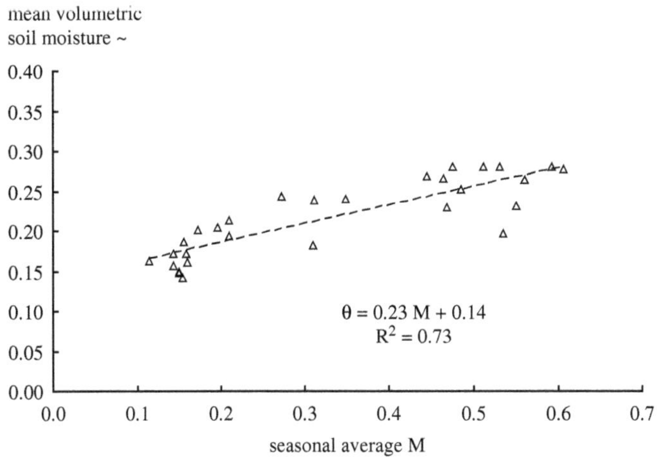

Fig. 3. Relationship between mean volumetric soil moisture (θ) and mean seasonal vegetation coverage (M) for the 29 catchments of the Loess Plateau.

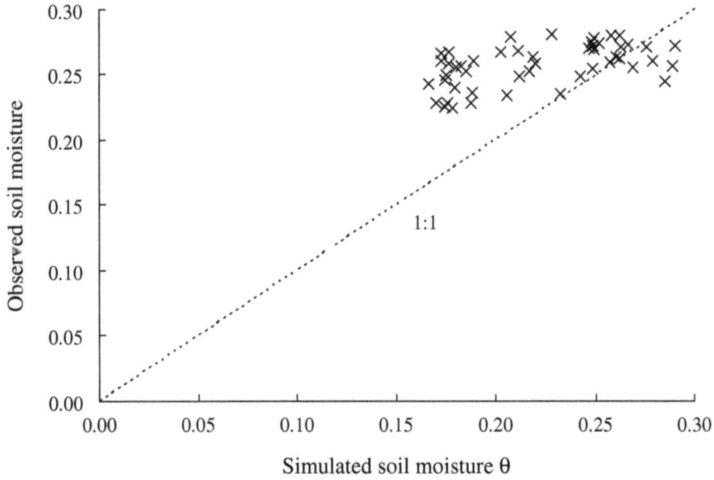

Fig. 4. Comparison between the simulated soil moisture and the observed soil moisture in the 51 catchments in the Loess Plateau region.

long-term seasonal average vegetation coverage and the soil moisture (Eq. (2)) in the Loess Plateau region based on the obtained meteorological, hydrological, and vegetation data. However, for different regions, the simplified form of Eq. (1) may be different due to their different climate and landscape conditions. The application of the water balance model should be based on the analysis of the water balance components. For example, Eagleson's simplification model as Eq. (2) can only apply to the region with very dry, non-seasonal climate and dense vegetated landscape, where almost the total precipitation during the growing season converts to the canopy moisture flux. In the present study, for the Loess Plateau region, the deep percolation and the capillary rise are neglected because of the deep groundwater; however, in some other regions with shallow groundwater conditions, these terms need to be considered. The relationship between the long-term seasonal average vegetation coverage and the soil moisture region is clearly positive in the Loess Plateau, which seems be reasonable; since in the arid or semi-arid regions, the growth of the vegetation coverage largely depends on the soil water content. In other regions, the relationship between the vegetation coverage and the soil moisture can also be explored from the water balance analysis based on local climate and landscape conditions.

5. Conclusion

The present study focuses on the relationship between vegetation coverage and water balance on the Eagleson's water balance model. A simplification of the Eagleson's water balance model is applied, and the mean seasonal vegetation coverage is proved to be correlated to the meteorological factors ($m_h/m_{tb}E_0$) in the Loess Plateau region. The relationship between the mean seasonal vegetation coverage and the soil moisture content is estimated through the long-term water balance analysis. It is found that there is an approximate linear spatial relation between the vegetation coverage and the soil moisture in the Loess Plateau region. This implies that the change of vegetation coverage due to the climate change and human activity in the Loess Plateau region could be potentially predicted through the change in soil moisture, which is expected to provide a useful reference for the water and soil management in the non-humid regions of China.

Acknowledgments

This research was supported by the joint research program of National Nature Science Foundation of China (NSFC, project no: 50721140161) and Japan Science and Technology (JST). The authors would like to express their appreciation for the grants received in this research.

Appendix

The physical meanings of the parameters in Eq. (1) are listed as following:

c = the permeability index of soil, [ ];
E_{ps} = the potential rate of evaporation from a wet, simple surface, [LT^{-1}];
E_{pv} = the potential canopy transpiration, [LT^{-1}];
G = the gravitational infiltration parameter, [ ];
$\overline{h_0}$ = the expected value of surface-retained precipitation in the growing seasons, [L];
$K(1)$ = the effective saturated hydraulic conductivity of soil, [LT^{-1}];
k_v^* = the potential canopy conductance, [ ];
M = the average vegetation coverage for the growing season, [ ];
m = the soil pore size distribution index, [ ];
m_h = the mean storm depth, [L];
m_{tb} = the mean duration between storms, [T];
$m_{tb'}$ = the mean inter-storm duration available for transpiration, [T];
$m_{tb''}$ = the mean inter-storm duration available for bare soil evaporation, [T];
m_v = the number of independent storms per season, [T^{-1}];
m_τ = the mean length of the growing season, [T];
P_τ = the precipitation of the growing season, [LT^{-1}];
s_o^c = the critical space-time average root-zone soil moisture concentration, [ ];
z_w = the depth to water table, [L];
β_s = the bare soil evaporation efficiency, [ ];
β_v = the canopy transpiration efficiency, [ ];
ΔS = the average carryover soil moisture storage, [L];
σ = the capillary infiltration parameter, [ ];
$\psi(1)$ = the saturated matrix potential of soil, [L].

References

1. J. Jiao, J. Tzanopoulos, P. Xofis, W. Bai, X. Ma and J. Mitchley, *Restor. Ecol.* **15**(3) (2007) 391.
2. D. Yang, C. Li, H. Hu, Z. Lei, S. Yang, T. Kusuda, T. Koike and K. Musiake, *Water Resour. Res.* W06502 (2004).
3. T. McVicar, L. Li, T. Niel, L. Zhang, R. Li, Q. Yang, X. Zhang, X. Mu, Z. Wen, W. Liu. Y. Zhao, Z. Liu and P. Gao, *Forest Ecol. Manage.* **251** (2007) 65.

4. X. Xu, H. Zhang and O. Zhang, *Environ. Sci. Pol.* **7** (2004) 79.
5. P. S. Eagleson, *Ecohydrology* (Cambridge University Press, 2002), p. 170.
6. D. Yang, F. Sun, Z. Liu, Z. Cong and Z. Lei, *Geophys. Res. Lett.* **33** (2006) L18402.
7. D. Yang, F. Sun, Z. Liu, Z. Cong, G. Ni and Z. Lei, *Water Resour. Res.* **43** (2007).
8. W. J. Shuttleworth, in *Handbook of Hydrology*, ed. D. R. Maidment (McGraw-Hill, New York, 1993), pp. 4.1–4.53.
9. R. Allen, Pereira, D. Raes and M. Smith, *Food Agric. Org. U.N.FAO Irrig. Drain. Pap.* Rome, Italy **56** (1998).
10. G. Gutman and A. Ignatov, *International Journal of Remote Sensing* **19** (1998) 1533.
11. B. Choudhury and J. Monteith, *Q. J. R. Meteorol. Soc.* **114** (1988) 373.
12. J. Noilhan and S. Planton, *Mon. Wea. Rev.* **177** (1989) 536.
13. S. Ortega-Farias, A. Olioso, S. Fuentes and H. Valdes, *Agric. Water Mang.* **82** (2006) 421.
14. M. De Groen, and H. Savenije, *Water Resour. Res.* **42** (2006) W12417.
15. R. Nathan and T. McMahon, *Water Resour. Res.* **26**(7) (1990) 1465.
16. T. Chapman, *Water Resour. Res.* **27**(7) (1991) 1783.

Advances in Geosciences
Vol. 17: Hydrological Science (2008)
Eds. Namsik Park *et al.*

THE EFFECTS OF TEMPORAL RESOLUTION ON ANNUAL INTERCEPTION ESTIMATION

AZINOOR AZIDA ABU BAKAR
Department of Civil and Environmental Engineering, Nagaoka University of Technology, Kamitomioka 1603-1, Nagaoka, Niigata, 940-2188 Japan
gnogda@yahoo.com

MINJIAO LU
Department of Civil and Environmental Engineering, Nagaoka University of Technology, Kamitomioka 1603-1, Nagaoka, Niigata, 940-2188 Japan

The objective of this study is to draw attention to the effect of temporal resolution on the estimation of annual losses by interception. Precipitation data with a different temporal resolution ranging from one hour to twenty four hours are processed as the input and the depth of the storage capacity varies from 0.1 to 10.0 mm. The physical interception was determined by the function of annual interception obtained from interception with high resolution. The calculated interception with the temporal resolution effect was determined by the slope function. The relationship between 'true' depth and the calculated depth was determined and the relationship between the annual calculated interceptions with the annual 'true value' of interception was derived. The result shows that temporal resolution affected the estimated annual interception value depending on the depth of storage capacity.

1. Introduction

The interception phenomenon is normally considered as one of the losses in hydrological cycle that comes into attention of many hydrologists for several decades. In general, interception models can be categorized into three groups: stochastic models (e.g. Calder[1]), empirical models (e.g. Horton[2]; Leonard[3]; and Aston[4]) and physical models (e.g. Rutter *et al.*[5,6]; Rutter and Morton[7]; Gash[8]; and Liu[9,10]). Most of the models have been developed according to the characteristics of the precipitation and canopy characteristics. For example, the canopy storage capacity of the forest is a key factor in estimating the amount of interception.[11]

Calibrated models of interception and evaporation loss; and direct measurement from the site are two main methods to estimate interception losses. Since the interception losses are geographically variable, there are no established standard field methods especially for the forest-based case studies. Most of the interception models obtained the data from the standard meteorological data, and from the literature, indirect methods and direct methods. But there are many flaws to the methods and the data that makes the model underestimates or overestimates the interception. Clearly, it is important to prepare a good data to determine the interception.

Few studies have been made on how the estimate interception value varies with temporal resolution and depth of the storage capacity with known annual precipitation. Therefore, this study was conducted to draw attention into the importance of considering the effect of temporal resolution in estimating the interception losses.

2. Study Area and Data Preparation

The Doki River basin, located in Kagawa Prefecture, Shikoku Island, western part of Japan is selected as the study area of this study. This is a small river basin just about 106.8 km^2 where most of the basin area is covered by the forest (85.5%), paddy field (5.6%) and vegetable field (6.2%),[12] and the influences of human activities are very limited. The basin receives annual rainfall of about 1600 mm and the evaporation is about 780 mm. For this case, snowmelt is considered not so significant because even the upper part of the Doki river basin has no snow during the winter.

The rainfall data was calculated using a rainfall data set of 245448 hourly rainfall data in the period between January 1978 and December 2005. The data was collected from 16 stations from two sources. The precipitation data was provided by the Japan Meteorological Agency and Ministry of Land, Infrastructure, Transport and Tourism Japan. The locations of the stations are shown in Fig. 1 and listed in Table 1.

To temporate the data, the hourly precipitation data was calculated as an average value for each resolution. In this process, the total precipitation during divisions is preserved for all temporal resolutions. For example, for time resolution 2 hours, the precipitation is divided into $P_{1,1}$ and $P_{1,2}$ in two sub-periods.

$$\frac{\Sigma P}{2} = P_{1,1} = P_{1,2}. \tag{1}$$

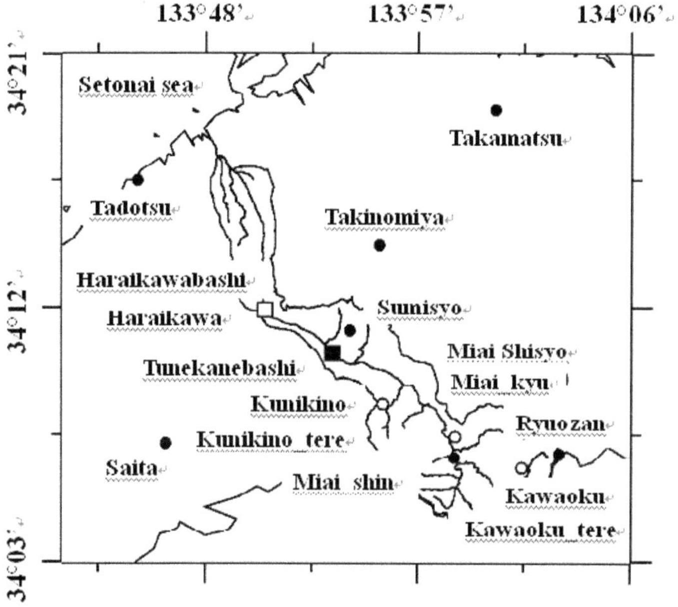

Fig. 1. Location of AMeDAS and observed stations at Kagawa Prefecture, Japan.

Table 1. Locations of the AMeDAS and observed stations in Kagawa Prefecture, Japan.

Station	Type of Data	Latitude (deg)	Longitude (deg)
Tadotsu	AMeDAS	34.273	133.775
Takamatsu	AMeDAS	34.313	134.057
Hiketa	AMeDAS	34.210	134.410
Uchinomi	AMeDAS	34.492	134.303
Takinomiya	AMeDAS	34.235	133.928
Saita	AMeDAS	34.117	133.775
Ryuozan	AMeDAS	34.112	134.053
Haraikawa	Observed	34.199	133.842
Kawaoku	Observed	34.105	134.024
Kawaoku_tere	Observed	34.105	134.024
Kunikino	Observed	34.143	133.926
Kunikino_tere	Observed	34.143	133.926
Miai_kyu	Observed	34.124	133.976
Miai_shin	Observed	34.111	133.976
Miai_tere	Observed	34.124	133.977
Sumisyo	Observed	34.187	133.902

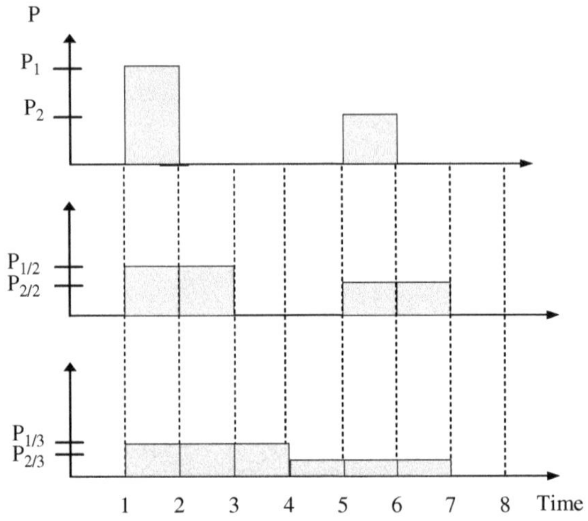

Fig. 2. The time step is fixed but rainfall is averaged according to the temporal resolution.

The time step of the sub-period is fixed as one hour but the averaging time for every rainfall event is changed. This is to avoid the computational error during calculation. Figure 2 illustrates the temporal resolution applied to prepare the data.

3. Methodology

3.1. *Interception concept*

Figure 3 shows the simple interception concept applied in this study. There are several assumptions made to the concept. It was assumed that there is no water drips from the canopy before the canopy capacity is filled; the rainfall pattern is represented by a series of discrete storms which are separated by sufficiently long intervals to allow the canopy to dry; the rainfall and the evaporation rates are constants during the storm, and under the conditions of canopy saturation the mean rainfall and evaporation rates are used; and evaporation from stems occurs only when the rainfall has ceased.

The potential evaporation, E_p is calculated by Priestley-Taylor equation. It is expressed by Eq. (2)

$$E_p = \frac{\alpha}{\lambda}\frac{\Delta}{\Delta + \gamma}R_n \tag{2}$$

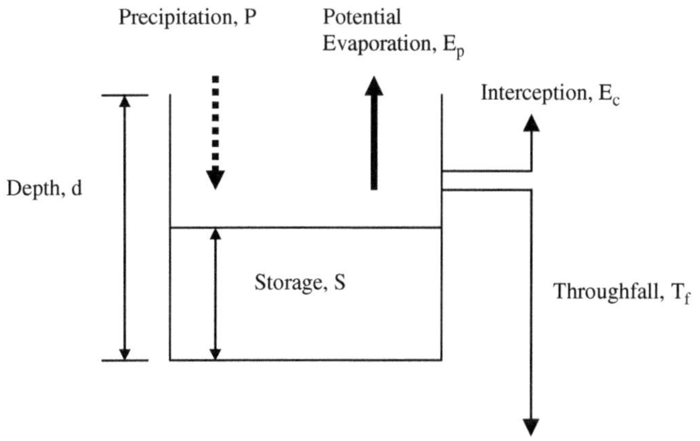

Fig. 3. Interception concept.

where α is the Priestley-Taylor constant (1.26); Δ is the slope of saturation vapor pressure function of temperature, $\frac{de_{sat}}{dT_a}$; γ is psychometric constant (0.677); and R_n is the net radiation.

But in this study, the potential evaporation is calculated as Eq. (3). This is because due to limited data, the air temperature was calculated by the ground surface temperature data at the catchment. Besides, it can be presumed that the quantity of direct solar radiation is same with the reflected solar radiation from the AMeDAS data. This value is already calibrated in previous study.[13]

$$E_p' = 0.4\,E_p. \tag{3}$$

For this study, the slope of saturation vapor pressure function of temperature is determined by the saturation vapor pressure, e_{sat} and temperature, T_a (^{o}C). e_{sat} was calculated by Eq. (4).

$$e_{sat} = 2.749 \times 10^8 \exp\left(\frac{-4278.6}{T_a + 242.8}\right). \tag{4}$$

The conditions considered to determine the interception are as follows:

$$T_f = \begin{cases} P - (d - S) & P > d - S \\ 0 & P \leq d - S \end{cases} \tag{5}$$

$$E_c = \begin{cases} S & E_p > S \\ E_p & E_p \leq S \end{cases} \tag{6}$$

where T_f is the throughfall, P is the precipitation, d is the depth of storage capacity, S is the storage, E_c is the interception and E_p is the potential evaporation.

The throughfall will occur when the precipitation is exceeding the storage capacity but no throughfall is allowed when the canopy is unsaturated. This will affect occurrence of the interception values. For evaporation greater than storage capacity, the interception value is the storage capacity but if the evaporation is equivalent or lesser than the storage capacity, the interception is the evaporation value.

3.2. *Determination of physical interception*

Temporal resolution ranging from 1 hour to 24 hours and various values of the depth of storage capacity that is between 0.1 to 10.0 mm were applied to derive the relationship between the interception and temporal resolution. The form of relationship is determined from a linear function. The range of the depth of storage capacity is wide enough so that it is applicable to all types of canopy. This is because there is some substantiation for high depth of canopy from both laboratory experiments,[14] and field based interception observations.[15,16] To determine the function of the annual interception, the value is taken at temporal resolution 0 hour because the temporal resolution is extremely high whereas longer averaging time indicated lower temporal resolution. The interception values at this point represent the physical annual interception of the site. This will be proved at Sec. 4.1. Each depth of storage capacity gives a relationship between the interception and the temporal resolution. From the relationship, the physical annual interception is determined and the slope for each temporal resolution is verified. The interception function is shown in Fig. 4 and summarized in Table 2.

When the depth is 0 which means there is no vegetation cover, there will be no interception process however when the depth is approaching infinity, all the rainfall will be intercepted. From this theory, the function of the interception can be expressed by Eq. (6)

$$E_{c,0} = P(1 - e^{-g(d)}) \tag{7}$$

where $g(d) = 0.109d^{0.58}$, $E_{c,0}$ is the annual physical interception value or annual 'true value' of interception (mm), P is the annual precipitation (mm), and d is the depth of storage capacity (mm). The value that is determined by Eq. (6) will be called 'true value' of interception here after.

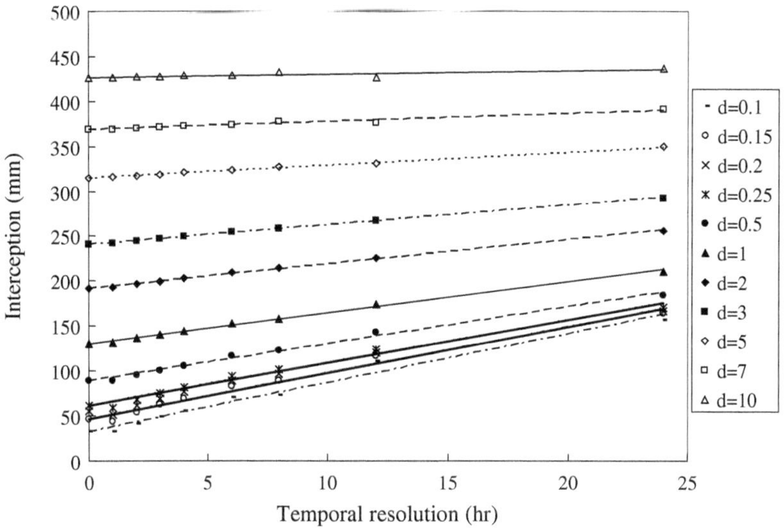

Fig. 4. Relationship between interceptions with temporal resolution for each depth.

Table 2. Function of temporal resolution with depth.

Depth (mm)	E_c (mm)
0.10	5.3831TR + 33.145
0.15	5.1315TR + 46.494
0.20	4.9067TR + 54.610
0.25	4.7336TR + 61.344
0.50	4.1249TR + 88.587
1.00	3.4268TR + 130.27
2.00	2.6909TR + 191.93
3.00	2.1884TR + 240.41
5.00	1.4419TR + 314.39
7.00	0.9065TR + 368.20
10.00	0.3722TR + 426.09

It can be seen that from Fig. 4, the interception loss is increase with the decreasing of temporal resolution where longer averaging time indicated lower temporal resolution. For example, when temporal resolution is 1 hour, the interception loss is lower than interception loss by temporal resolution 24. This is because when the temporal resolution applied to the rainfall data, it affects the duration of the storm and the storm break period. The storm duration will be longer and the break period becomes shorter. As a result, it may reduce the rainfall intensity and it also changes the amount of interception loss too. Thus in this study, the highest interception value is given by the depth of storage capacity 10 mm, and low interception obtained when the depth of storage capacity is 0.1 mm. One of the characteristics of the interception process is that a typical storm can saturate the canopy very quickly because of its low water holding capacity. Therefore for a different rainfall event but with the same total rainfall amount, the heavy rainfall may produce less interception loss comparing to the light rainfall.

3.3. *Determination of calculated interception with temporal resolution effect*

Figure 5 shows the slope relationship to determine the value of calculated annual interception which is affected by the temporal resolution. The slope values were taken from the function of temporal resolution with various depths (Table 2). The assumption made was there will be no interception when depth is 0, but when the depth is approaching infinity, the interception will occur. The value is determined by the slope function. The relationship of each depth with the temporal resolution is expressed by Eq. (7)

$$f(d) = A.d^m.\exp(-d^n) \tag{8}$$

where $A = 9.2$, $m = 0.08$, $n = 0.45$ and d is the depth of storage capacity.

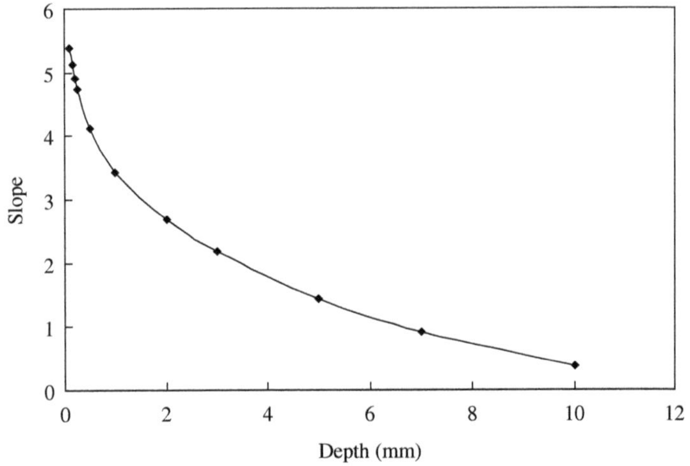

Fig. 5. Slope function.

According to Eqs. (6) and (7), the calculated annual interception function, E_c is determined by the annual precipitation P, depth of storage capacity d, and the temporal resolution TR, which is suitable for the set of circumstances. The interception value is represented by Eq. (8).

$$E_c = E_{c,o} + f(d) \cdot TR. \tag{9}$$

From these two relationships, it shows that the calculated annual interception is affected by the temporal resolution which means the temporal resolution factor should be considered when estimating the annual interception.

4. Results and Discussions

4.1. *'True value' of interception versus measured interception*

In order to determine the accuracy of Eq. (7), the measured value of interception studies from others are compared to the interception value established from the equation. Five locations from six studies were used to make the comparison namely southern France,[17] northern Queensland, Australia,[18,19] southeastern US,[20] and eastern Japan.[21] All these locations give 16 different cases as shown in Table 3.

The percentage of the 'true value' and the measured interception is shown in Table 3 and the values are illustrated in Fig. 6. From the results, there are 3 cases with the 'true value' of interception are higher than measured interception. The cases are Case 6, 12 and 13 where the equation overestimated the interception value for about 1.67–23.08%. In this case, the highest overestimated is the pine plantation case. This is because it was affected by the stemflow percentage of the precipitation. The study shows that the percent of incident precipitation for stemflow for the pine plantation is slightly higher (1.96%) than other forest community (0.54–0.655%). There are cases where the 'true value' of interception is lower than the measured interception value (11.36–56.11%). Hutchinson Creek with complex Mesophyll vine forest give the highest different value (56.11%). This is due to the dry season interception fractions where Hutchinson Creek received precipitation below 10 mm (40%) in the wet season, compare to dry season (85%). As a result the interception fraction increased when the dry season rainfall rates are lower than wet season rates.

Figure 6 shows that in most of the cases, the 'true value' of interception harmonized with the measured interception. This proved that at temporal

Table 3. 'True value' of interception and measured interception value for each case.

Study	Case	Description	Annual Precipitation (mm)	Interception (%) Measured	Interception (%) 'True value'
[17]	1	Control plot	908	30.90	17.28
	2	Thinned plot	908	20.20	13.78
[18]	3	Oliver creek - Complex Mesophyll Vine Forest	3068	33.87	20.48
	4	Hutchinson Creek - Complex Mesophyll Vine Forest	2680	34.25	15.04
	5	Mount Lewis 1- Simple Notophyll Vine Forest	3316	33.32	20.48
	6	Mount Lewis 2 - Simple Notophyll Vine Forest	2609	17.75	18.30
	7	Upper Barron - Complex Notophyll Vine Forest	1831	22.83	17.63
[19]	8	Storm based rainfall	4037	24.75	20.18
	9	Daily total rainfall	4037	24.75	20.18
[20]	10	Wetland	830	16.07	10.21
	11	Pine	830	20.23	14.91
	12	Pine plantation	830	15.34	13.78
	13	Hardwood	830	15.23	12.41
	14	Mixed	830	15.35	13.25
[21]	15	1991	1826	11.83	10.21
	16	1993	1642	14.62	10.21

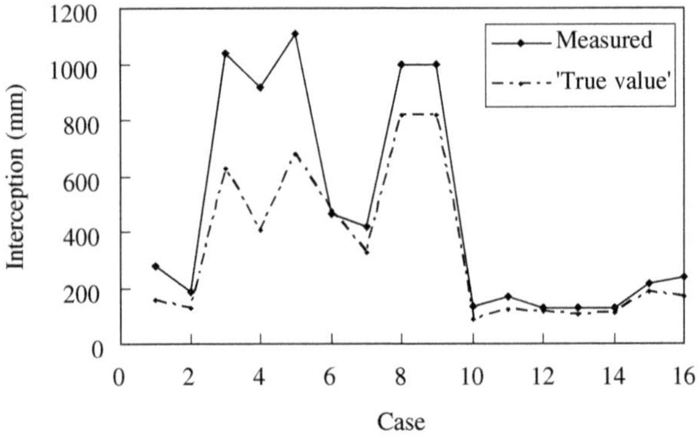

Fig. 6. Measured and 'true value' of interception for each case.

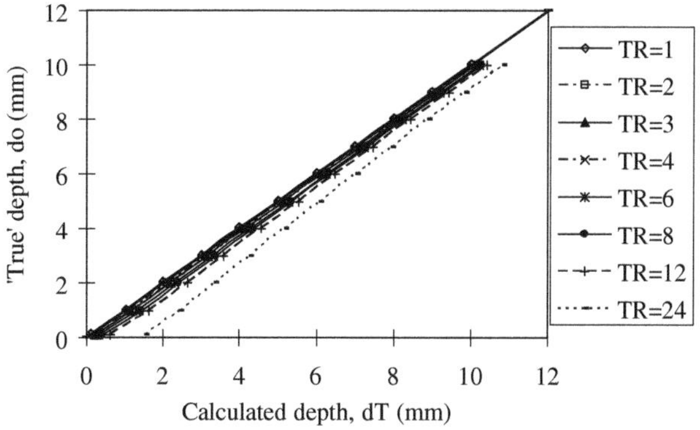

Fig. 7. The relationship between 'true' depth and calculated depth for each temporal resolution.

resolution 0, the interception value represents the 'true value' of interception at the site.

4.2. *Relationship between 'true' depth and calculated depth*

In order to assess the effect of temporal resolution on 'true' depth, d_o and the calculated depth, d_T, the relationship between these two depth with various temporal resolution is determined. The depth values were determined by applying Eq. (9) and the relationship is shown in Fig. 7. For each temporal resolution, there are 2 depth values with the same interception value. For example, for temporal resolution 1 hour, the d_0 0.1 mm gives d_T 0.13 mm but with the same d_0 value, the d_T is 0.16 mm, 0.19 mm, 0.23 mm, 0.31 mm, 0.40 mm, 0.61 mm and 1.5 mm for temporal resolution 2, 3, 4, 6, 8, 12 and 24 hours respectively. This proved that the magnitude of difference was because of the temporal resolution factor.

5. Conclusion

The results presented here proved that the calculated annual interception is affected by the temporal resolution of the precipitation data. The relationship of the temporal resolution with the calculated annual interception is presented here:

$$E_c = E_{c,0} + f(d) \cdot TR$$

where $E_{c,0} = P(1 - e^{-g(d)})$, $g(d) = 0.1092d^{0.5824}$, $f(d) = A.d^m.\exp(-d^n)$, $A = 9.2$, $m = 0.08$, $n = 0.45$; P is the annual precipitation (mm), d is the depth of storage capacity (mm), and TR is the temporal resolution (hr).

References

1. L. R. Calder, *J. Hydrol.* **89** (1986) 65–71.
2. R. H. Horton, Mon. *Wea. Rev.* **47** (1919) 603–623.
3. R. E. Leonard, W. E. Sopper and H. W. Lull (Editors), *Forest Hydrology. Pergamon, Oxford* (1965) 131–136.
4. A. R. Aston, *J. Hydrol.* **42** (1979) 383–396.
5. A. J. Rutter, A. J. Morton and P. C. Robins, *J. Appl. Ecol.* **12** (1975) 367–380.
6. A. J. Rutter, K. A. Kershwa, P. C. Robins and A. J. Morton, *Agric. Meteorol.* **9** (1971) 367–384.
7. A. J. Rutter and A. J. Morton, *J. Appl. Ecol.* **14** (1977) 567–588.
8. J. H. C. Gash, *Quart. J. R. Met. Soc.* **105** (1979) 43–55.
9. J. Liu, *Ecol. Model* **42** (1988) 111–123.
10. S. Liu, *Sci. Silvae Sin.* **28** (1992) 445–449.
11. J. Schellekens, *Phys. Chem. Earth (b)* **25**, 7–8 (2000) 659–664.
12. I. Makino, S. Ogawa and G. Saito, *Proceedings of 20th Asian Conference on Remote Sensing* (1999).
13. T. Yamamoto and M. Lu, *Journal of Japan Society of Hydrology & Water Resources*, Submitted (2008).
14. S. R. Herwitz, *Journal of Hydrology* **77** (1985) 237–252.
15. R. H. Crockford and D. P. Richardson, *Hydrol. Process* **14** (2000) 2903–2920.
16. K. Fleischbein, W. Wilcke, R. Glooer, J. Boy, C. Valarezo, W. Zech and K. Knoblich, *Hydrological Processess* **19** (2005) 1355–1371.
17. J. Limousin, S. Rambal, J. Ourcival and R. Joffree, *Journal of Hydrology* **357** (2008) 57–66.
18. J. Wallace and D. McJannet, *Journal of Hydrology* **348** (2008) 480–495.
19. J. Wallace and D. McJannet, *Journal of Hydrology* **329** (2006) 477–488.
20. M. L. Bryant, S. Bhat and J. M. Jacobs, *Journal of Hydrology* **312** (2005) 95–108.
21. S. Murakami, Y. Tsuboyama, T. Shimizu, M. Fujieda and S. Noguchi, *Journal of Hydrology* **227** (2000) 114–127.

Advances in Geosciences
Vol. 17: Hydrological Science (2008)
Eds. Namsik Park et al.

EXPERIENCES ON WATER-RELATED DISASTER MANAGEMENT IN CHINA

JINPING LIU
UNESCAP/WMO Typhoon Committee Secretariat,
Avenida de 5 de Outubro
Coloane, Macao, China
jpliu@typhooncommittee.org

XIAOHE DU
Department of Flood Control and Disaster Reduction,
China Institute of Water Resources and Hydropower Research,
Yuyuantan Science and Technology Garden Haidian district,
Beijing,100038, China
duxiaohe1981@yahoo.com.cn

The paper discusses recent experiences in water-related disaster management in China, including recent trends of socio-economic impacts and investment for better management. The paper noted important achievement in the reduction of the number of annual deaths by water-related disasters, but pointed out the rapid increase in economic damage. It noted important increase in investment in recent years, particularly on flood management. In the context of the recent trends, which saw rapid increase of damage in the urban areas, the paper discusses recent increase in the attention of the Government to improve the planning and implementation of urban flood management in the country. Finally, the paper discusses opportunities for enhanced regional cooperation on urban flood management among the Typhoon Committee Members and through the framework of UNESCAP.

1. Condition of Geography and Hydrology in China

China is situated at the eastern part of Asia, west of the Pacific Ocean, with wide territory and numerous rivers. China is characterized by high in west part and low east part, gradually descending in elevation from the west towards the east to form three steps. The total land area is about 9.6 million km^2. China is one of the most typical monsoon climate countries in the world, with two-third area influenced by monsoon. Southeast monsoon can reach Eastern China and southwest monsoon covers coastland of Southwestern and Southern China. Most part of China is characterized by humid and rainy, rain-heat synchronous in summer. Northwestern China is dry and rainless, stretching into the inlands of Eurasia.

The total length of river in China is about 430,000 km, one of the most numerous rivers in the world. There are more than 20 rivers longer than 1,000 km, 50,000 rivers larger than 100 km^2 in catchment area, 1,500 rivers larger than 1,000 km^2 and 79 rivers larger than 10,000 km^2. The total area of Yangtze, Yellow, Huai, Hai, Songhua, Liao and Pearl River Basin is about 4,300,000 km^2, amount to 45.6% of the territory. Most of these rivers are distributed in the east and south part of China, characterized by rainy and humid. On the contrary, there are seldom rivers in the northwest part of China.

China has uneven precipitation distribution timely (monthly and yearly) and spatially. It is so different from southeast coastal area, where average annual precipitation can reach 1,600 mm, to inland area, where that is less than 200 mm.

China is dominated by mountainous area, two third of total territory, including mountain, plateau and hill, which is mainly located in west of China, with low population density. Compared with elevation of land, water level is relatively low. On the other hand, East China is mainly plain region, characterized by high population density and boomed economy. Compared with low elevation of land, water level is relatively high, so it is the concentration region of flood, with huge flood loss. At same time, the coastal area is often struck by typhoon and rainstorm.

2. The Trend of Water-Related Damage in China

China has been frequently hit by large floods and suffered from flood disasters. As mentioned above, the critical issue is that about 8% of the land area located in the mid- and down stream parts of the seven major rivers of the country are prone to floods. However, in these areas lives 50% of the total population of the country, and they contribute over 2/3 of total agricultural and industrial product value. According to the historical records, large flood events occur once in every two years.

According to the historical record, almost 1,100 large flood disaster events occurred since 206 BC in China, which means that the large flood events occur once in every two years in average. At the beginning of the twentieth century, the major rivers in China were struck by a number of disastrous floods. In 1931, the disastrous floods hit the Huai River basin and the Yangtze River basin. Over 51 million people were affected, of which more than 400,000 people died. In 1975, a catastrophic local flood caused typhoon caused almost 30,000 people died.

In recent two decades, there is no river in the seven major rivers (Yangtze, Yellow, Huai, Hai, Songhua, Liao, Pearl) escaping from big flood. Flood is still the top trouble for Chinese in all natural disasters. About 63% of the total loss by natural disaster is contributed by flood since 1990's, which amounts to 1.7% of corresponding GDP. In 1998, a nationwide flood caused 255 billion RBM direct economic losses which amounts to 4.1% of GDP and 4150 people died.

On the other hand, drought disasters almost occur every year in China while flood disasters hit. In recent 20 years, drought caused 28.4 billion kilograms of grain losses, 33.1 billion RMB of economic-crop losses, 24.03 million people and 1.68 million livestock suffering from scarce drinking water every year in average. In 2007, the drought damage reached 37.4 billion kilogram of gram, 42.2 billion RMB of economic-crop, and 27.56 million people and 20.6 livestock were suffered from scarce drinking water.

With the development of society and economy and the changing of the concepts on flood control and drought relief, China has been perfecting its system of water-related disaster prevention and reduction gradually, including construction and non-construction measures. However, extreme water-related disaster events, such as rainstorm, flood, strong typhoon, mudflow, debris flow and landslide and etc, happened frequently because of climate change, urbanization and etc. The damage caused by the water-related disaster is showing itself a trend of increasing of economic losses as well as decreasing of life losses.

Figure 1 shows that the magnitude of the yearly life losses caused by flood disaster is decreased from near 10,000 in 1950's close to 1,000 in 21 century, and meanwhile, the magnitude of the yearly direct economic losses caused by flood is contrarily increased from tens billions RMB yuan in the early of 1990's to more than hundreds billions RMB yuan at present.

Figure 2 shows that the yearly impacted farmland by drought disaster is increased by 10 times since 1950's and the magnitude of the yearly grain losses caused by drought disaster is increased from billions kilogram in 1950's to tens billions kilogram at present.

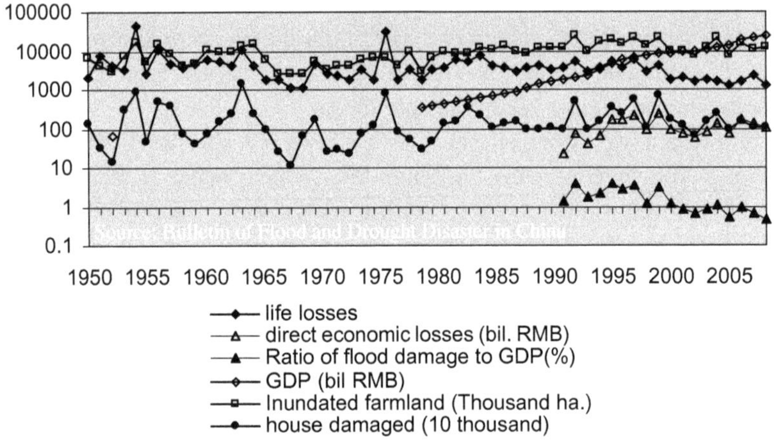

Fig. 1. The damage caused by flood disaster in China.

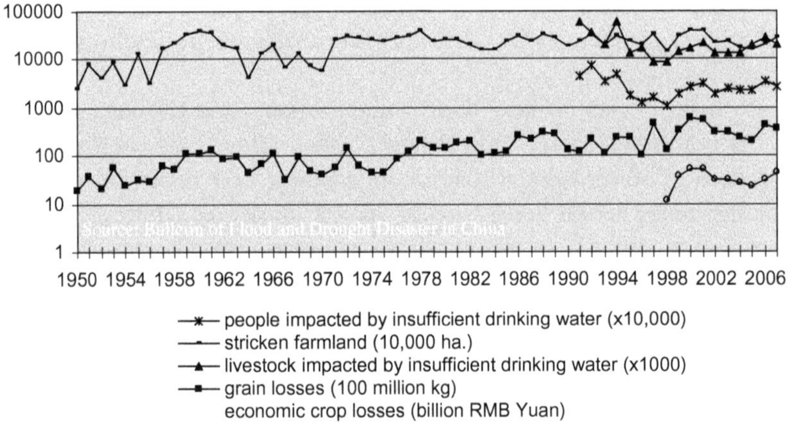

Fig. 2. The damage caused by drought disaster in China.

3. Experiences on Investment for Water-Related Disaster Management

Abundant experiences on water-related disaster management have been accumulated in the long period of flood control and flood fighting, especially in recent decades. China's government updated its knowledge and concept on flood control and drought relief with economic and social development.

As a result, the strategy is transferred to reasonable and sustainable way. Now, to achieve sustainable development goal, scientific development and human-nature harmonious development are established as guideline for water-related disaster management, which is adjusted from flood control strategy. It means that construction and non-construction measures such as engineering, legislation, command, economy, technique and education measures will coordinated for water-related disaster management on basis of integrated water resources management, comprehensive consideration of river training, water resources conservation, allocation and environmental protection.

According to strategy adjustment of water-related disaster prevention and mitigation, many other measures are also necessary to be taken into account for integrated water-related disaster management system as well as construction measures: (1) to regulate social development to accommodate natural condition, aiming to reduce flood loss; (2) to establish flood mitigation guarantee system and responsibility-share mechanism of water-related disaster mitigation and relief; (3) to enhance public awareness of China's water-related disaster especially flood issue and improve public involvement in flood mitigation and disaster relief; (4) to take comprehensive consideration of flood management, water resources utilization and environmental protection; (5) to complete legal system for flood management and to strengthen capacity on legal enforcement; (6) to establish flood insurance and post-disaster reconstruction mechanism; (7) to speedup construction of flood mitigation and draught relief command system and decision-making supporting system; (8) to broaden financial support for flood management and etc.

The Governments at all levels pay more close attention on the capacity building for water-related disaster prevention, preparedness and reduction, especially on flood disaster management. In recent years, the yearly total investment for flood disaster management is increasing year by year which makes the country enhance and promote the capacity of construction and non-construction measures for water-related disaster reduction. Statistics shows that the total completed water-related fixed assets investment reached 278.3 billion RMB Yuan since 1998 to 2006 (see Fig. 3), of which 170.6 billion RMB Yuan has been put into construction of flood control projects. There are some experiences on the investment for water-related disaster management in China Water sector summarized here.

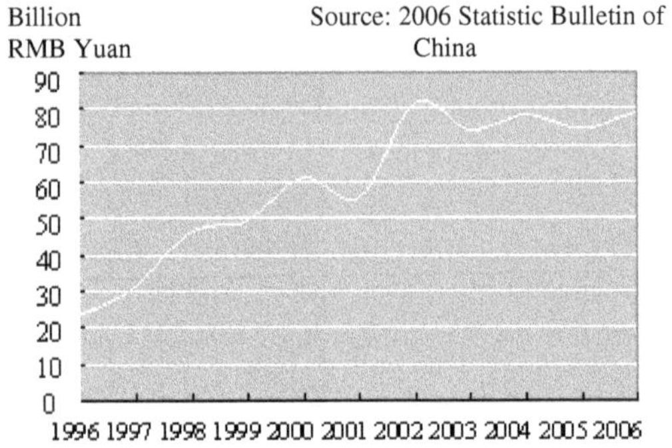

Fig. 3. Completed water-related fixed assets investment from 1996 to 2006 in China.

3.1. *Planning and early-stage work for disaster prevention and reduction*

The most important foundation for investment from Central Government and Provincial Governments is planning and early-stage work including carrying out the possibility study, analyzing the result of investment and so on. The investment for disaster prevention and reduction from the provincial-above governments mainly lies on the planning and early-stage work. Once the proposals of big projects pass the approval by National Development and Reform Commission (NDRC), the investment plan would be authorized by the Central Government.

In 2006 (the first year of 11th five-year plan of National economy), a total investment of 39.4 billion Yuan was allocated to carry out the planning and early-stage work for the proposals of sixteen new key water projects in the 11th Five-Year Plan period.

In the aspect of water-related disaster reduction, the State Council approved in 2006 the National 11th Five-Year Plan (2006–2010) for Rural Drinking Water Safety with a total investment of about 40 billion RMB Yuan and the National Plan for Mountain Flood Control and Prevention with a total investment of about 170 billion RMB Yuan. These two projects will be carried out in next 5 years.

3.2. *Stable investment channels*

In China, there are three main stable investment channels for water-related disaster (flood and drought disaster) prevention, reduction and management.

The first one is the Water Construction Funds founded in 1997. In order to support and push the development of some special undertakings, the central and provincial governments levy the stated financing with special purposes from citizen, corporations and other organizations according to the correlative laws and regulations. The Central Government allocates a certain proportion of the total of the Central Governmental Funds on water construction every year to form Water Construction Funds, among which 30% is used to flood disaster management including flood control project's operation and maintenance, emergency treatment, devastated equipments and facilities repair, flood control scheme revision, flood fighting training and flood awareness publicizing and etc.

The second one is the Extraordinary Floods and Droughts Funds from Central Government. The investment from this part will be distributed to the regions suffered from large catastrophic flood and severe drought disaster, depend on how severe the flood and drought damage is. For example, about 2.3 billion RMB Yuan from the Funds for Extraordinary Floods and Droughts was arranged on the emergency treatment for national-wide flood in 1998 and about 1.26 billion RMB Yuan in 2007.

Figure 4 shows the investments on water-related disaster prevention and mitigation from the Central Water Construction Funds and the Central Extraordinary Floods and Droughts Funds in China from 1997 to 2007.

The third channel is the investment from special projects mentioned above. For example, about 1.8 billion RMB Yuan was invested into the National Flood Control and Drought Relief Command System (NFCDRCS) since 2002 to promote the capacity of flood monitoring, communication, forecasting and warning and decision-making supporting.

3.3. *Multi-investment sources*

Water-related disaster reduction is the business concerned the whole society. All stakeholders should be involved in the investing for disaster prevention, reduction and mitigation. The capitals of investment are raised from various financial sources including central government, local governments, domestic loan, foreign investment, private investment and other investment. The local residents contribute for disaster reduction as the form of labour work.

For an example, the total investment of about 170 billion RMB Yuan for the National Plan for Mountain Flood Control and Prevention will be shared 50% by central government and local government, separately. Among the total capital of investments for flood control in 1997~2002 period,

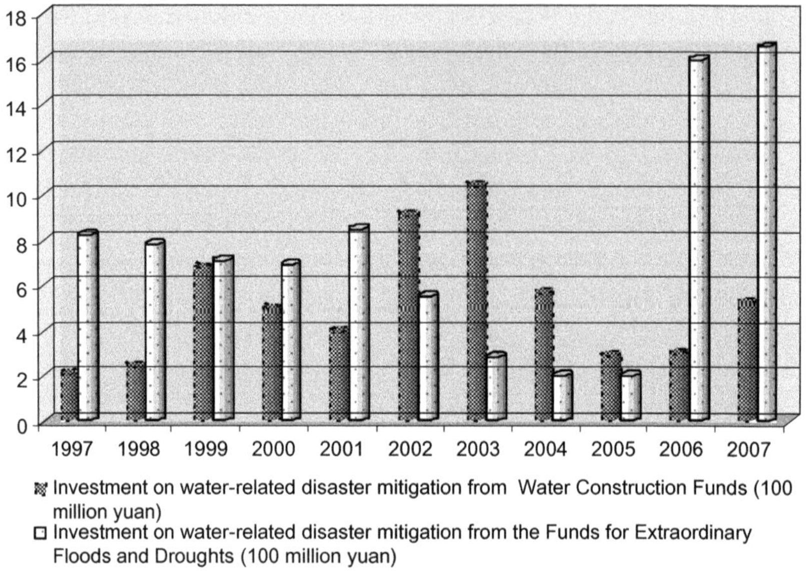

Fig. 4. The investments on water-related disaster management from the Funds of Central Government in China.

the investment from Central Government accounts the proportion of 49%, Provincial Governments account 30% and local governments account 21% including labour force expended by local residents.

In 2006, the planned total investment of fixed assets in the water sector was 93.27 billion Yuan, of which 30.84 billion Yuan from the Central Government, 44.15 billion Yuan from the local governments, 1.16 billion Yuan of foreign investment, 11.20 billion Yuan of domestic loans, 2.89 billion Yuan from enterprises and private sector, and 3.03 billion Yuan from other financial sources (see Fig. 5).

For another example, among the total investment for hydrology sector including capacity building of flood monitoring, forecasting and warning and etc in 2006, 13% was from central government, 58% from provincial governments, 6% from city and county governments and 23% from other special projects' investment.

3.4. *The appropriate regional investment policy*

China has a prominent characteristics, namely the economic development is uneven in all over the country. Comparing to the more developed

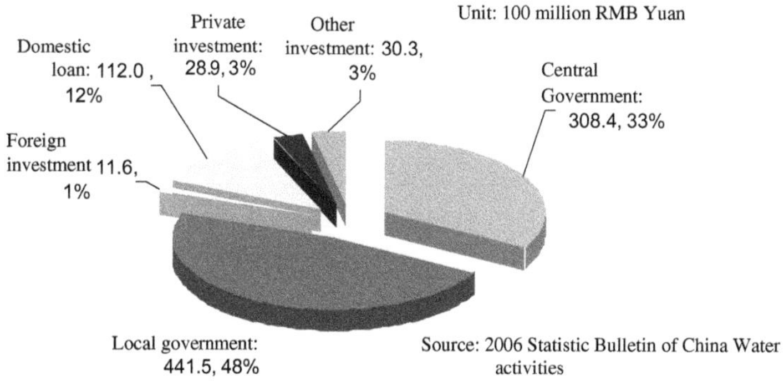

Fig. 5. Components of Fixed Assets Investment Plan on water infrastructure in China in 2006.

eastern regions, northwest region has relatively low developed economy. The investment from Central Government should consider this China characteristics and lean to those regions so that the low developed northwest economic regions can promote their ability of water-related disaster mitigation and management to protect the people's life and property.

China has a principle that the investment from the Central Government is inclined to the western region and flood prone areas, while the investment in the eastern part mainly depends on their own provincial and local government. The percentage of local self-raising funds and private investment in the eastern region of China is much higher than others in water project financing, while the percentage of government investment and bank loan is relatively higher in water project financing in the western region.

For instance, the proportion of investment for phase two of the project on National Flood Control and Drought Relief Commanding System to be carried out in next years, Central Government will invest 50% to the provinces in east part, 80% for the provinces in middle and west part and 100% for Tibet.

The principle of investment in different regions mentioned above is incarnated in all water-related sections. Table 1 shows the proportions of the different investment sources on hydrology capacity building in eastern region and western region in 2006.

3.5. *More attention to urban flood management*

Economic growth and urbanization are inextricably linked. Economic growth often implies the conversion of rural land to urban uses, massive

Table 1. The investment sources on hydrology in eastern region and western region in 2006.

	Ratio of the investments from different sources to the total investment	
Investment sources	Eastern 5 Provinces	Western 5 Provinces
Central Government	4.19%	29.47%
Provincial Government	29.02%	43.2%
City and county government	31.06%	2.99%
Other projects	35.74%	24.34%

Source: 2006 Statistic Bulletin of China Water activities.

inflows of capital transformed and more people concentrated in urban areas. An important effect of urbanization is to increase the high sensitivity of urban watersheds to rainfall which causes flash flood phenomena and high frangibility to flood disaster which causes much larger potential flood-related damage especially in the Asia and the Pacific region influenced by typhoon.

For an instance, severe flood and waterlogging disaster was caused by heavy rainfall in Jinan, capital city of Shandong province, China on July 18, 2007, which caused 37 peoples died and 171 injured. In flood season of 2004, a short-time heavy rainfall in Beijing almost broke down the public transportation system.

China has pay attention to the plan of the urban flood management. More investment has been involved in urban flood control. For an example, in Nanjing city of Jiangsu province in east of China, the total investment for whole water sector is about 3 billion RBM Yuan in the period of 10th five-year plan and the total investment plan reaches 5.7 billion Yuan for flood control and waterlogging draining projects in 11th five-year plan period (2006~2010) and beyond. Table 2 shows the components of investment on flood control and waterlog draining in Nanjing up to 2020.

Considering the universality of urban flood problem, as the proposal by China, UNESCAP/WMO Typhoon Committee launched the project on flood-related disaster risk management in urban areas at its Integrated Workshop on Social-economic Impacts of Extreme Typhoon-related Events held in Bangkok, September 2007. The purposes to launch this project are to exchange the experiences on management and mitigation of floods and typhoon-related disaster in urban area between TC Members; to share the technology of urban flood monitoring and methodology of urban

Table 2. The components of investment on flood control and waterlogging draining in Nanjing Unit: billion yuan

		Sources of investment for 2006–2010			
Total planned investment for 2006–2020	Total investment for 2006–2010	Provincial & above investment	Municipal investment	County investment	The planned investment for 2011–2020
5.7	2.3 (40.4%)	1.05 (45.7%)	0.62 (27%)	0.63 (27.4%)	3.4 (59.6%)

Source: the Report of Nanjing 11th Five-year Plan for Social-economic Development.

flood forecasting and prediction, early warning and disaster assessment between TC Members and to promote management of urban flood and other typhoon–related disasters in TC area.

4. Conclusion

Even though China has been perfecting its flood control system in recent decades, the water-related disaster is still main nature disaster and the water related damage is still severe because of the special condition of geography and hydrology, climate change and urbanization and others. Statistics shows that the damage caused by the water-related disaster is showing itself a trend of increasing of socio-economic impacts as well as decreasing of life losses.

In recent years, China has been adjusting the strategy of water-related disaster management and recognizes that many other measures are also necessary to be taken into account for integrated water-related disaster management system as well as construction measures. To broaden financial support for flood management is one of the most important measures to insure reduction and mitigation of flood impacts on the society and economy. China increased the investment on water-related disaster mitigation in recent years, particularly on flood management.

China also recognized that economic growth is inextricably linked with urbanization and as a result the urban flood damage is rapidly increasing. China has started to improve the planning and implementation of urban flood management in the country. China is willing to enhance the regional cooperation on urban flood management among the Typhoon Committee Members through the framework of UNESCAP.

References

1. Headquarter of State Flood Control and Drought Relief of China, Ministry of Water Resources of China, Bulletin of Flood and Drought Disasters in China, 2007.
2. Ministry of Water Resources of China, 2006 Statistic Bulletin of China Water Activities, 2006.
3. Ministry of Water Resources of China, Gazette of the Ministry of Water Resources of China, 2007.01.
4. Asia Development Bank, Urbanization and Sustainability in Asia, 2006.
5. National Bureau of Statistics of China, 2006 Bulletin of Social and Economic Development in China, http://www.stats.gov.cn/tjgb/.

Advances in Geosciences
Vol. 17: Hydrological Science (2008)
Eds. Namsik Park *et al.*

HYDROLOGICAL MODEL COMPARISON AND COMBINATION FOR FLOOD FORECAST IN THE THREE GORGES REGION, CHINA*

HUAN MA
Department of Hydraulic Engineering,
Tsinghua University, Haidian district,
Beijing 100084, China
mah06@mails.tsinghua.edu.cn

DAWEN YANG
Department of Hydraulic Engineering,
Tsinghua University, Haidian district,
Beijing 100084, China

Lumped conceptual models are traditional methods for flood forecast, and with the development of GIS and RS, physically-based distributed models became one of the most popular tools for flood forecasting in the past two or three decades. However, in order to deal with the complexity of hydrological system, many hydrological models have become too complex to understand and too data-hungry to be widely applied. In this study, a physically-based distributed hydrological model (GBHM) and a conceptual hydrological model (Xinanjiang model) are examined for short-term flood simulation in the Three Gorges Region. Through an inter-comparison of the two models at six catchments in the study area, this study aims to establish a combined method for flood forecasting. A new model is developed on the framework of GBHM, which simplified the runoff calculation by using conceptual methods instead of the complex physical descriptions in the original GBHM model. The new model combines the advantages of the GBHM and the Xinanjiang model, and the results showed that the model performance is similar to the two original models and can be used for flood forecasting, although some refinement is still necessary.

1. Introduction

The hydrological models for flood forecasting can be classified into black-box model, conceptual model and physically-based model, depending on

*This work is supported by the National '948' project of China and the National Key Technology R&D Program of China.

their descriptions of the hydrological process. They can also be classified into lumped model and distributed model, depending on whether or not the model considers the spatial variation of the catchment characteristics and meteorological data. Among the different kinds of hydrological models, conceptual lumped model is a widely used traditional method, and physically-based distributed model has been developing fast since the 1980s. In general, conceptual lumped model is simple to understand and use, while physically-based distributed model is much more complex but can give a detailed and potentially more correct description of hydrological processes, and have the advantage of using spatial information supported by GIS and remote sensing techniques.[1] The development of physically-based distributed model is a progress in hydrological modeling, but there are some obstacles to its further development. On the one hand, this kind of model is supported by detailed catchment characteristics and weather data, but in practice some parameters have to be estimated by model calibration, due to lack of detailed data. Since the 1990s, the study on uncertainties from the parameters as well as model structure and input data became a hot topic for physically-based distributed models.[2,3] On the other hand, most physically-based distributed models are based on well known small-scale theories such as Darcy's law and the Richards equation, and there have been difficulties in their application to larger watersheds.[4] A possible way is to develop a new kind of model which can capture the aggregate behavior of complex hydrologic systems while avoiding complex model structure and elaborate mathematical schemes.

There are different types of hydrological models that can be chosen for flood simulation, but none of them is robust, because different models may have different performance in different area and parameter calibration influences the results greatly. Some multi-model combination methods have been developed to get better simulation results on the basis of different models, such as Bayesian model averaging (BMA) method and so on,[5] which are mainly used in conceptual models.

In this paper, we made a comparison between a lumped conceptual model and a physically-based hydrological model by applying them to six small catchments in the Three Gorges Region, and the Xinanjiang model[6] and the GBHM (Geomorphology-Based distributed Hydrological Model)[7] are selected respectively as examples of these two kinds of models. Basing on an inter-comparison between the two models, we modified the GBHM by calculating the runoff in each grid using conceptual methods like the Xinanjiang model instead of the complex physical equations in the original GBHM model.

2. Study Area and Dataset Description

The Three Gorges Region is part of the upper Yangtze River basin, located between the two stations of Cuntan and Yichang on the main Yangtze River. The region covers an area of about 55,097 km^2, and the length is about 658 km. The average annual precipitation in the Three Gorges region is about 1,200 mm, mainly concentrated in summer (June, July and August), resulting in frequent floods, and flood predictability in this region is discussed a lot. Six small catchments in this region are selected as the study areas in this study, as shown in Fig. 1. The data used in this study are listed in the Table 1.

The weather data is used in calculating potential evaporation, and for each vegetation type, the leaf-area-index (LAI) is calculated from the monthly Normalized Difference Vegetation Index (NDVI).

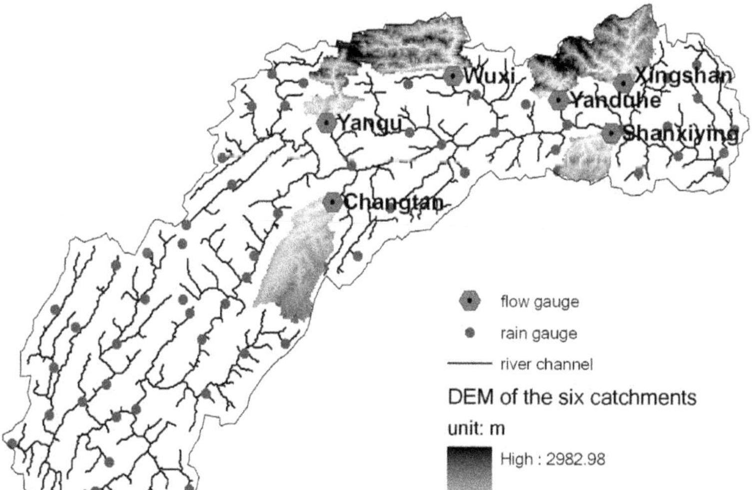

Fig. 1. Study area: The six small catchments in the Three Gorges region.

Table 1. Description of data used in the study.

Data type	Data contents	Data source
DEM	100 m resolution digital elevation data	global topography database
Land use	Reclassified into 7 types in this study, including water body, forest, urban area, irrigation filed, grassland, upland and shrub	USGS global land cover characteristics database
Soil	Totally 237 types in China, there are mainly 5 types in this study catchment, 10km resolution	FAO/UNESCO, IGBP-DIS
Weather	35 stations, daily record of precipitation/temperature/ humidity/winds peed/ sunshine time	China Meteorological Administration
Precipitation	78 rainfall gauges, hourly record of rainfall or snowfall from 2004 to 2007	the Bureau of Hydrology, Ministry of Water Resources
NDVI	Global dataset of monthly NDVI, 10 km resolution	the DAAC of GSFC/NASA
River discharge	6 flow gauges at the outlets of the six catchment, 4 records in a day (at 2:00/8:00/14:00/20:00)	the Bureau of Hydrology, Ministry of Water Resources

3. Brief Description of the Hydrological Models

3.1. *The geomorphology-based hydrological model (GBHM)*

The Geomorphology-based Hydrological Model (GBHM) was developed for hydrological simulation especially considering the catchment geomorphological properties[8,9] and was revised for application to large river basins.[10] Representations of the catchment landform characteristics and description of the hydrological processes are two basic aspects of a hydrological model. The GBHM represents the topography and river network based on the catchment geomorphology area function and width function.[11] The catchment is discretized into a number of flow-intervals along the flow direction, and each flow-interval is represented as a series of hill-slope elements.[8,9] Figure 2 shows the discretization of watershed in GBHM.

As mentioned above, the resolution of DEM is 100 m, but the grid-based distributed models usually cannot run at such high resolution due

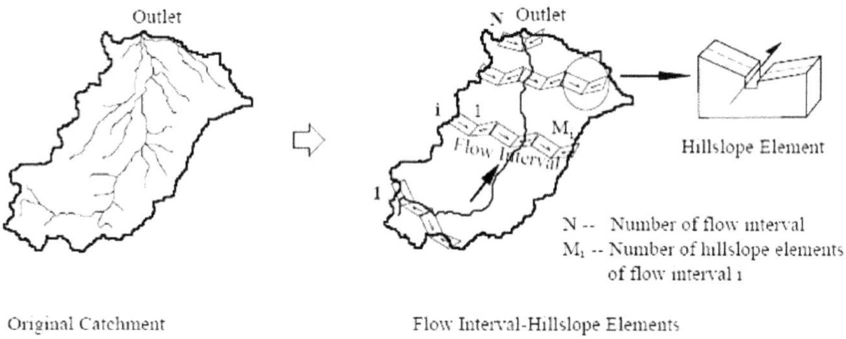

Fig. 2. Discretization of watershed in GBHM.

to the heavy computational requirement, especially for large catchments. In this study, the grid size of hydrological simulation is 1 km, and the spatial heterogeneity inside a 1km grid is considered using the sub-grid parameterization method.[10] The catchment is firstly divided into a number of sub-catchments numbered by the Pfafstetter scheme, and each sub-catchment is then divided into a number of flow-intervals. The calculation grids within the same flow-interval have the same flowing distance to the outlet of the sub-catchment. Each 1 km grid is generalized to a hill-slope element, but the slope parameters (length and gradient) of the element are the average values of the slope parameters calculated from the corresponding 100m DEM. The runoff is calculated basing on the simulation of hydrological process in each hill-slope element, and then added to the total runoff in each flow-interval.

For the hydrological modeling in the hill-slope element, evapotranspiration is estimated from potential evaporation, which is calculated by the Penman equation.[12] Soil moisture movement in the unsaturated soil is described by Richards equation.[13] Subsurface flow process is considered as a function of the hydraulic conductivity and the slope. Surface runoff is treated as sheet flow and solved by Manning's equation.[11] The exchange between groundwater and river water is considered as steady flow and is calculated using Darcy's law.[11]

The kinematic wave method is used to simulate the confluence process in the rivers, get the discharge at the outlet of each sub-catchment from the discharge of the flow-intervals.

There are mainly three kinds of parameters used in the model, which are vegetation and land surface parameters, soil water parameters and river channel parameters. The parameters have clear physical meanings, and generally can be determined by measurements or experimentations. In this study, some of the parameters are estimated by model calibration due to lack of data, including the parameters related to river channel as well as groundwater.

3.2. *The Xinanjiang model*

The three component Xinanjiang model is a conceptual rainfall-runoff model developed by ZhaoRenjun,[6] which has been widely used for flood forecasting in the humid and semi-humid areas in China in the past few decades. It is based on the concept of storage-excess runoff, which means runoff only yields where the soil moisture is saturated. The storage capacity curve is used to describe the spatial variability of the soil-moisture storage capacity in the study area. In this model, soil evaporation is composed of upper evaporation, lower layer evaporation as well as deep layer evaporation. Runoff is divided into surface runoff, subsurface runoff and groundwater runoff, and their contributions to flow discharge are respectively described by Unit Hydrograph, subsurface and ground water depletion coefficients. The schematic of the Xinanjiang model is illustrated in Fig. 3.

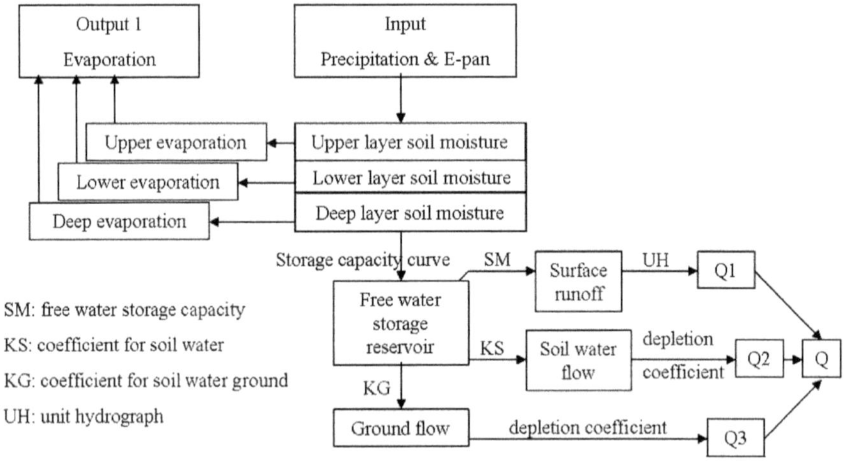

Fig. 3. Schematic of the Xinanjiang model.

4. Model Application and Result Analysis

4.1. *Model applications*

The two models are applied to the six catchments in the Three Gorges region to simulate two series of floods in the year of 2004 and 2005, respectively as calibration periods and validation periods. For reducing the influence of initial state variables, a warm-up period is used in both models. In the GBHM the rainfall and weather data are interpolated to each 1 km grid considering the distance and direction of the station to the target grid, and for the Xinanjiang model the rainfall in each catchment is the average value of the distributed rainfall calculated by the GBHM, so that the total rainfall input is equal in the two models. The Nash-Sutcliffe efficiency coefficient (R^2), the relative error of flood volume (RE) and the coefficient of persistence (CP)[14] are used as evaluation criteria, and the results are listed in Table 2. The observed and simulated discharges at two selected stations of Yanqu and Yanduhe are shown in Fig. 4.

We also applied the two models to continuously simulate the discharge from 2004 to 2007. For each catchment, we selected the four biggest floods in each year (totally 96 floods in the four years) to analyze. Among the results given by the GBHM and the Xinanjiang model, there are 15 and 18 floods with the Nash-Sutcliffe efficiency coefficient bigger than 0.7, 28 and 22 floods with the relative error smaller than 15%, respectively.

Table 2. Results of the two models in individual flood simulation in the six catchments.

				By GBHM			By Xinanjiang		
Gauge	River	Area (km^2)	Flood event	RE(%)	R^2	CP	RE(%)	R^2	CP
Yanqu	Tangxi	1152	04-7-15∼7-20	−1	0.81	0.89	−4	0.90	0.94
			05-7-08∼7-13	−18	0.64	0.83	−20	0.75	0.88
Changtan	Modaoxi	2135	04-6-11∼6-16	−7	0.86	0.92	−8	0.82	0.89
			05-8-26∼8-31	19.7	0.63	0.74	23	0.77	0.82
Yanduhe	Yanduhe	637	04-6-02∼6-07	−14	0.90	0.94	4	0.92	0.95
			05-7-08∼7-13	−10	0.83	0.88	17	0.49	0.66
Shanxiying	Qinggang	576	04-7-15∼7-20	8	0.93	0.95	−15	0.93	0.95
			05-7-08∼7-13	9	0.55	0.71	26	0.55	0.71
Wuxi	Daning	2001	04-7-15∼7-20	4	0.82	0.89	0	0.82	0.89
			05-7-08∼7-13	−8	0.86	0.92	−21	0.87	0.92
Xingshan	Xiangxi	1900	04-7-14∼7-19	−20	0.77	0.85	−8	0.72	0.82
			05-8-19∼8-24	8	0.78	0.80	−22	0.42	0.48

The flood enveents in the year of 2004 and 2005 are calibration and validation periods, respectively.

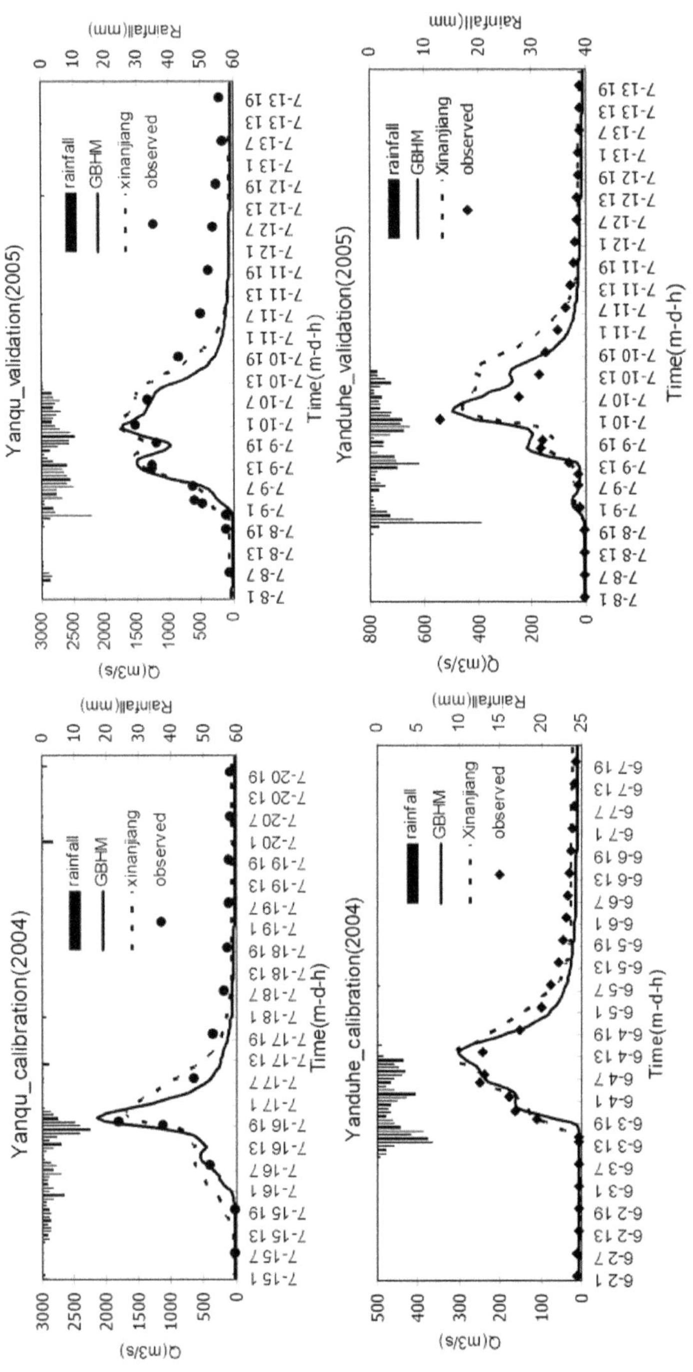

Fig. 4. Simulations results of flood events by the GBHM and the Xinanjiang Model at the stations of Yanqu and Yanduhe.

4.2. Results analysis and model comparison

As shown in the Table 2, out of the 12 flood events, there are respectively 9 and 6 floods with the relative error less than 15% for the GBHM and the Xinanjiang Model, and 9 floods with the Nash-Sutcliffe efficiency coefficient bigger than 0.7 for both. For most of the flood events, the two models can give a good simulation. For the Xinanjiang model, there is a phenomenon that the results for the calibration periods are usually much better than the validation periods. In the Xinanjiang Model, unit hydrograph is estimated from the calibration periods, and it may not represent the hydrological process very well for the validation period, which can be one of the main reasons for the phenomenon.

For the continuous simulation, both of the models failed to simulate all the floods very well. It is difficult for a model to give good hourly simulation of long period flow, and the errors come from the uncertainties of model structure, parameters and input data. Firstly, evaporation calculation is very important for long period flow simulation, but almost negligible for short flood, so the uncertainty of the parameters related to evaporation influences the long period simulation more than short flood simulation. Secondly, equifinality for different parameters influences model calibration, especially for the short flood, and the parameter group calibrated from a single flood may not best reflect the catchment characteristics. At last, accumulation of the errors is a big problem especially for long period simulation. In addition to the uncertainty of the model themselves, uncertainty of input data is also a problem. For example, in July 2005 at the Yanduhe station, a flood happened while there is almost no rainfall, and input data uncertainty can be a main reason in such cases. From the discussion above, it is known that long period simulation gives better evaluation of model performance than short floods.

In this paper, we do not want to decide which model is 'better', but to discuss their advantages and disadvantages as conceptual model and physically-based model, and try to make a combination. The Xinanjiang model is much easier to understand and use, while the GBHM gives detailed description about the catchment characteristics and the hydrological processes. Although there are some problems in using the micro-scale equations in the watershed scale simulation, the framework of physically-based distributed model provided the possibility of studying the real function of watersheds. Further development of the detailed models may rely on getting better knowledge about the catchment characteristic and discovering the macro-scale 'laws' that govern the hydrological responses of watersheds.[4]

4.3. *Flood simulation by the modified GBHM*

Based on the discussions above, the study aims to combine the advantages of the two kinds of models, by establishing a model basing on the framework of distributed model, but using some conceptual method to make the calculation simpler (easier and faster). We developed a new model based on the framework of the GBHM, and runoff calculation in the new model is simplified the by using conceptual methods instead of the complex physical descriptions in the original GBHM. We applied the new model to the six catchments to make continuous simulations from 2004 to 2007, using the same parameters calculated in the GBHM and the Xinanjiang model. The results of the modified GBHM as well as the two original models for the 96 floods are listed in Table 3.

Among the 96 floods simulated by the new model, there are 12 floods with Nash-Sutcliffe efficiency coefficient bigger than 0.7 and 17 floods with relative error smaller than 15%. Compared to the results by the two original models, the performance of the modified GBHM is also not very satisfying. The runoff related parameters are calibrated in the Xinanjiang model, and these parameters are directly used in the new model. The parameters suitable for the watershed scale conceptual model may not be suitable for the runoff calculation in each grid, which can be one of the reasons. Further refinement is necessary for the new model.

Table 3. Results of the three models in long term flow simulation in the six catchments.

		Number of simulated flood events that agree with the criteria		
Criteria	Station	By GBHM	By Xinanjiang	By Modified model
$R^2 > 0.7$	Yanqu	2	6	3
	Shanxiying	5	4	2
	Wuxi	2	3	5
	Yanduhe	3	3	0
	Changtan	1	2	0
	Xingshan	2	0	2
	Sum	15	18	12
RE < 15%	Yanqu	4	3	0
	Shanxiying	6	8	2
	Wuxi	5	4	8
	Yanduhe	6	4	3
	Changtan	3	2	1
	Xingshan	4	1	3
	Sum	28	22	17

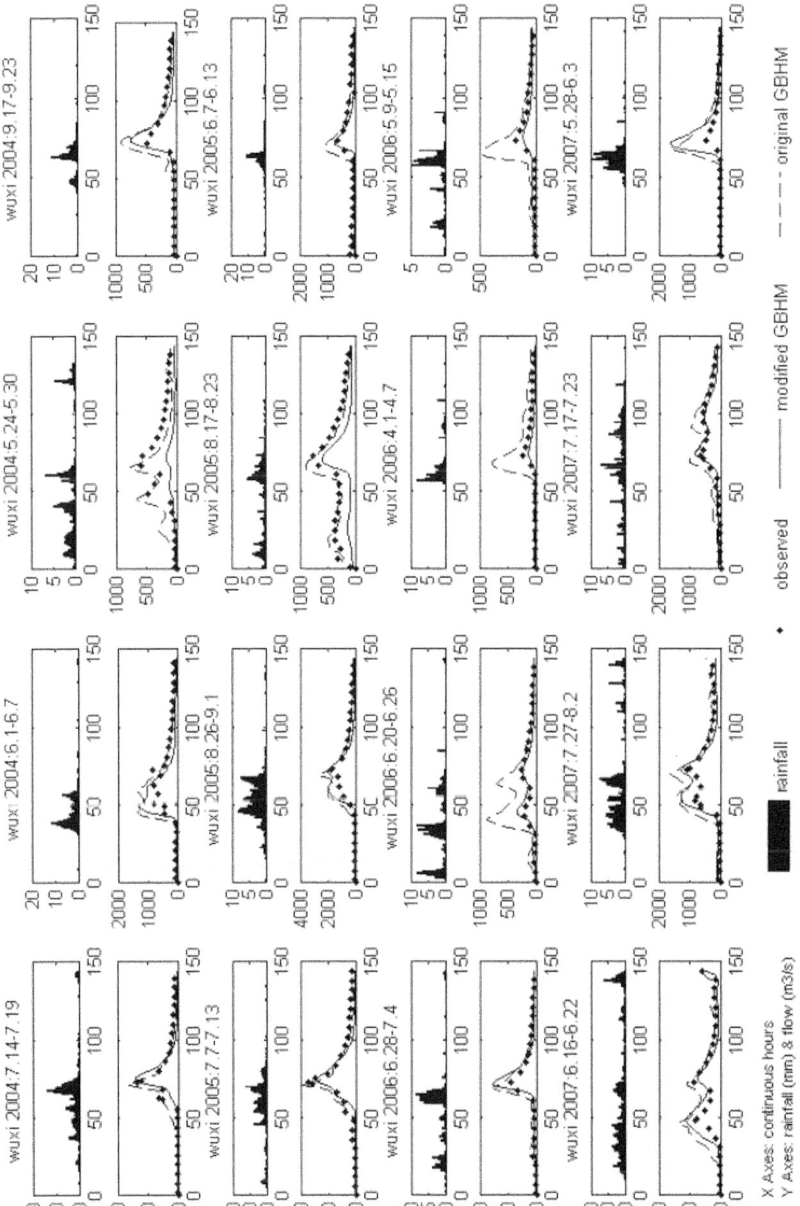

Fig. 5. Results of the four biggest floods at Wuxi station by the modified and original GBHM.

For certain station, for example the Wuxi station, the modified model has a better performance than the two original models. The simulated discharges by the modified and original GBHM of the four biggest floods in each year at the Wuxi station are plotted in Fig. 5.

5. Conclusion

We applied the GBHM and the Xinanjiang model to six catchments in the Three Gorges Region, and simulated 12 flood events in 2004 and 2005. Out of the 12 flood events, there are 9 and 6 floods with the relative error less than 15% for the GBHM and the Xinanjiang model respectively, and 9 floods with the Nash-Sutcliffe efficiency coefficient bigger than 0.7 for both. We conclude from the results that for most of the flood events, the two models can give a good simulation, and the GBHM has a better performance. For the Xinanjiang model, unit hydrograph is estimated from the calibration period, so it may not represent the hydrological process very well for the validation period.

The two models are also used to continuously simulate the discharge from 2004 to 2007 at hourly time step. Among the total 96 flood events simulated by the GBHM and the Xinanjiang model, there are 15 and 18 floods with the Nash-Sutcliffe efficiency coefficient bigger than 0.7, 28 and 22 floods with the relative error smaller than 15%, respectively. It is more difficult for a model to give good hourly simulation of long period flow, and the errors come from the uncertainties of model structure, parameters and input data. From this aspect, long period simulation can give better evaluation of model performance than short floods.

To combine the advantages of the GBHM and the Xinanjiang model, a new model is developed basing on the framework of the GBHM, and the runoff generation is calculated by using conceptual methods as the Xinanjiang model instead of the complex methods in the original GBHM. Comparing to original GBHM, the modified model is much simpler, and doesn't need the detailed information of land surface especially the soil. Comparing to the Xinanjiang model, it can consider the spatial distribution of the rainfall, which is the biggest obstacle for lumped models. The new model is also applied to continuously simulate the discharge from 2004 to 2007, and among the 96 flood events, there are 12 floods with Nash-Sutcliffe efficiency coefficient bigger than 0.7 and 17 floods with relative error smaller than 15%. The results showed that the model performance is similar to the tow original models and can be used for flood forecasting, although some refinement is still necessary.

Acknowledgments

The research was supported by the national '948' project (contract no. 200760) funded by the Ministry of Water Resources and by the National Key Technology R&D Program of China (project no. 2006BAB05B04). The reviewers' comments are gratefully acknowledged.

References

1. J. Refsgaard, *Terminology, Modeling Protocol and Classification of Hydrological Model Codes* (Kluwer Academic Publisher, 1996).
2. N. K. Ajami, Q. Y. Duan and S. Sorooshian, An integrated hydrologic Bayesian multimodel combination framework: Confronting input, parameter, and model structural uncertainty in hydrologic prediction, *Water Resources Research* **43** (2007), W01403, doi:10.1029/2005WR004745.
3. K. J. Beven, and A. M. Binley, The future of distributed models: Model calibration and uncertainty prediction, *Hydrological Process* **6** (1992) 279.
4. J. J. McDonnell, M. Sivapalan, K. Vache, S. Dunn, G. Grant, R. Haggerty, C. Hinz, R. Hooper, J. Kirchner, M. L. Roderick, J. Selker and M. Weiler, Moving beyond heterogeneity and process complexity: A new vision for watershed hydrology, *Water Resources Research* **43** (2007), W07301, doi:10.1029/2006WR005467.
5. J. A. Hoeting, D. Madigan, A. E. Raftery and C. T. Volinsky, Bayesian model averaging, *A tutorial Stat. Sci.* **14** (1999) 382.
6. R. J. Zhao, *Water Hydrological Model-Xinanjiang Model and Shanbei Model* (Water and Power Press, Beijing, 1984).
7. D. Yang, S. Herath and K. Musiake, A Hillslope-based hydrological model using catchment area and width function, *Hydrological Sciences Journal* **47** (2002) 49.
8. D. W. Yang, S. Herath and K. Musiake, Development of a geomorphology-based hydrological model for large catchments, *Annual Journal of Hydraulic Engineering JSCE* **42** (1998a) 169.
9. D. Yang, Distributed hydrological model using hillslope discretization based on catchment area function development and applications, in Engineering, Vol. Doctor, University of Tokyo, Tokyo (1998b).
10. D. Yang, T. Koike and H. Tanizawa, Application of a distributed hydrological model and weather radar observations for flood management in the upper Tone River of Japan, *Hydrological Processes* **18** (2004) 3119.
11. D. Yang, S. Herath and K. Musiake, Comparison of different distributed hydrological models for characterization of catchment spatial variability, *Hydrological Processes* **14** (2000) 403.

12. H. Penman, Natural evaporation from open water, bare soil and grass, *Proc. Phil. Trans. Roy. Soc., Series A* **193** (1948) 120–145.
13. L. Richards, Capillary conduction of liquids through porous mediums, *Physics* **1** (1931) 318–333.
14. P. Kitanidis and R. Bras, Real time forecasting with a conceptual hydrologic model 2: applications and results. *Water Resources Research* **16**, 6 (1980) 1034–1044.

Advances in Geosciences
Vol. 17: Hydrological Science (2008)
Eds. Namsik Park *et al.*

MONITORING THE 2004 FLOOD IN THE MEGHNA RIVER BASIN IN BANGLADESH USING GROUND DATA, RADARSAT IMAGERY AND GIS

ROXANA HOQUE[*,†], DAICHI NAKAYAMA[‡], HIROSHI MATSUYAMA AND JUN MATSUMOTO
Department of Geography,
Tokyo Metropolitan University, Minami-Ohsawa 1-1, Hachioji,
Tokyo 192-0397, Japan
[*] *roxanna_bang@yahoo.com*
[†] *mullah_usa@yahoo.com*
[‡] *daichi@comp.metro-u.ac.jp*

This study investigated an unusual flood event of 2004 in Bangladesh. The unusual early flood occurred in the 2nd week of April, and flood situation became worst when the greater Ganges-Brahmaputra-Meghna (GBM) rivers were coincidently synchronized their peaks on 24th of July.

Flood hazard becomes a regular phenomenon in Bangladesh. In order to investigate the flood of 2004, we conducted a comparative analysis between major historical flood of 1998 with 2004 flood, using remote sensing, GIS and ground data.

Five RADARSAT yearly maximum inundation images from 2000 to 2004 and three time series images of 2004 were used to identify wide flooding area around Bhairab Bazar, including northeastern Bangladesh. The result indicates that on 24th of July 2004, inundated areas in the northeastern region were about 29,900.72 km^2. This result as well corresponded with the GBM rivers discharge, water level and rainfall anomaly maps.

In 2004, about 5027.0 mm of annual rainfall found around Bhairab Bazar and 14816.8 mm at Cherrapunji in India. The annual rainfall anomaly maps from 2000 to 2005 illustrate that exceptionally high rainfall contributes severe flooding of northeastern Bangladesh. In 2004, the water level of the Meghna River at Bhairab Bazar station found that higher than that in 1998. In 2004, inundation period of Bhairab Bazar was longer than that in Bahadurabad of Brahmaputra River.

From the flood damage maps and hazard maps, it was clarified that study area was very much vulnerable for flood disasters. It is crucial for Bangladesh to be able to monitor and assess the flood disasters using remote sensing and GIS techniques, to reduce the cost of general survey and hazard mapping services.

1. Introduction

Bangladesh is situated in the Asian tropical monsoon system lying within 88.05–92.74°E and 20.67–26.63°N. The topography of Bangladesh is very flat. Bangladesh is one of the most flood prone countries in the world and also one of the most densely populated countries Islam and Dhar, [1998]. The countries' economy is basically dominated by agriculture. About 80% of the country is deltaic floodplains with about 230 major rivers including 57 transboundary rivers. Floods are the most destructive and pervasive natural disaster in Bangladesh Dewan *et al.* [2004]. Every year one fourth to one third of the country is inundated during monsoon season by overflowing rivers. Flood disrupts people's live, damage infrastructures and road networks in urban area. In rural areas it damages crops, causes death to livestock, and people become isolated due to the unavailability of communication mode.

Matsumoto [1996] reported that in the northeastern regions the mean annual rainfall ranges roughly from 1,000 to 3,000 mm. Our own results show that the annual rainfall exceeded 5027.0 mm in 2004 at the Meghna river basin at Bhairab Bazar station, which is higher than previous studies. The heavy rainfall in 2004 caused severe flooding in the Meghna basin in the districts of Bhairab Bazar, Sunamganj, Sylhet, Habiganj, Moulvibazar, Kishoreganj and Netrokona.

To understand the hydrological characteristics of 2004 flood, the hydrological behavior of these three major rivers (GBM) was analyzed. In order to investigate the 2004 flood, we conducted a comparative analysis between severe floods of 1998 and 2004.

The RADARSAT remote sensing can be effective tools to determine flood inundation areas. The SIR-B radar penetrated the latent monsoonal cloud cover to reveal hydrological conditions on the ground during severe weather conditions Imhoff *et al.* [1986]. The use of SAR system as a solution to the cloud cover problem has shown great promise Imhoff *et al.* [1987]. Therefore, this study makes an effort to use active sensor the RADARSAT satellite data to identify inundation area and develop a methodology for flood monitoring mapping and assessing. In order to explicitly delineate flood, unique flood hazard maps that have not been utilize in previous studies are also presented.

1.1. *Previous study*

Numbers of previous studies have been devoted to flood monitoring in Bangladesh using remote sensing, GIS and ground data, such as Smith

[1997], Imhoff *et al.* [1986]. Wilson and Rashid [2005], Chowdhury and Ward [2004], Rasid and Pramanik [1993], Oberstadles *et al.* [1997], Dewan *et al.* [2006], Hofer and Messerli [2006], Matsumoto [1996] and Rogers [1993] etc.

However, the majority of previous studies focused mainly on frequent monitoring and mapping of the ground surface using variety of satellite image that were limited to the sensors available at the time. Several previous studies have dealt with the mapping of inundation extant around Dhaka City. We specifically investigate the flood from 2000 to 2004. Such data drawn from satellite research activity for inundation mapping around northeastern Bangladesh have not been analyzed in previous studies.

1.2. *The purpose of this study*

The objective of this study is flood monitoring, mapping and assessing around Bhairab Bazar, northeastern part of Bangladesh. Specifically highlight the feature of unusual flood in 2004 in comparison with that in 1998. This study also intended to develop very detailed flood inundation map and flood hazard map. To accomplish these purposes, we investigated the Ganges-Brahmaputra-Meghna (GBM) Rivers, hydrological characteristics such as discharge, water level, and rainfall from 1998 to 2006, and the RADARSAT satellite images from 2000 to 2004.

2. Data Sources and Description

The active sensor microwave RADARSAT remote sensing data, spatial resolution is 50 m by 50 m, were used for the purpose of flood monitoring, mapping and frequency analysis of floods from 2000 to 2004, and time series images in 2004. We created maximum water extent map, estimated inundation area and compared the results with ground data around northeastern Bangladesh. These data were derived from the Center for Environmental and Geographic Information Services (CEGIS), Dhaka, Bangladesh.

To depict the hydrological characteristics of the 2004 flood, we analyzed three major rivers systems. From the discharge, water level and rainfall data of 95 stations from 1998 to 2006, we made comparative analysis of 1998 and 2004 floods. Data were received from Bangladesh Water Development Board (BWDB), Flood Forecasting and Warning Center (FFWC) and CEGIS.

Using precipitation data from 2000 to 2004, we created distribution maps of rainfall and rainfall anomaly. The precipitation data used were the 51 gauging stations, received from BWDB, FFWC and NOAA. Using GIS data and social and infrastructure flood damage data in 2004, we created a set of damage maps and hazard maps. Social and infrastructure flood damage data were received from the Ministry of Agriculture, Dhaka, Bangladesh.

3. Results and Discussion

3.1. *Seasonal and annual changes of rainfall around Bhairab Bazar from 1998 to 2006*

On average, more than 70% of annual rainfall in Bangladesh is concentrated in the 4 months from June to September Ohsawa *et al.* [2000]. Seasonal and annual changes of the monsoon rainfall around Bhairab Bazar from 1998 to 2006 (Fig. 1) indicate that onset of the summer monsoon generally occurs in the end of May or beginning of June. In 2004, during the pre-monsoon period, unusual rainfall occurred in April, and rainfall exceeded the monthly average from June to September. The highest rainfall was found in August, and total annual rainfall found to be 5027.0 mm. Likewise in 1998, the highest rainfall was found in July, and total annual rainfall was found to be 2230.0 mm. The monthly rainfall scenario at Bhairab Bazar in 1998 and 2004 (Table 1) shows that in 2004, annual precipitation at Bhairab Bazar was larger than that in 1998. Precipitation in May, June and August in 2004 was larger than previous monthly maximum. The results suggested that precipitation around Bhairab Bazar was a principal factor in the flood in 2004.

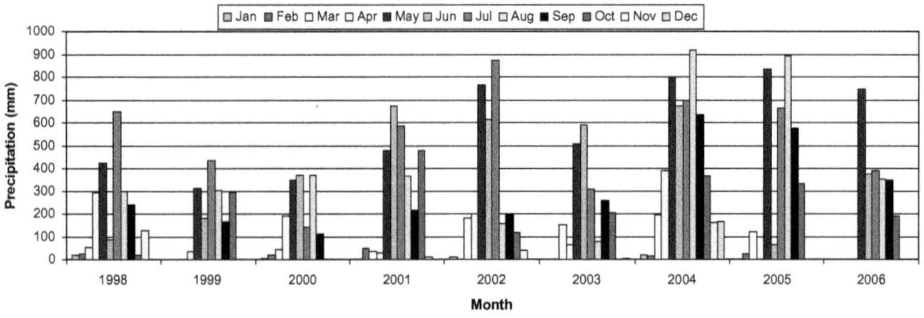

Fig. 1. Seasonal and annual changes of rainfall around Bhairab Bazar for 1998 to 2006.

Table 1. Rainfall scenarios for Bhairab Bazar in 1998 and 2004 (Unit: mm).

Month	Monthly Maximum	Monthly Average	Observed in 1998	Observed in 2004
January	21.5	3.3	19.0	21.5
February	120.2	29.4	22.0	12.2
March	193.5	50.5	52.0	193.5
April	388.8	167.7	295.0	388.8
May	801.0	379.9	424.0	801
June	674.6	327.4	87.0	674.6
July	695.4	395.4	647.0	695.4
August	622.0	314.5	298.0	916.2
September	634.8	293.9	239.0	634.8
October	365.1	153.7	19.0	365.1
November	159.0	25.5	128.0	159
December	164.9	12.7	0.0	164.9
Annual Total	5440.8	2153.9	2230.0	5027.0

Monthly maximum and average are calculated from 1970 to 2000.
Source: BWDB, Dhaka, Bangladesh.

3.2. *Rainfall comparisons between upper reach Cherrapumji in India and Bhairab Bazar from 1998 to 2004*

The Cherrapunji is well known as the world's heaviest rainfall zone. It is located on the southern slope of the Meghalaya Plateau and about 10 km away from the northeastern border of Bangladesh and India Pant and Kumar [1997]. The extensive rainfall in the Meghalaya Plateau is very important, perhaps even decisive, for flooding processes in Bangladesh Matsumoto *et al.* [1996]. The monsoon rainfall at Cherrapunji is on average larger than 8,000 mm Ohsawa *et al.* [2000].

The rainfall situations at the Cherrapunji and Bhairab Bazar are presented in Figs. 2(a) and (b). There was a huge amount of rainfall in 1998. The annual rainfall in Cherrapunji and Bhairab Bazar was found to

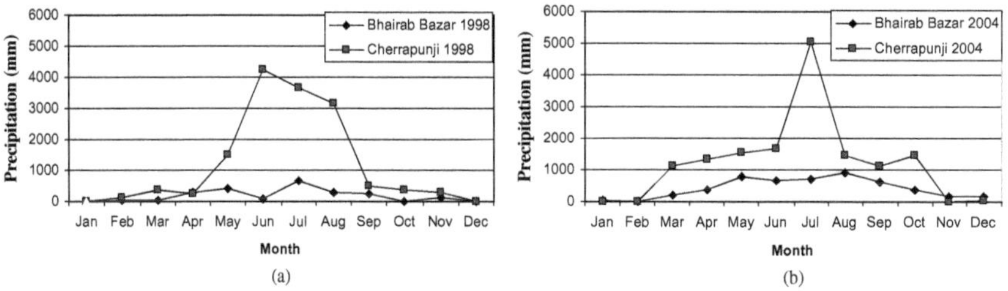

Fig. 2. Rainfall comparison between Bhairab Bazar and Cherrapunji in 1998 and 2004.

be 14536.9 mm and 2230.0 mm respectively. In 2004, the highest rainfall was found in July and annual total rainfall was found to be 14816.8 mm in Cherrapunji. In contrast, the highest rainfall was found in August and annual total rainfall was about 5027.0 mm in Bhairab Bazar. This result suggested that rainfall in 2004 was considerably higher than that in 1998 at both stations.

3.3. *Annual summer monsoon rainfall anomaly maps from 2000 to 2005*

The Asian summer monsoon in South Asia, there remains many unsolved problems. Distribution of excessive rainfall causes rain water to flood, and devastating floods have occurred frequently, causing misery to vast areas of the country Hoque *et al.* [2007].

Based on the precipitation data, annual rainfall anomaly maps from the average of the years 2000 to 2005 were created. On the legend, the gray and black part of the map illustrate 2,000 mm rainfall, while the ash color part of the map illustrates −2,000 mm rainfall (Fig. 3). The rainfall anomaly map for the year 2000 shows that Bangladesh had less rainfall and in the northern part there was little rainfall. In 2001, the western and southern parts had little rainfall. The rainfall anomaly map of 2002 illustrates significant rainfall, while in 2003 rainfall anomaly map shows less rainfall throughout Bangladesh. The rainfall anomaly map for 2004 illustrates extremely high rainfall. The 2005 rainfall anomaly map indicates again less rainfall in Bangladesh. Similar results were found by the RADARSAT imageries in 2002 and 2004 (Fig. 6) inundation maps. From these data we conclude that the relationship between the annual rainfall anomaly maps and RADARSAT inundation maps corresponds quite closely.

3.4. *Seasonal and annual changes of water level around Bhairab Bazar and GBM Rivers*

Seasonal and annual changes of water level of Meghna River around Bhairab Bazar from 1998 to 2006 (Fig. 4) indicated that in 2004 water level was significantly higher than that in 1998 event. In 2004, excessive early monsoon rainfall occurred from April (Fig. 1 and Table 1). The water levels around Bhairab Bazar in these months in 2004 were higher than those in other years. In 2004 flood around northeastern region were principally

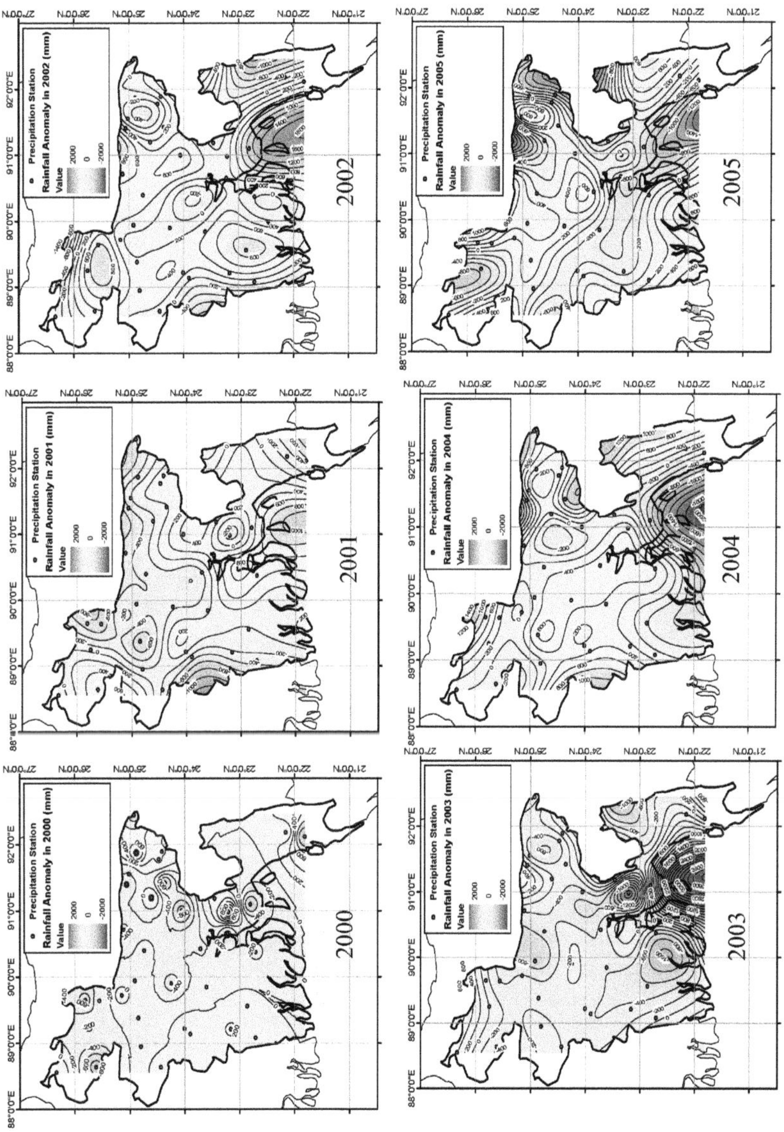

Fig. 3. Annual summer monsoon rainfall anomaly maps from 2000 to 2005 (Black part of the map indicates 2,000 mm rainfall and ash part of the map indicates −2,000 mm rainfall).

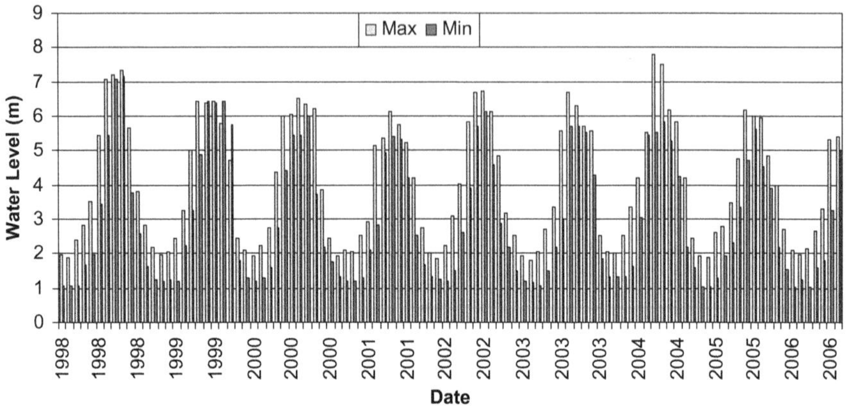

Fig. 4. Water level of Meghna River around Bhairab Bazar from 1998 to 2006.

responsible to much rainfall and high water level. Generally, the flood peaks of Meghna River were found in July and August.

Water levels of three major river stations of the Ganges (Harding Bridge), Brahmaputra (Bahadurabad), and Meghna (Bhairab Bazar) from 2001 to 2006 (Fig. 5) shows that their peaks appear in July and August almost every year. Generally, the Brahmaputra starts rising in March and April, due to snowmelt in the Himalayas, and usually attains its peak in July. It rises again and reaches the annual peak in late July to early August from heavy monsoon rainfall. The Ganges begins rising in the end of June to the beginning of July and attains the peak in the end of August to beginning of September. The Meghna reaches its peaks between June and August. It was revealed that GBM Rivers were coincidently synchronized their peaks on 24th July, resulting severe flooding.

3.5. *Water level and discharge of three major rivers (GBM) in 1998*

Water level of three major Rivers (GBM) is very unique. It was found that the Meghna and Brahmaputra raised water level from June and decreased from October. It was found that discharge increased sharply from June and extremely high peak discharge was found at Bahadurabad on 12th September in 1998. The Meghna attained its peak earlier than the Brahmaputra, probably due to its shorter length Rahman *et al.* [2005]. Therefore, a much greater flood was inevitable in 1998 (Figure not shown).

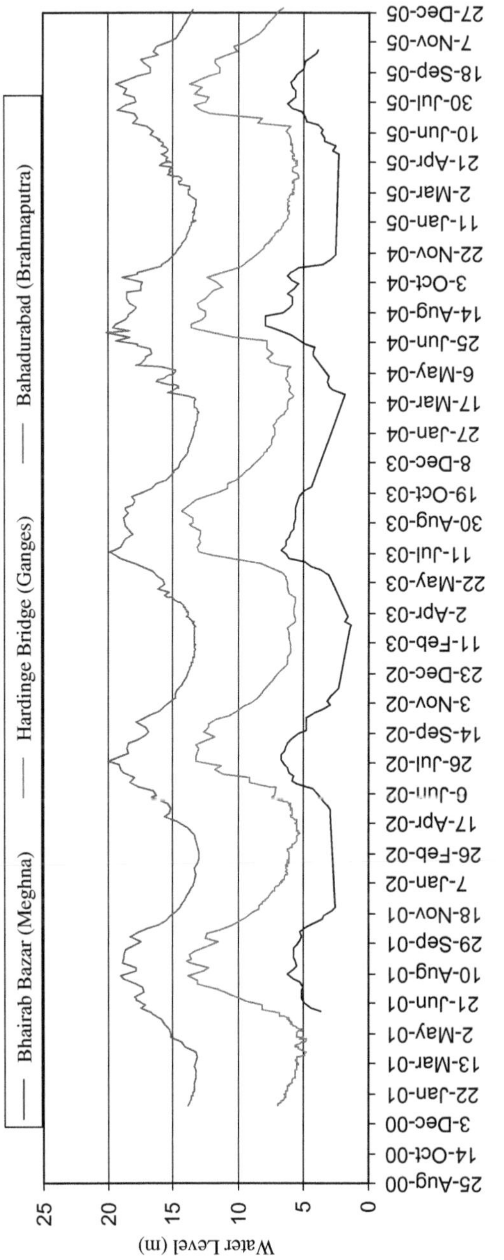

Fig. 5. Water level of three major rivers (GBM) 2001 to 2006.

Table 2. Three major river's Maximum water level (W. L) and duration over danger water levels (D. L) from 1998 to 2004.

	Bahadurabad (Brahmaputra)		Hardinge bridge (Ganges)		Bhairab Bazar (Meghna)	
Year	Max^m WL(m) Water level	Duration over D.L	Max^m.WL(m) Water level	Duration over D.L	Max^m WL(m) Water level	Duration over D.L
1998	20.36 (Aug)	68 days	15.9 (Sep)	27 days	7.33 (Aug)	68 days
1999	19.82 (28Aug)	8 days	14.15 (Sep)	0 days	6.43 (Jul)	20 days
2000	20.16 (6Aug)	10 days	14.18 (25Sep)	0 days	6.50 (19Aug)	20 days
2001	19.33 (4Aug)	0 days	13.90 (31Aug)	0 days	5.98 (5 Aug)	0 days
2002	20.09 (28Jul)	8 days	13.53 (lAug)	0 days	6.73 (3 Aug)	38 days
2003	19.87 (13Jul)	14 days	14.28 (20Sep)	4 days	6.70 (18Jul)	27 days
2004	20.18 (24 Jul)	16 days	13.68 (24 July)	0 days	7.87 (24Jul)	40 days

3.6. *Comparison between the major historical floods of 1998 and 2004 flood*

The 1998 flood was not only of very high magnitude, but also of very long duration of more than two months. The 1998 flood was found prolonged and severe. In 2004, major parts of the study area found severely vulnerable to floods, causing tremendous human sufferings and considerable damages. In 1998, 70% of the country was flooded, while 40% in 2004 Rahman *et al.* [2005]. In 2004 maximum water levels around GBM Rivers coincided on 24th of July, resulting severe flood. In 1998, Bhairab Bazar and Bahadurabad, duration over danger water level found to be same (68 days), while in 2004 duration over danger water level is longer in Bhairab Bazar (40 days) than in Bahadurabad (16 days) (Table 2). This point indicates that Bhairab Bazar station is more severe than Bahadurabad station in 2004. As far as inundation area in the whole country is concerned 1998 flood was much greater, while 2004 flood was also severe with many respects around northeastern region.

3.7. *Satellite image date interpretation of flood in 2004*

3.7.1. *Detection of the flood inundation area around the northeaster region based on RADARSAT images from 2000 to 2004*

Microwave remote sensing the RADARSAT data are particularly useful for hydrological studies in monsoon countries where inclement weather and cloud cover precluded imaging by optical sensors in the monsoon season

Imhoff *et al.* [1987]. Precise mapping of the maximum flood extent is also required for detecting deficiencies in existing flood control measures and arbitrating damage claims later. Remote sensing has proved useful in ecological, hydrological and geomorphological river studies Smith [1997]. The RADARSAT satellite images can be taken at specified locations on specified dates and very useful for inundation mapping Imhoff *et al.* [1986].

Inundation areas are clearly seen on microwave remote sensing technique by the RADARSAT data Badji and Dautrebande [1997]. It is very difficult to obtain such kind of figures based on the optical remote sensing such as NOAA or LANDSAT, due to high amount of rainfall and cloud cover.

Figure 6 illustrates progress of monsoon flood around northeastern Bangladesh from 2000 to 2004. It was monitored by active sensor RADARSAT remote sensing imaging. The RADARSAT images spatial resolution is 50 m by 50 m, were classified by CEGIS. RADARSAT images focus only on two classes, maximum extent of water body in black and other surface conditions in ash color. These map clearly depicted the extent of flood from 2000 to 2004. The RADARSAT classified images were digitized; they were processed for creating a maximum water extent map and for estimating inundation areas.

Visual interpretation of the RADARSAT images for 2000 suggests that the extent of the inundation area was extremely large. The Meghna river basin at the Bhairab Bazar station illustrates over bank flooding. The district of Bhairab Bazar, Sunamganj, Mymensingh Kishoreganj and Netrakona district where completely submerged, while southern side of the district Habiganj Maulvibazar Sherpur showed incomplete inundation. Based on digital analysis of RADARSAT 2000 image, our estimates of the maximum extent of inundation area was approximately 23, 678.55 km^2.

The RADARSAT images of the 2001 data depicted that there was also over bank flooding in the Meghna river basin and estimates of the maximum extent of inundation area was roughly 20,956.14 km^2. Therefore, based on the satellite image interpretation and the ground data, it can be deduced that the 2001 flood was not so significant. The 2002 image data demonstrated that the six districts of the northeastern region were completely submersed. Estimate of the maximum extent of the inundation area was about 27,478.04 km^2. For 2002, the satellite data interpretation and the ground data verification, it can be deduced that, the 2002 flood was extremely severe. The 2003 image data suggested that the Meghna River basin and Bhairab Bazar, Netrokona, Sunamganj, Moulavi Bazar and

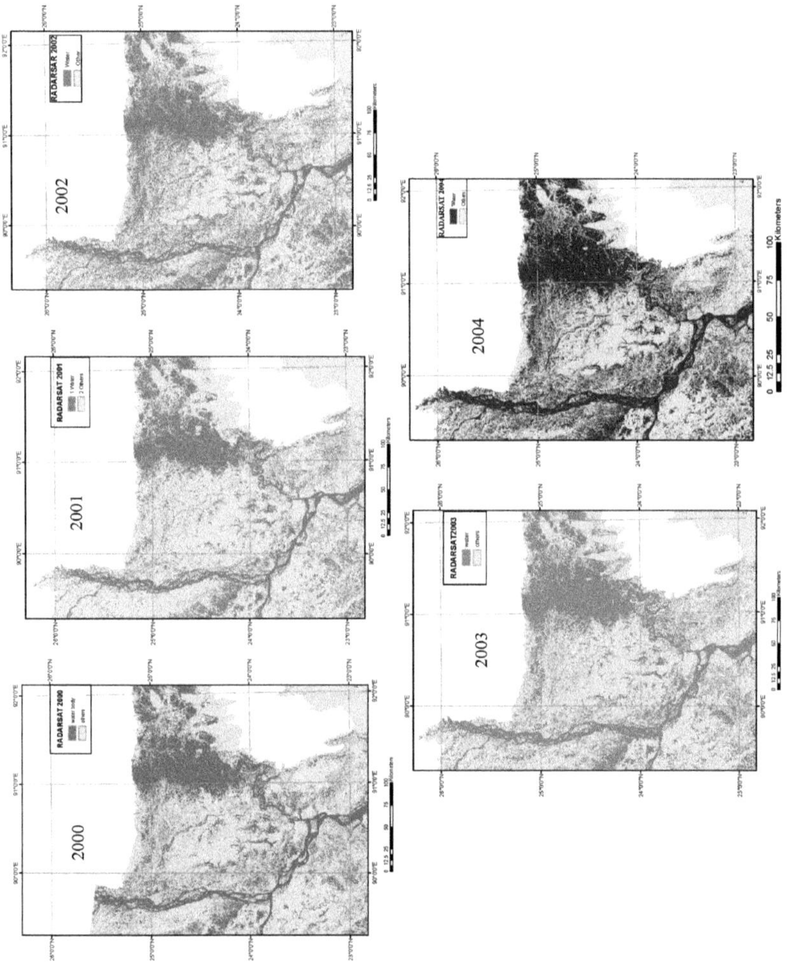

Fig. 6. RADARSAT inundation map illustrates flooded area around Bhairab Bazar northeastern part of Bangladesh from 2000 to 2004. Black part of the map indicates water and white part are indicates others surface conditions.

Habiganj districts were severely flooded. Estimate of the maximum extent of inundation area was about 25,417.91 km^2. As a consequence, satellite data and the ground data indicate that the 2003 flood was also severe.

Visual interpretation of the 2004 imagery also demonstrated that the inundation area was extremely large and our estimate the maximum extent of the inundation area was approximately 29,900.72 km^2. RADARSAT images focused on the northeastern region the Meghna river basins showed that the Sunamjang, Habiganj and Maulvibazar districts and southern side; Sherpur, Mymensingh and Netrakona districts were completely submerged. RADARSAT imagery also suggested that the surrounding dry areas had been submerged in both 2002 and the 2004 floods. It was revealed that in 2004 inundated area was larger than in other years. The visual interpretation of RADARSAT imagery showed good comparison with the ground data interpretation. The RADARSAT imaging system is capable of allowing for flood delineation. From the images major river lines can be easily located. However, problems arose for example, flood affected family housing areas such as buildings, bridges, culverts, agricultural lands, roads, trees and embankments could not be detected by RADARSAT image data.

3.8. *Time series of RADARSAT inundation maps around northeastern region for 2004*

The spatial distribution of flood inundation maps obtained by the RADARSAT data is adequate to precisely facilitate flood mapping. Time series analysis of RADARSAT inundation maps (Fig. 7) allowed for visualization of the progress of monsoon flooding in northeastern Bangladesh for 2004.

Visual interpretation of RADARSAT images on 5th of June in 2004 illustrated that, rain water and river water gradually increased and filled the Meghna River including the northeastern Bangladesh. The RADARSAT image on the 24th of July in 2004 illustrated extensive flood water prevailed, and flooding around northeastern Bangladesh was extremely severe. The visual assessment of RADARSAT images on the 9th of September in 2004 showed that the flood water was gradually decreasing, but large parts of flood water still presented. It is to be noted that it takes long time to dry out this area. It is thus quite possible to create maximum inundation flood maps by overlaying the time-series of the inundation images Otsubo *et al.* [2000].

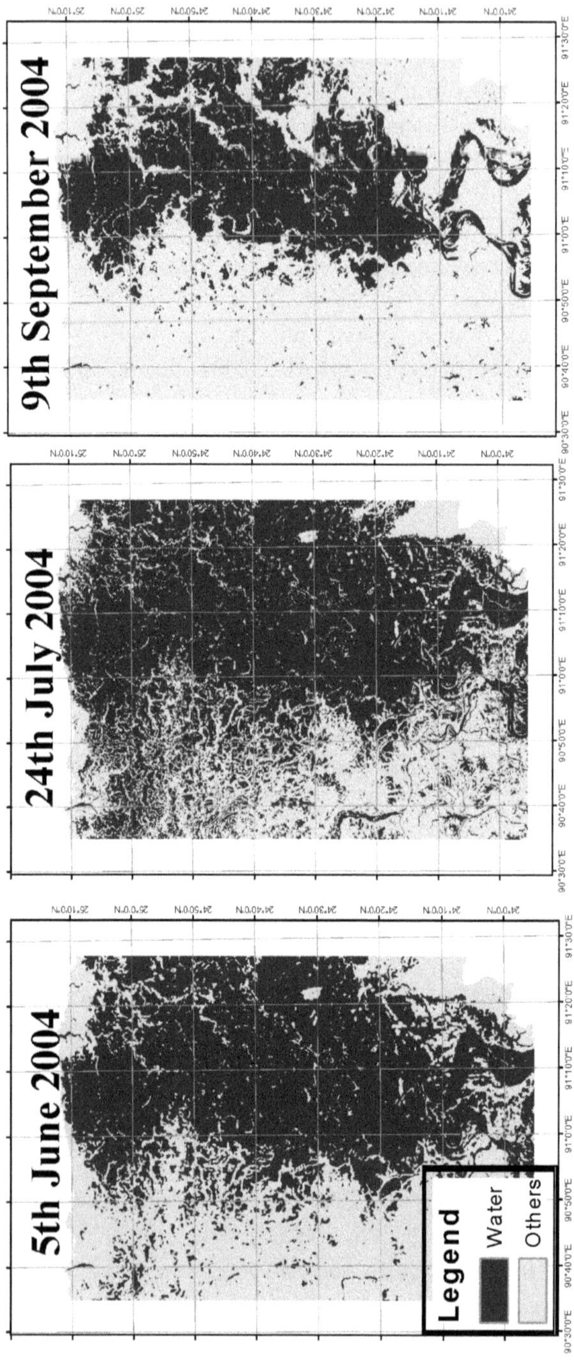

Fig. 7. Time series of RADARSAT inundation map illustrates flooded area around Bhairab Bazar northeastern part of Bangladesh from on June, July and September in 2004. Black part of the map indicates water and white part are indicates others surface condition.

3.9. *Flood damage maps and hazard maps northeastern region for 2004*

Due to lack of sufficient flood damage data, we presents six flood damage maps in six districts, such as, Bhairab Bazar, Netrokona, Brahmanbaria, Sunamganj, Moulavi Bazar and Habiganj, based on 18 days data from 14th to 31st of July, in 2004 around northeastern part of Bangladesh.

Figure 8(a) displays flood affected families. It was found that 1,590,425 families in total were affected by severe flood. On the map black parts indicate number of the flood affected families in the Habiganj district is between 116,769 to 176,548. This map indicates that a number of families suffered from severe floods, and a number of families moved from the inundation area. Nevertheless, large number of families continued to live in their homes in the inundation area, in spite of the fact that this area is not habitable during flooding.

Figure 8(b) shows crop damage. A total of crop damage was 177,842 hectare in fully and 90,160 hectare in partially damaged. The black parts of the map indicate the crop damage area is 7,551–11,855 hectare around the Bhairab Bazar, Brahmanbaria and Sunamganj districts. In fact, at the beginning of crop plantation and harvesting period, unusual heavy rainfall occurred from April, as a result agriculture land was submerged. The Boro rice crops were completely damaged.

Composites of the five RADARSAT inundation maps from 2000 to 2004, the GIS data and damage data in 2004, allowed the creation of unique flood hazard maps. Five year hazard zones can be classified based on the number of years affected by floods. Hazard maps covered the northeastern region, while damage data focused only in six districts: Bhairab Bazar, Netrokona, Sunamganj, Moulavi Bazar, Habiganj and Brahmanbaria. The Hazard maps illustrate that the very vulnerable and risky areas are Bhairab Bazar, Sunamganj, Habiganj, Netrokona, and Moulavi Bazar.

The black part of the map (Fig. 9) illustrates the five years flood hazard zone and ash part illustrates no hazard zone. On the hazard map, white circles indicate the severity of flood damage and risk zone area (large to small) along with numbers showing serially. Six hazard maps are presented here, which focus on the severity of the flood hazard around northeastern Bangladesh in 2004.

Figure 9(a) display homes that were fully damaged. It was found that a total of 276,516 homes were fully damaged. On the hazard map one white circle illustrates 25,000 homes are completely damaged around Bhairab Bazar. It was also observed that there were no hazard zones around Brahmanbaria district. One white circle indicates that for a no hazard zone

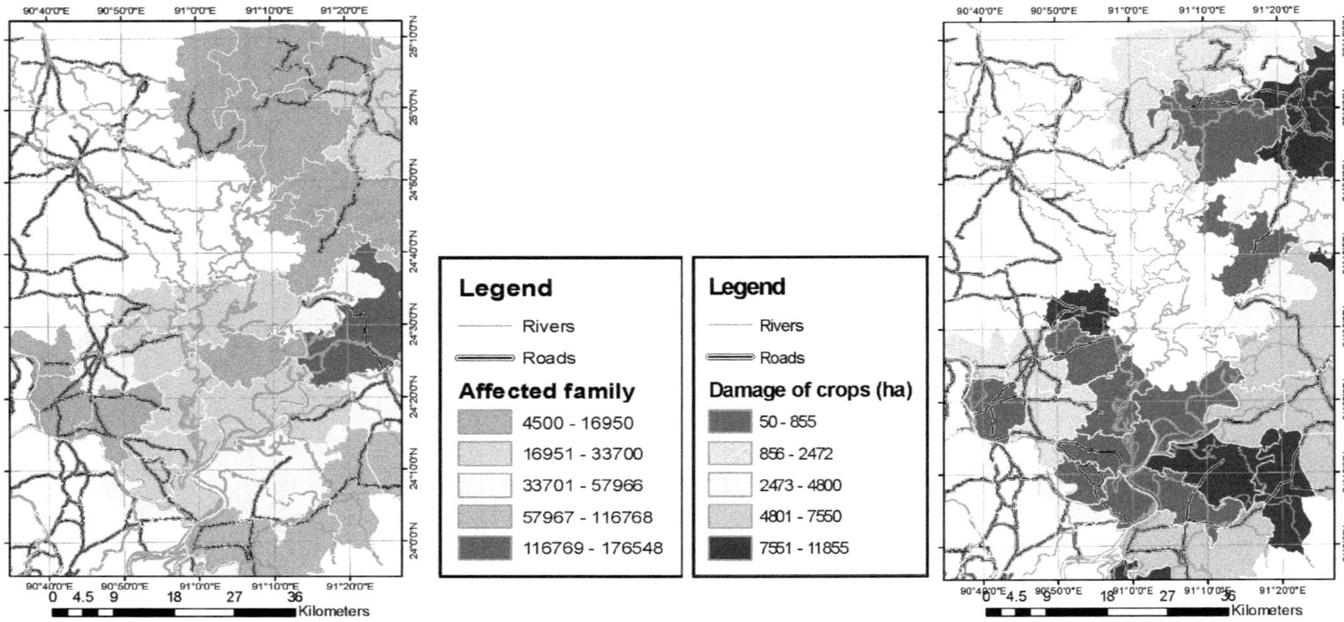

Fig. 8. Illustrate (a) Flood affected family (b) Damage of crops (ha).

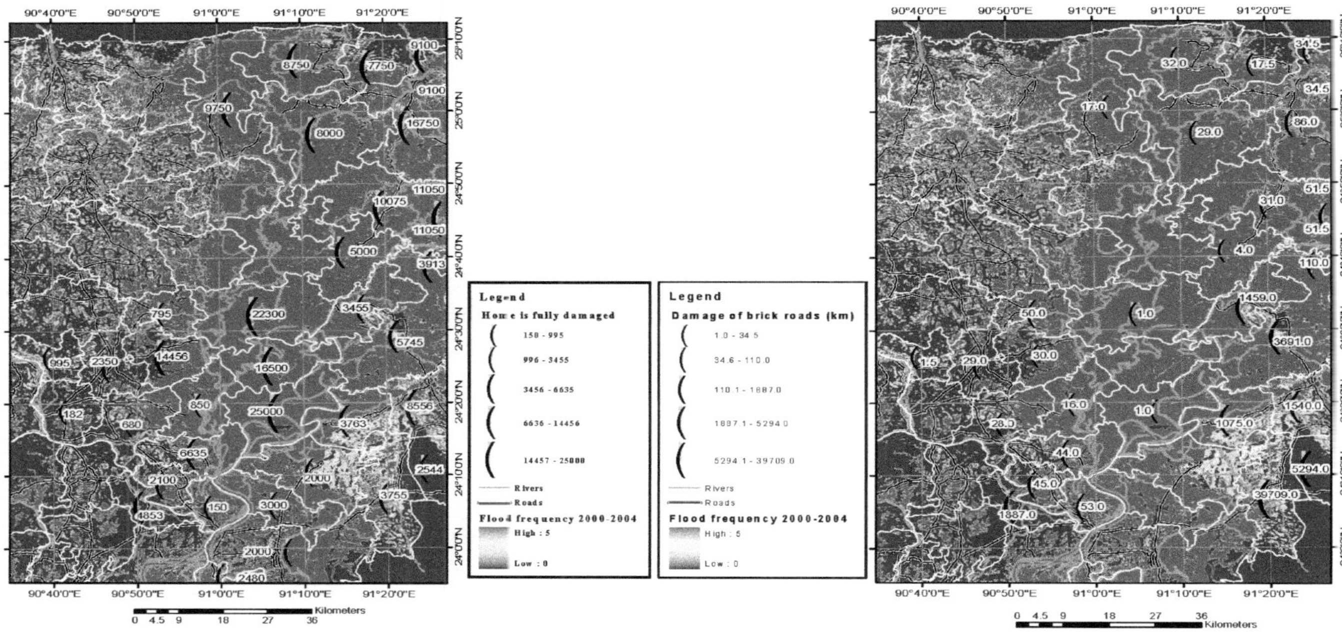

Fig. 9. Flood hazard maps in 2004 (a) Home is fully damage (b) Damage of brick roads.

2,544 homes are completely damaged. It can be deduced that the hazard zone was increasing daily. The flood affected people left their damaged homes and took shelter with their relatives or in other area. Those villagers, who did not leave their home, managed to live by making machangs.

Figure 9(b) shows the damage of brick roads. It was found that a total of 4,489,044 km^2 brick roads were completely destroyed in 2004. It can be deduced that a number of roads were destroyed around the Hobiganj district. No hazard zone was in the Brahmanbaria district two white circles indicate that 5,294 km^2 brick roads completely destroyed. Damaged roads are a very common feature in northeastern Bangladesh. It is very difficult and complicated to commute to other cities during flooding conditions. The government of Bangladesh, therefore, tries to repair these roads during the winter season.

4. Discussion and Conclusions

This study explored the widest flood prone areas around northeastern part of Bangladesh as determined from RADARSAT images. Flood monitoring, mapping and estimates RADRARSAT data can be done effectively. Though just a tool, remote sensing and GIS can play a great role in identifying flood prone area and estimation precisely, mainly in developing countries like Bangladesh where there are numerous inadequacies in the availability and access to the data and information required for effective management of this highly sensitive environmental condition.

Analysis of inter-annual rainfall variations, anomalous rainfall was found around northeastern region. In 2004, significant precipitation was observed from May to August. Upper reach catchment in India at Cherrapunji, much rainfall was observed. In 2004, water level around the Meghna River Bhairab Bazar station was found significantly high. The maximum water level around GBM Rivers coincided on 24th July, resulting severe flood. These all characteristics were well captured the maximum water extent image by RADARSAT in 2004.

The damage map shows the six district's damage from GIS data and damage data. The hazard map illustrates the five years duration hazard zone, using RADARSAT satellite data, GIS data and damage data.

Acknowledgment

The authors express their sincere thanks to the Center for Environmental and Geographic Information Services (CEGIS), Flood Forecasting and

Warning Center (FFWC) and Bangladesh Water Development Board (BWDB) for provide satellite data and ground data. The authors are wish to thank Dr. Fumie Murata of Kochi University, for provide precipitation data. The authors are wishes to thank Dr. Masumi Zaiki (Seikei University) for her constructive suggestions.

References

1. A. M. Dewan, K. Yeboah and M. Nishigaki, *J. Japan Soc. of Hydrol. and Water Resource* **19** (2006) 44.
2. B. Moussa and D. Sylvia, *Hydrol. Process* **11** (1997) 1441.
3. B. A. Wilson and H. Rashid, *The Canadian Geographer* **49** (2005) 100.
4. A. M. Dewan, M. Nishigaki and M. Komatsu, *J. Faculty of Environmental Science and Technology, Okayama University* **9** (2004) 99.
5. G. B. Pant and K. R. Kumar, *Climates of South Asia* (John Wiley, New York, 1997).
6. J. Matsumoto, M. R. Rahman, T. Hayashi and N. Monji, *Bull. Dept. Geogr. Univ. Tokyo* **28** (1996) 25.
7. L. C. Smith. *Hydrol. Process* **11** (1997) 1427.
8. M. R. Chowdhury and Y. Sato, *J. Japan Association of Hydrol. Sci.* **26** (1996) 241.
9. M. L. Imhoff, C. H. Vermillion, M. Story and F. Polcyn, *The Science of the Total Environment* **56** (1986) 277.
10. R. Oberstadles, H. Honsch and D. Huth, *Hydrol. Process* **11** (1997) 1415.
11. R. Hoque, D. Nakayama and H, Matsuyama, *Int. Conference on Water and Flood Management* **2** (2007) 431.
12. H. E. Rashid, *Geography of Bangladesh* (University Press Ltd, Dhaka, 1991).
13. R. Rahman, A. Haque, S. Khan, A. Salehin and S. K. Bala, Dhaka, (Bangladesh Univ. Engi. and Tech. 2005).
14. T. Hofer and B. Messerli, *Floods in Bangladesh* (United Nations University, Tokyo 2006).
15. T. Ohsawa, T. Hayashi and Y. Mitsuta, *J. Geophys. Res.* **105** (2000) 445.
16. Y. Otsubo, T. Ito and H. Iida, *J. Japanese Rem. Sen. Soc.* **20** (2000) 80.

Advances in Geosciences
Vol. 17: Hydrological Science (2008)
Eds. Namsik Park *et al.*

RECENTLY FLOOD EVENTS AND GROUNDWATER RELATIONSHIP IN LOWER NAMKAM BASIN THAILAND

SOMBAT CHUENCHOOKLIN
Department of Civil Engineering, Naresuan University, 99 Moo 9 Thapho, Muang District, Phitsanulok Province 65000, Thailand
sombatc@nu.ac.th

TSUTOMU ICHIKAWA
Department of Environmental Conservation Sciences, School of Industrial Engineering,
Tokai University, 9-1-1 Toroku, Kumamoto 862-8652, Japan
tichi@ktmail.tokai-u.jp

PHATTAPORN MEKPRUKSAWONG
Project Management Office, Royal Irrigation Department, Samsen Road, Dusit, Bangkok, 10300, Thailand
phatta05@yahoo.com

The lower Namkam River Basin was chosen for study in this research. The experimental site is located at a tributary basin of the Mekong River Basin in Nakhon Phanom Province, Thailand. In this area, the ground water salinity is influenced by some saline soil layers beneath ground surface. Recently, the regulation structures in the Namkam River and its tributaries were built and operated for downstream water utilization and flood control through the Royal Irrigation Department (RID). However, the flood has occurred and become a problem in the lowland agricultural areas year by year. In contrast, the surface water is insufficient for the cultivated areas during the dry season. The Gumbel distribution and a river system model (HEC-RAS) were used to mimic the levels of water surface and flood events with the historical data of flood in the period of 1964–2007. The field measurements with automatic observation of groundwater level and salinisation were applied for the model simulation of 390 square kilometers. The results from the model simulation and from the observation reveal that the groundwater levels in the shallow aquifers significantly fell after the end of May. This event caused by high amounts of groundwater withdrawal during drought periods. On the other hand, the groundwater levels in the shallow aquifers increased in the beginning of October resulted from high recharge from flooding during wet seasons. However, the groundwater levels in the confined aquifers remained less impacted than in the shallow aquifers. Thus, the water tables in the deeper aquifers are higher than in the shallow aquifers during the end of wet season. This reason leads to salinisation of the groundwater and thus cannot be used for crops. This research project is a part of model developments for the balance of conjunctive surface runoff and groundwater use at high efficiency in irrigation management in this study area.

1. Introduction

The Namkam River, one of the Mekong River's tributaries, is located in Sakhon Nakhon and Nakhon Phanom provinces in Northeast region of Thailand with a catchment area of 3,440 square kilometers as shown in Fig. 1. The study area, the Lower Namkam Basin in Nakhon Phanom Province covers an area of 1,570 square kilometers. Apart of this area located in Na Kae, Renu Nakhon, and That Phanom districts was chosen for study with emphasise on floods and groundwater levels change. Lam Nam Kam (Namkam) is the main river used as the water source for agriculture in this area but there is a high variation of flow between rainy and dry season. The direction of water flow is from the west in Sakhon Nakhon to the east and meets the Mekong River in Nakhon Phanom. Nam Bang (Nambang) is the largest tributary river of Namkam which flows from the north in Pla Pak district to the south in Renu Nakhon and Na Kae districts.

The major existing water resources development projects in this basin include a big dam, a lake and 5-regulated structures. The Nam Phung Dam (with a storage capacity of 155.8 million cubic meters: MCM) and the Nong Han Lake (266.9 MCM) are operated for downstream water utilization and flood control through the Electricity Generating Authority of Thailand (EGAT), the Department of Fisheries (DOF), and the Royal Irrigation Department (RID). Suratsawadee is a gate structure in order to release water from the Nong Han Lake to the Namkam River and the Mekong River via 4-regulated structures: Nongbueng (97 km), Nakham (71 km), Nakhu (46 km) and Namkam (1.7 km). Those gates are situated in the Namkam River and were built in 2000, 1997, 1999, and 2009, respectively (RID, 1995; RID, 2006). The tributary Nambang is not influenced by those structures, except for a medium reservoir of 11 MCM storage (Huai Kaen) and 2-regulators: Tabtao (34 km), and Nabua (20 km), respectively. Moreover, 14-small reservoirs upstream in Phu Phan Range in the south of the Namkam basin were built.

Those water resource projects are used for flood control and drought remedy in the upland agricultural area. However, flood problems is still remain in the lowland agricultural area downstream of Nakhu regulator

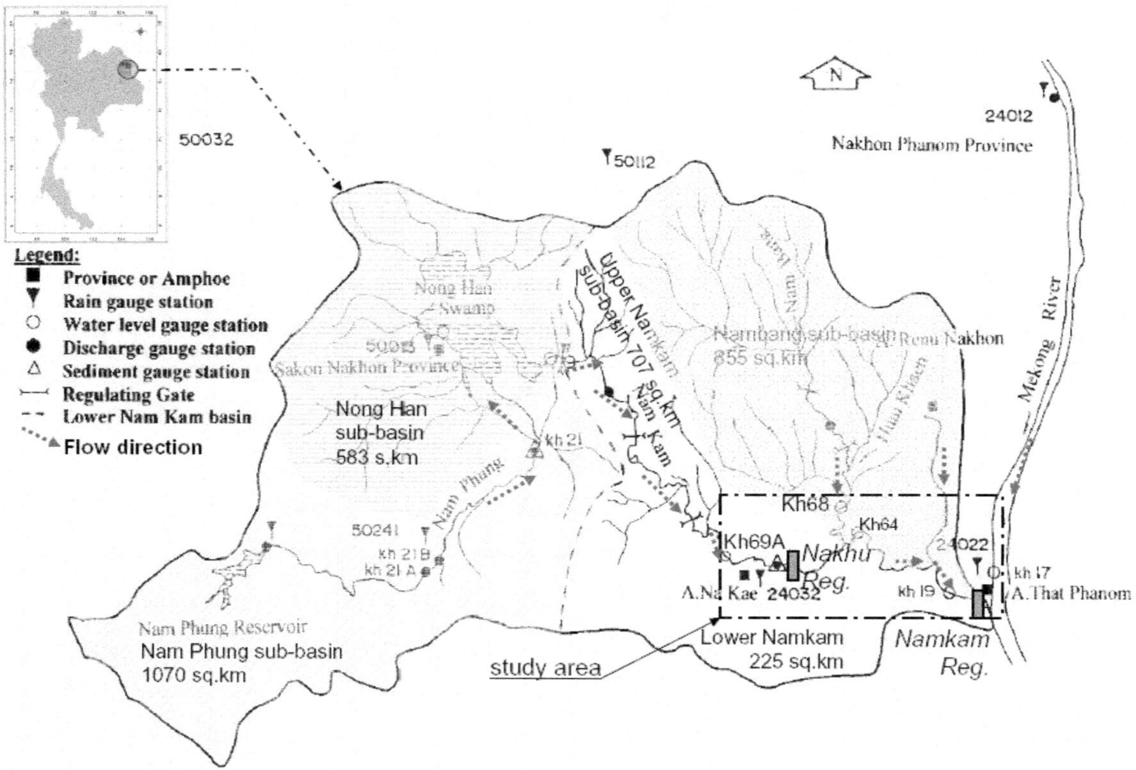

Fig. 1. Map shows Namkam River Basin, streams layouts, location of the study area, the hydrological observation stations, and regulated-structures in study area.

(46 km) to the river mouth (0 km) caused by heavy rainfall, large upstream surface runoffs, and backwater from the Mekong River. In contrast, during the dry season there is insufficient surface water for the cultivated areas. Therefore, many shallow groundwater tube-wells were drilled and pumped as the water supply system for irrigation and drinking purposes (Ichikawa, 2007).

Observations of groundwater quality in this area reported that there are some problems of groundwater salinity and might cause deteriorative quality for irrigation purposes (RID, 2006; KKU, 2001). It is evident that there are some traditional salt making spots in the north of the study area i.e. Phra Song Noi and Bo Dok Son villages. Moreover, the geological map showed that there is Mahasarakham Formation (KTms) which is composed by thick layers of mudstone, shale, and siltstone inter-bedded with rock salt which are significantly influence groundwater salinity as shown in Figs. 2 and 3. Improper water management from those water sources and uncontrolled groundwater utilization will result in severe deterioration of the groundwater (Ichikawa, 2007; Mekpruksawong, 2008). This research reported in this paper was undertaken to analyze the balance of conjunctive surface runoff and groundwater use using the interpretation data from automatic observation stations installed since 2005. This paper explains recently flood conditions and groundwater levels change which is one part of the overall research goals and will be useful for further development of the conjunctive model.

2. Materials and Method

The observation daily recorded data of rainfall and river water level (WSL) along the Namkam River, Nam Bang River and Mekong River, conducted by RID, the Department of Water Resources (DWR) and Thailand Meteorology Department (TMD) were gathered and analyzed. The data from stations: Kh69a (57.3 km), Kh69 (49.0 km), Kh64 (30.6 km), Kh64a (21.5 km), Kh19 (5.5 km), and Kh17 (0.0 km) with measured from the river mouth at That Phanom and Na Kae districts are shown in Figs. 1 and 2, respectively. The frequency analysis based on Gumbel distribution or extreme value type-I (Chow, 1988) was applied for computing of occurrence year of maximum runoff and rainfall as either daily or season from the past to present period given by:

$$X_T = X + K_T S \tag{1}$$

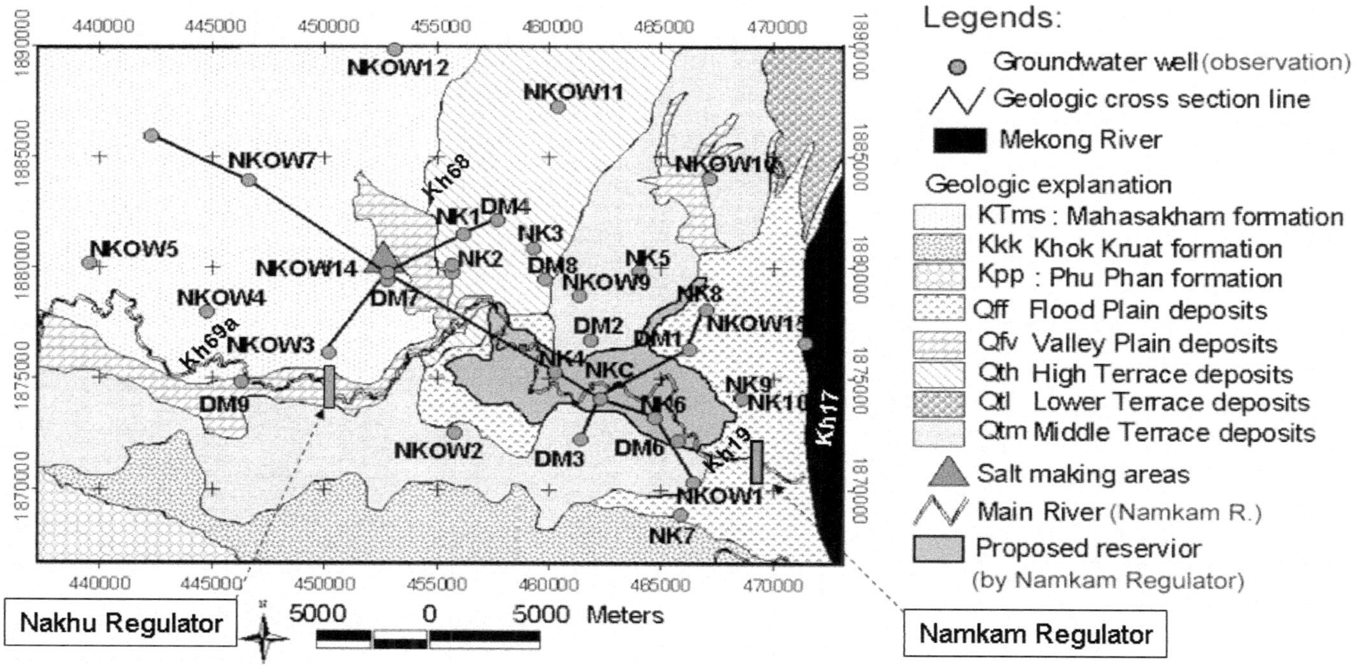

Fig. 2. Map shows the interpreted geological explanations, cross-sectional lines, and the location of observation wells in the study area.

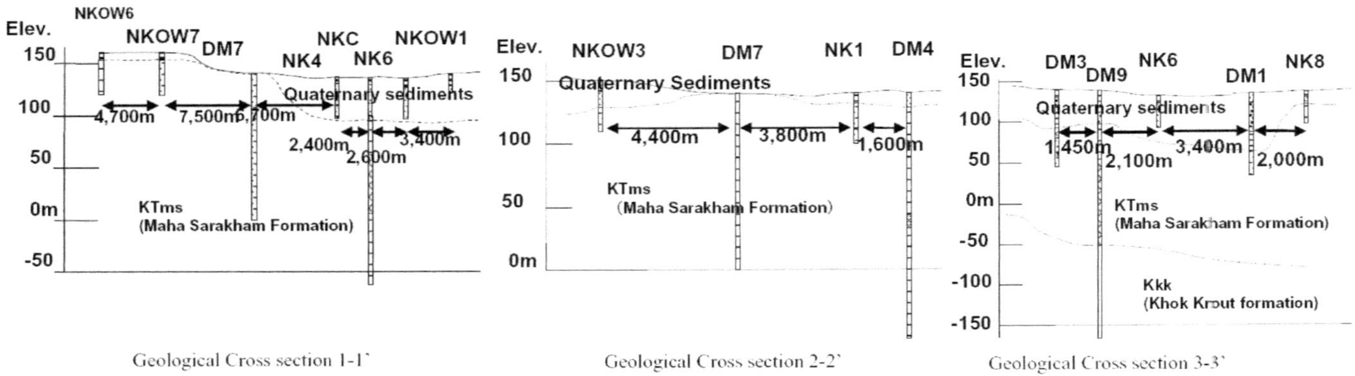

Fig. 3. Cross-sectional profiles of geological maps interpreted by well's boring-log data.

where X_T = either discharge or rainfall at each occurrence year (T), S = standard deviation, K_T = frequency factor for each occurrence year (T), K_T can be found as:

$$K_T = (-\sqrt{6}/\pi)\{0.5772 + \ln[\ln(T/(T-1))]\} \tag{2}$$

However, most of observation stations were not automatically recorded and there is missing data relating to the influence by backwater from the Mekong River at the Namkam river mouth and from the construction of cross regulators. Thus the rating curve profiles of water levels and discharges at each gauging station are uncompleted. Moreover, recording stopped at sites Kh19, Kh64a, Kh64, and Kh69 a number of years ago. The remaining water level recorders, Kh17 and Kh68 are still recording the river depths at Mekong River and Nambang River, respectively. Kh69 was moved in 2001 from 49 km to 57 km (8 km upstream of previous site) and renamed Kh69a and remains operational to this day. Therefore, linear interpolation and river levels correlations from those previous observation stations were used for the extension of the missing data or to generate data for the new stations.

The study of water surface profiles using mathematical simulation model for river analysis system: HEC-RAS (USACE, 2002) in one dimensional (1-d) steady and unsteady states was applied for the simulation of water surface levels (WSL) along the river reach (from 57 km to 0 km along the Namkam River) and to verify hydraulic parameters. Daily discharge (translation from rating curve) at Kh69a (57 km) was used as the upstream boundary and as the initial conditions in the model. Daily WSL at Kh17 was transferred to WSL at the Namkam river mount (0 km) and used as the downstream boundary condition. The simulation results of daily WSL of some cross-sectional profiles between upstream and downstream in the reach were produced by the model and verified using historical data. If WSL flows across a natural levee or flood protection dike and is higher than the natural ground surface elevation (NGL), it was recorded as a flood depth (d) over NGL. Therefore, subsurface flow and groundwater recharge at each grid points can be computed using flood depth and therefore give field infiltration capacity since 2005.

Field infiltration measurements in floodplain using ring infiltrometers based on Horton's equation (Chow, 1988) were tested at 15 points during the wet season and 55 points in the dry season since 2005, respectively. The spatial analysis was applied to generate point to aerial infiltration capacities as contour levels. The result of infiltration (F) can be converted to the amount of infiltration capacity over head of water (H) or flood depth (d) and therefore related to the seepage coefficient (Ac = F/H). Therefore, seepage volume could be estimated using existing flood depth, Ac, inundated area (A), and period of flood (t). The amount of groundwater recharge from floods calculated from the change of the water table (GWL) in phreatic aquifer through Ac and ground properties could be computed using further development of the numerical conjunctive surface and groundwater flow model. The boundary of the study area was based on maximum flood extent boundary topographic map based on a grid of 200m by 200m with a total of 390 square kilometers (sq.km) as shown in Figs. 2 and 9.

The automatically groundwater level (GWL) recorders were installed at 9- observation wells since 2006 as shown in Fig. 2. The first group of deep wells with groundwater levels deeper than 80 meters (m) are DM1, DM2, DM4 and DM8 are located in confined aquifers. DM6, NK2, NK3, NK8 and NK9 are located in phreatic groundwater or shallow aquifer. The electric conductivity (EC) was also automatically recorded for the change of groundwater salinity at DM2, DM6, DM8, and NK3. An automatic river levels recorder (WSL) and a rain-gage were installed at the Namkam Bridge (0.47 km) and at the head-works of the Namkam regulator (2.0 km), respectively. These measurements recorded data 30-60 minutes' time intervals.

3. Results and Discussion

The data of daily WSL and discharges (Q) during wet seasons (June–October) since 1964 at every river gauging stations in Namkam (Kh69a), Nambang (Kh68) and Mekong (Kh17) have been analyzed and converted into runoff volumes (million cu.m: MCM) and transformed into event of occurrences (T) as shown in Table 1 and Fig. 4. It was found that the trend of T since 2000 is greater than previous due to higher annual rainfall as shown in Fig. 5.

Due to a lack of recorded data since 1999 of gate opening and WSL between 46 km to 0 km, the WSL during flood periods (1 July-31October) in 2005, 2006, and 2007 were simulated for each station using HEC-RAS. The results shown in Figs. 6 and 7 were overlaid with existing topographical map shown in Fig. 8. The max flood levels (FSL), inundated area (% of study area), and return periods were summarized in Table 2.

Table 1. Comparison on flood events occurred in wet season since 1964.

year	Annual runoff, MCM	Q max: Kh69a	WSL max: Kh69a	WSL max: Kh17	% wet season runoff	*T* wet-season
1964	840.5	143.0	144.43	137.53	81.3	1.3
etc.	etc.	etc.	etc.	etc.	etc.	etc.
2001	2248.4	249.7	145.71	138.92	60.7	10.3
2002	2657.4	259.7	145.68	138.44	65.1	30.8
2003	1408.8	196.7	145.21	136.72	45.6	1.2
2004	1625.0	291.0	145.98	138.49	58.0	2.9
2005	2112.8	322.4	146.16	139.30	78.2	24.4
2006	1145.3	131.2	144.01	136.76	55.1	1.1
2007	1106.4	223.5	145.59	136.47	84.3	2.8
Mean	1003.3	186.8	145.00	137.70	87.1	2.1

Note: T = return period in years, Q max = max discharge in cu.m/s, WSL max in m(MSL)

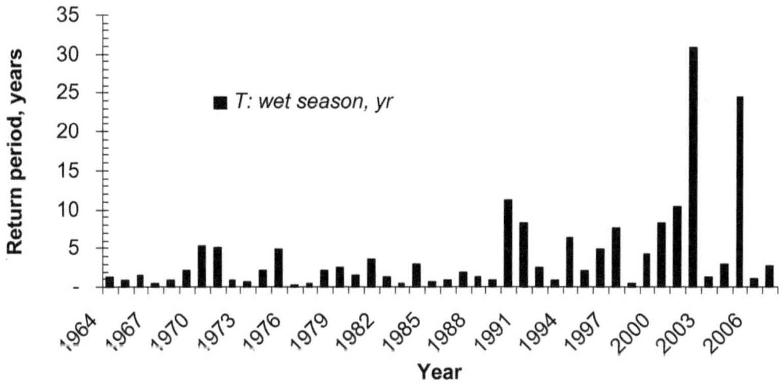

Fig. 4. The occurrence years of annual runoff volume occurred in wet season since 1964.

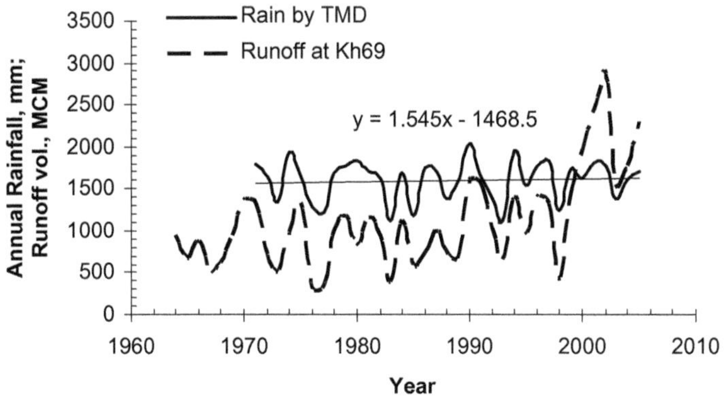

Fig. 5. The annual rainfall and runoff at Kh69a were compared using historical data since 1964.

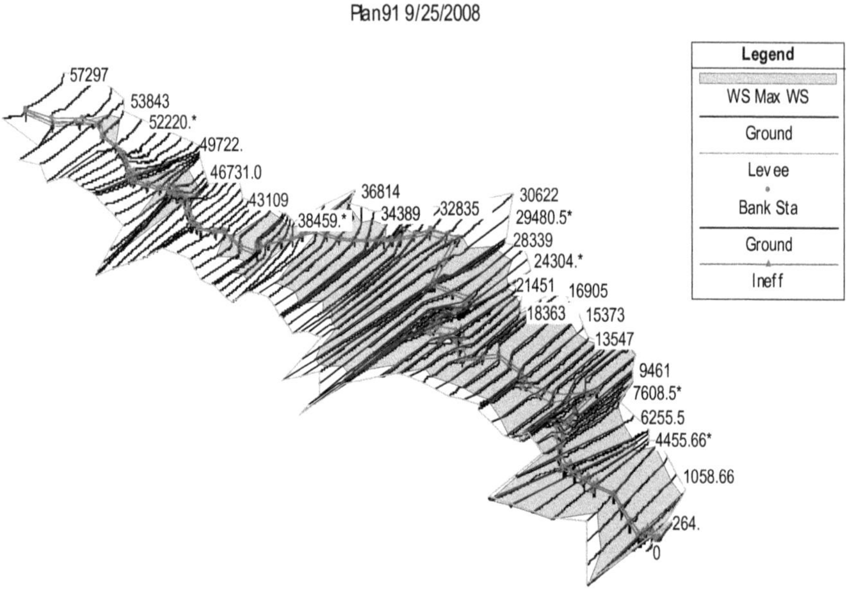

Fig. 6. 3-D view of flood extents (WS) as an example in 2005 simulated by HEC-RAS for each bank station from upstream (sta. 57297 m) to downstream end (sta. 0 m) of the river.

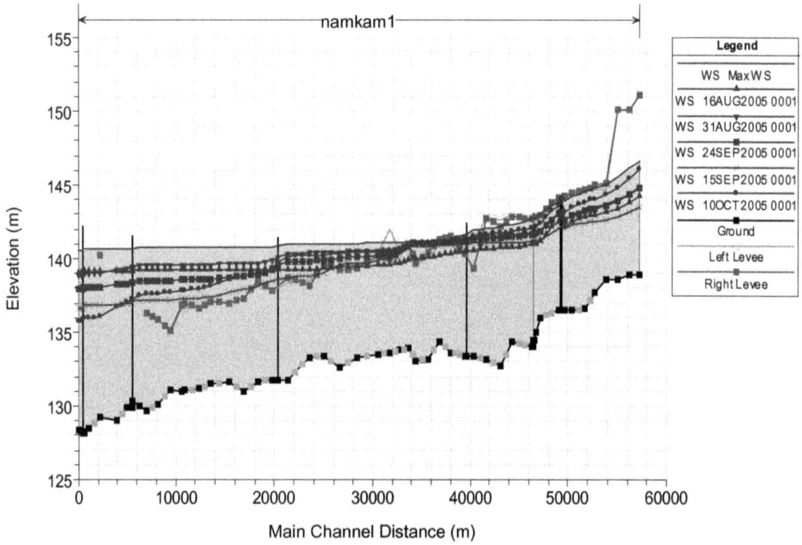

Fig. 7. 2-D view of daily flood elevations (WS) as an example in 2005 simulated by HEC-RAS unsteady flow module for each river station from upstream to downstream end of the river.

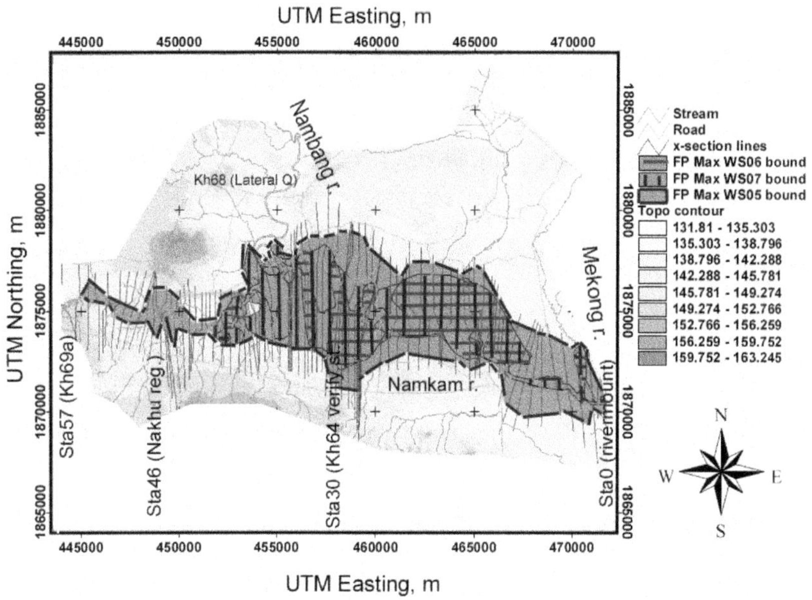

Fig. 8. Result of max flood extents (FP Max WS), in 2005 2006, and 2007 which are overlaid with the existing topographic map.

Table 2. Comparison on flood events occurred during wet season 2005 to 2007.

Year	Mean WSL, m(MSL)	Date of max FSL	Max FSL, m(MSL)	% flood area per study area	Inundated days	Return period: T, years*
2005	138.88	25 August	141.48	23.15	110	24.4
2006	138.25	22 August	139.67	9.61	65	1.1
2007	138.44	13 October	140.20	12.73	85	2.8

Note. *based on Fig. 4 and simulated WSL by HEC-RAS at 0 −57 km, with study area of 390 sq.km.

The result of averaged infiltrations (F) during flood season was 6 mm per day (mm/d) and averaged seepage coefficients (Ac) was 32.4 mm/d per 1 m of flood depth as shown in Fig. 9. This will be useful for further computing sub-surface flow and groundwater recharge.

The results of observed GWL and EC at DM6 in 2006 are shown in Fig. 10. It could be summarized that EC would increase when GWL reached lower level. Saline water would be raised up to NGL when GWL in confined aquifer became the same level of phreatic aquifer lending to groundwater saline.

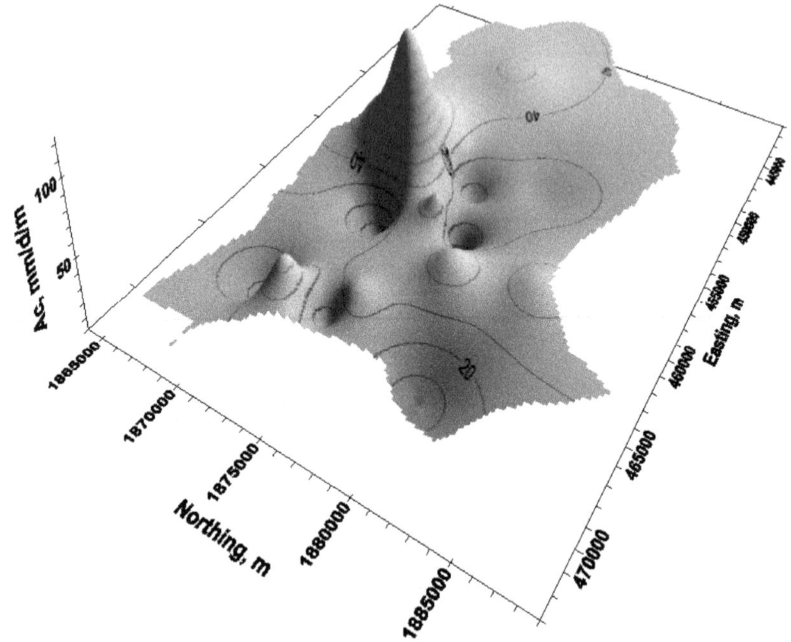

Fig. 9. 3-D view of aerial seepage coefficients (Ac) resulted from filed infiltration measurements in the study area during wet season 2005–2007.

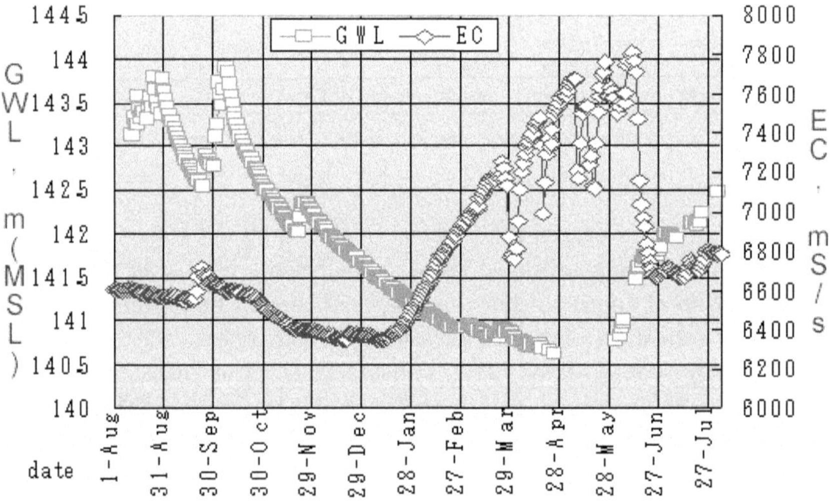

Fig. 10. Hourly recorded of GWL and EC values at DM6 from Aug 2006 to July 2007.

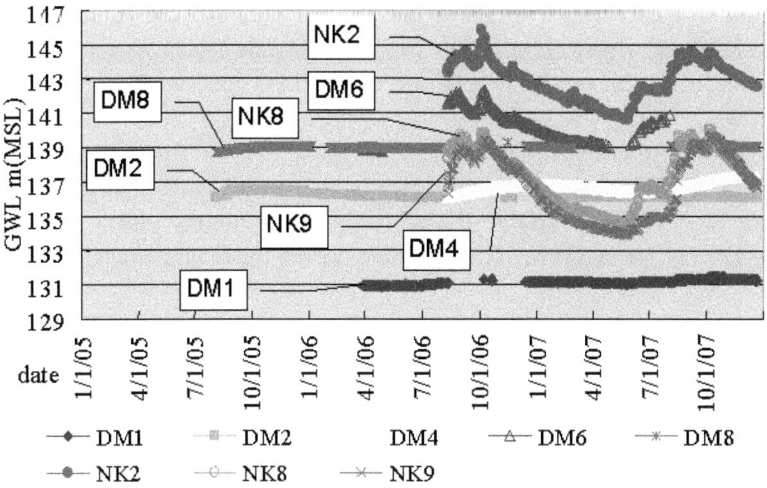

Fig. 11. Hourly recorded of groundwater levels (GWL) from 8-observation wells since 2005.

The daily GWL at shallow observation wells shown in Fig. 11 fell significantly after the end of May and increased in the beginning of October. These GWL fallers were influenced by high amounts of groundwater withdrawal during drought periods. In contrast, the increasing of GWL result from high recharge from flooding during wet seasons in June to September. However, GWL in the confined aquifers remained less impacted than in the shallow aquifers. Thus water tables in the deeper aquifers are higher level than shallow aquifers during the end of the wet season. This reason leads to salinisation of the groundwater and thus can not be used for crops.

4. Conclusion and Recommendation

The above results showed that both river levels and runoff volume during recently events since 1996 have been increasing compared to the previous periods. This is due to increasing annual rainfall and is also influenced by backwater in the Mekong River. Due to a lack of WSL data and rating curves in the middle reach downstream of Kh69a, the influence of the backwater effect from the Mekong River should be further investigated by using field observation and development of a mathematical model, particular during flood periods. The continuing research of water utilization in both wet and dry seasons should be carried out as well.

The regulated river levels in the ponds in floodplains of the Namkam and Nambang rivers should be carefully controlled with awareness of the dispersion of saline water from those salt making spots and other potential sources.

Further research on flood risk mapping, hydrodynamic modeling of GWL and WSL, best cropping patterns, and water management scenarios should be developed and distributed to the local administration officers and farmers in order to aware of dispersion of saline water and change right cash crop.

Acknowledgments

The authors wish to thank the Royal Irrigation Department, Department of Groundwater Resources, Department of Water Resources, and related agencies for supporting this research. Tokai University (School of Industrial Engineering) is highly appreciated to the main contribution of research funds.

References

1. Royal Irrigation Department (RID), The feasibility and environmental impact study of Lower Nam Kam irrigation project, Asdecon Corporation Company: Vol. 1 in Thai, 1995.
2. Royal Irrigation Department (RID), The environment mitigation and rehabilitation for salt rock geology and hydrogeology of Lower Nam Kam Irrigation Project, Chiang Mai University in Thai, 2006.
3. Khon Kaen University (KKU), *Environmental Impact Study of the Lower Nam Kam Irrigation Project Report*, Thailand, 2001.
4. Chow, V.T., Maidment, D.R. and Mays, L.W., *Applied Hydrology*, 1988.
5. Ichikawa, T., *et al.*, *Proc. Intl. Symposium 4th INWEPF*, Thailand, 2007.
6. Mekpruksawong, P., *et al.*, *Intl. Conference in HydroChange*, Japan, 2008.
7. U.S. Army Corps of Engineers (USACE), HEC-RAS River Analysis System: Hydraulics Reference Manual Version 3.1, Retrieved from http://www.hec.usace.army.mil. Public distribution unlimited (2002).

Advances in Geosciences
Vol. 17: Hydrological Science (2008)
Eds. Namsik Park *et al.*

IMPACT OF ERROR COVARIANCE ON FLOOD FORECASTING USING ENSEMBLE KALMAN FILTER

SRI SRIKANTHAN
Water Division, Bureau of Meteorology, GPO Box 1289, Melbourne, Victoria 3001, Australia
r.srikanthan@bom.gov.au

GEORGE KUCZERA
School of Civil Engineering, University of Newcastle, University Drive, Callaghan, NSW 2308, Australia

The flood forecasting process involves a number of steps. The first step is to obtain the observed rainfall up to the time of forecast and the future rainfall during a flood event. The rainfall is then transformed into discharge using a combined water balance and runoff-routing model to obtain the forecast discharge. The forecast discharge and the model states are adjusted when new discharge observations become available. There are uncertainties associated with rainfall measurement/forecasting, model (conceptualisation and parameters) and flow measurements. All these uncertainties contribute to the uncertainty in the resulting flood forecasts. Ensemble Kalman Filter (EnKF) enables all these uncertainties to be combined in a systematic way and it has been used by a number of researchers in the past. In this paper, the EnKF with state updating is used with the Probability Distributed Moisture model to forecast four flood events in the Gudgenby River, Australia. In many applications of EnKF in flood forecasting, the error covariances are either assumed *a priori* or found by trial and error procedure. The sensitivity of the error covariances on the performance of the EnKF is investigated. Imposing uncertainty on observed rainfall (which was used as a proxy on forecast rainfall) was found to have little effect on the forecast uncertainty. However, the uncertainties in the discharge and the stores affected the forecast performance considerably.

1. Introduction

There is a growing interest in knowing the uncertainty in flood forecasting and the resulting flood warnings. This is borne out of the fact that the processes involved in flood forecasting have inherent uncertainties in them. The procedure used in flood forecasting consists of a number of steps.

The first step is rainfall measurement and forecasting rainfall during a flood event. The rainfall is then transformed into flow using a combined water balance and runoff-routing model. There are uncertainties associated with rainfall measurement/forecasting, model conceptualisation and flow measurements. All these uncertainties contribute to the uncertainty in the resulting flood forecasts. Knowing the uncertainty in the flood forecasts and expressing that in the flood warnings will help decision makers prepare appropriate response plans.

The interest in quantifying uncertainty in real-time flood forecasting has led to a number of publications on the treatment of uncertainty caused by model parameters. Sequential data assimilation techniques provide a means of explicitly taking account of input, model and output uncertainties. One of the earliest data assimilation techniques is the Kalman filter developed for linear systems. For use with nonlinear models, it was later extended resulting in the extended Kalman filter (EKF). These two filters have been widely used in hydrologic modelling. If the nonlinearities in the model are strong, the linearization becomes very inaccurate. This has led to the development of the ensemble Kalman filter (EnKF) where the errors are allowed to evolve with the nonlinear model equations by performing an ensemble of model runs (Burgers *et al.*, 1998).

EnKF and its several variations were used with the PDM model to forecast a number of flood events in eight Australian rivers and the performance of EnKF was evaluated (Srikanthan *et al.*, 2008). In applying the EnKF, the main shortcoming is the *a priori* specification of the error covariances. The error variance for rainfall (Weerts and El Serafy, 2006) and discharge (Georgakakos, 1986) were obtained from the literature. The variances of the states and parameters were obtained by trial and error. Using these error variances, it was found that the state updating performed the best. Since the error variances were selected subjectively, the validity of the resulting uncertainty estimates of the forecast discharges remains questionable. Hence, the effect of error covariances on the performance of the EnKF and the forecast uncertainty is investigated in this paper.

2. Probability Distributed Moisture Model

The PDM model (Fig. 1) is a conceptual rainfall-runoff model which transforms rainfall and evaporation data to flow at the catchment outlet (Moore, 1985, 2007; Moore and Bell, 2002). The runoff production at a point in the catchment is controlled by the absorption capacity of the soil (treated together with canopy interception and surface detention) to take up water. This is conceptualised as a simple store with a given storage capacity. By considering that different points in a catchment have differing storage capacities and that spatial variation of capacity can be described by a probability distribution, it is possible to formulate a simple runoff production model which integrates the point runoffs to yield the catchment surface runoff into surface storage (S_2). The recharge from the soil moisture store (S_1) passes into subsurface storage (S_3). The outflow from surface (q_s) and subsurface (q_b.) storages forms the model output (q). A complete description of PDM model is given in Moore (1985, 2007). In the PDM model formulation, the surface runoff is calculated from the previous values of the surface runoff and net rainfall. In order to update the model states using ensemble Kalman filtering, the surface storage (S_2) is replaced by a cascade of two linear reservoirs (S_{21} and S_{22}) in which the storages are expressed explicitly to obtain the surface runoff.

3. Ensemble Kalman Filter

A brief exposition of the application of the EnKF is presented. There are two key steps. The first step involves assimilating real-time data on rainfall and discharge to update the states in the PDM model. The updated states represent the best estimates of the current hydrologic state of the catchment. The forecast step uses the updated states along with forecasted rainfall to generate ensembles of future discharge time series.

3.1. *Data assimilation using state updating*

An ensemble of state vectors is propagated through the PDM model such that each state vector represents one realisation of generated model state. The state forecast is made for each ensemble member as follows:

$$x_{t+1}^{i-} = f(x_t^{i+}, u_t^i, \theta) + \omega_t^i, \quad i = 1, \ldots n, \tag{1}$$

where x_{t+1}^{i-} is the ith ensemble forecasted state at time $t+1$, $x_t^{i+}$the ith updated ensemble state at time t, θ the model parameters and u_t^i the input to the model. In addition to representing the additive process noise $\omega_t^i \sim N(0, \sum_t^s)$, the EnKF represents the multiplicative model errors through forcing data perturbations. The input data perturbations are made by adding the ζ_t^i noise with covariance $\sum_t^u$ to the input data at each time step:

$$u_t^i = u_t + \zeta_t^i, \quad \zeta_t^i \sim N\left(0, \sum_t^u\right). \tag{2}$$

It was recognized that in order for the EnKF to maintain sufficient spread in the ensemble and prevent filter divergence, the observations should be treated as random variables by generating an observation ensemble with mean equal to the actual observation at each time and a predefined covariance (Burgers *et al.*, 1998). Thus the forecasted states x_{t+1}^{i-} are updated using Kalman gain K_{t+1}^x as follows:

$$x_{t+1}^{i+} = x_{t+1}^{i-} + K_{t+1}^x (y_{t+1}^i - \hat{y}_{t+1}^i) \tag{3}$$

where $\hat{y}_{t+1}^i$ is the ith predictive variable at time $t+1$ given by:

$$\hat{y}_{t+1}^i = h(x_{t+1}^{i-}, u_t^i, \theta) \tag{4}$$

y_{t+1}^i is the ith replicate of observation generated by adding the noise η_{t+1}^i with zero mean and covariance $\sum_{t+1}^y$ to the actual observation y_{t+1} as follows:

$$y_{t+1}^i = y_{t+1} + \eta_{t+1}^i, \quad \eta_{t+1}^i \sim N\left(0, \sum_{t+1}^y\right) \tag{5}$$

If the measurements are a nonlinear combination of state variables, the Kalman gain in adaptation to the ensemble based approach can be shown as (Moradkhani *et al.*, 2005):

$$K_{t+1}^x = \sum_{t+1}^{xy-} \left[\sum_{t+1}^{yy} + \sum_{t+1}^{y}\right]^{-1} \tag{6}$$

where $\sum_{t+1}^{yy}$ is the forecast error covariance matrix of the prediction $\hat{y}_{t+1}^i$, and $\sum_{t+1}^{xy-}$ is the forecast cross covariance of the state variables x_{t+1}^{i-} and prediction $\hat{y}_{t+1}^i$ and $\sum_{t+1}^y$ is the covariance of the measurements.

3.2. *Uncertainties in input and output*

The spread of the ensemble members is determined by the specified error in the model structure, the input forcing and discharge data. Realistic assumptions for errors in input forcing and response are essential for proper assimilation of data by filtering.

For areal average rainfall derived in operational flood forecasting systems with a limited number of rain gauges, the uncertainties can be up to 50% (Willems, 2001). In this study, a preliminary estimate of input error term based on Weerts and El Serafy (2006) is used:

$$P_{true} = P_{input} + \delta P \tag{7}$$

where $\delta P \sim N(0, (\alpha P_{input} + 0.2)^2)$ with $\alpha = 0.15$.

The uncertainty in the discharge measurement can be obtained from the derivation of rating curve using stream gauging data for a given gauging station. However, in this study, a preliminary estimate of discharge measurement error is chosen similar to that of Georgakakos (1986) who assumed a standard deviation of 0.1 times the measured discharge as given below.

$$Q_{true} = Q_{measured} + \delta Q \tag{8}$$

where $\delta Q \sim N\left(0, (\beta Q_{measured})^2\right)$ with $\beta = 0.1$.

3.3. *Forecasting*

Suppose t represents the current time. The assimilation step has made use of observed rainfall and discharge data up to and including time t to produce an ensemble of PDM model states $x_t^{i+}, i = 1, \ldots, n$. The forecast step involves propagating the model states into the future $\{t+1, t+2, t+3, \ldots\}$ using Eq. (1) along with ensembles of forecasted rainfall. At each future time step $t + l$, the model state is updated for each ensemble i using forecasted rainfall u_{t+l}^i and a random sample of process noise $\omega_{t+l}^i \sim N(0, \sum_{t+l}^s)$. This produces ensembles of future discharge time series from which forecasts and prediction limits can be derived.

In this study the focus is on the rainfall-runoff model updating. As a result, we will use the observed rainfall rather than forecasted rainfall. However, because the observed rainfall is subject to sampling and measurement error, Eq. (2) is used to generate ensembles of possible true rainfall that occurred for the forecast period $\{t + l, l = 1, \ldots, m\}$ where m is the maximum forecast lead time, taken here to be 12 hours.

4. Application to Gudgenby River

This section describes the calibration of the PDM model first and then the application of EnKF. The error variances for the states were obtained by trial and error. For the input and output, these were obtained from the values quoted in the literature. Finally, the effect of these error variances on the forecast uncertainty is investigated.

4.1. *Calibration of PDM model*

The PDM model was calibrated using daily and hourly data from Gudgenby River by the Shuffled Complex Evolutionary (SCE) algorithm. The PDM model was first run using daily rainfall data until the beginning of the flood event and then with hourly rainfall data. The root mean square error and the coefficient of efficiency of the calibration are 15.0 m^3/s and 0.87 respectively using only the hourly data for which the observed discharge is greater than 30 m^3/s to avoid the small discharge values influencing the statistics.

4.2. *Application of EnKF with state updating*

The variance of noises introduced to the input forcing and flow measurements are proportional to their magnitudes as stated in (7) and (8). For EnKF with state updating, the four stores in the PDM model (S_1, S_3, S_{21}, S_{22}) were considered as state variables and were updated sequentially as new measurements became available. The standard deviation of the four state variables was selected by trial and error and is summarized in Table 1. To obtain the lead time forecasts, a perfect knowledge of the future rainfall was assumed to avoid the error in forecasting rainfall. Ensembles of forecast rainfalls were obtained by perturbing the observed rainfall as in Eq. (7). States were not perturbed during forecasting discharges as this produced large oscillation. In reality, the uncertainty in the rainfall forecasts will add to the other uncertainties in the forecasting process. Forecasts (defined as the expected ensemble discharge) were made at 1, 3, 6, 9 and 12-hour lead times at every forecast time.

The adequacy of the EnKF was evaluated by using the root mean square (RMSE) and the coefficient of efficiency (Nash and Sutcliffe, 1970)

Table 1. Process noise standard deviation of the state variables.

State variable	S_1	S_{21}	S_{22}	S_3
Standard deviation	1.0	0.3	0.15	0.02

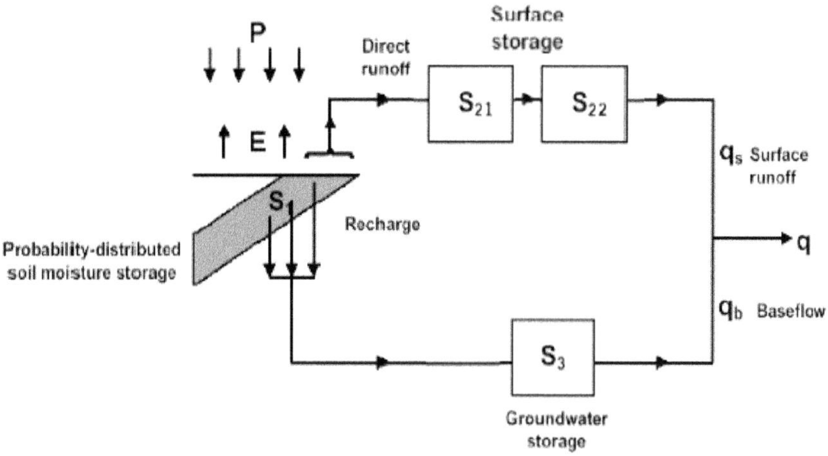

Fig. 1. The PDM rainfall-runoff model.

Table 2. RMSE and coefficient of efficiency of lead time forecasts.

	RMSE					Coefficient of efficiency				
Method	1	3	6	9	12	1	3	6	9	12
No update	15.0	15.0	15.0	15.0	15.0	0.87	0.87	0.87	0.87	0.87
EnKF	8.2	10.3	12.6	13.7	14.5	0.95	0.92	0.88	0.85	0.83

for 1, 3, 6, 9 and 12 hour forecasts. The average values of the root mean square of error and coefficient of efficiency for the four events for lead times of 1, 3, 6, 9, 12 hours are given in Table 2. These statistics were determined only for the period where the observed discharge is greater than 30 m^3/s to avoid the small discharge values influencing the statistics. In flood forecasting, small discharges are not important.

It can be seen from Table 2 that the EnKF is effective in reducing the RMSE for all the lead times. However, the coefficient of efficiency is marginally worse for lead times of 9 and 12 hours.

5. Effect of Error Variance of Forecast Uncertainty

The forecast uncertainty given by the EnKF is a function of the error variances used. As noted above, these were obtained either from the literature or by trial and error to minimise the forecast error. In this section,

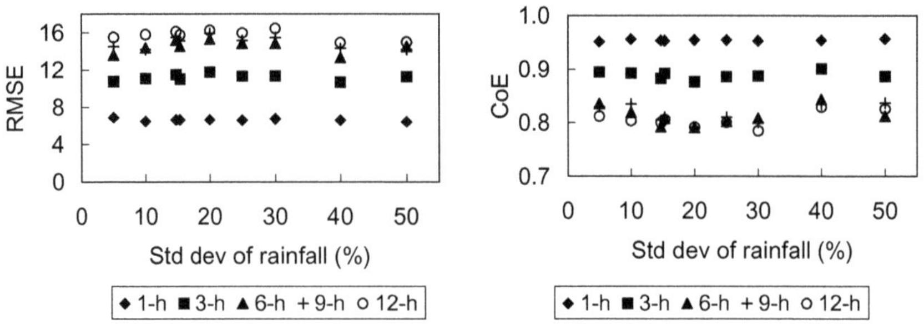

Fig. 2. RMSE and coefficient of efficiency as a function of the standard deviation of the rainfall noise with β fixed at 10%.

the effect of the error variance on the quality and uncertainty of the forecast is investigated.

5.1. *Effect of input error variance on forecast*

The value of α in Eq. (7) was taken earlier as 15%. The value of α is varied from 5 to 50% in steps of 5% while keeping the standard deviation of the discharge error β fixed at 10% and EnKF was run with state updating for the Gudgenby River. The standard deviations of the states were also fixed at the values given in Table 2. RMSE and coefficient of efficiency are shown in Fig. 2.

There is little variation in the values of RMSE, coefficient of efficiency or the standard deviation of the 1-h forecast for different values of α. This means that the rainfall uncertainty has a little influence on the forecast performance of the EnKF.

This insensitivity to rainfall sampling/measurement error is unexpected. However, one needs to recognise that this analysis does not reflect what would be encountered in real-time flood forecasting. Here the future observed rainfall is known and the errors are random and independent perturbations about the locus of the observed rainfall. In actuality the observed future rainfall is unknown. The best we can do is to generate ensemble forecasts of the future rainfall which are likely to exhibit more persistence and variability than rainfall series used here.

The independence of the rainfall errors may be a particularly important factor with regard to the insensitivity of the forecast discharge error. The hydrographs for the Gudgenby catchment suggest a response time of the order of 7 hours. It is therefore possible that hourly independent fluctuations

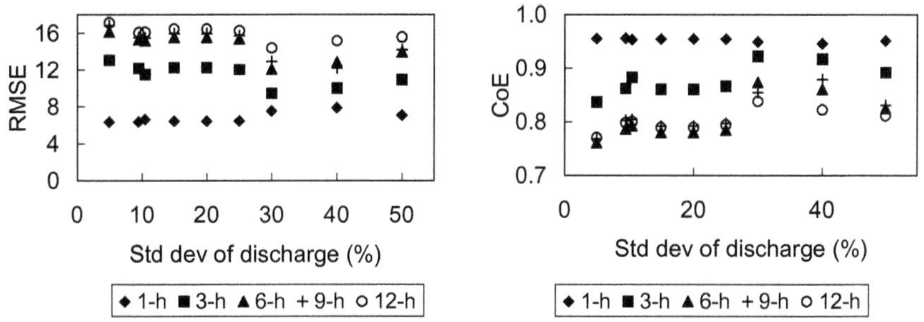

Fig. 3. RMSE and coefficient of efficiency as a function of the standard deviation of the discharge noise with α fixed at 15%.

in rainfall errors occur at sufficiently small time scales that the catchment largely disregards these fluctuations and primarily responds to the known low frequency component of the rainfall time series.

5.2. *Effect of discharge error variance on forecast*

The value of β was varied from 5 to 50 % while keeping α fixed at 15 % and EnKF was run as before with state updating for the Gudgenby River. The RMSE and coefficient of efficiency for lead time forecasts are presented in Fig. 3.

The RMSE increases slightly with β for 1-h lead time while it tends to decrease with β for other lead times. The coefficient of efficiency remains the same for 1-h lead time and increases slightly for other lead times with β. It appears that larger values of β give smaller RMSE and higher coefficient of efficiency values. Values of β greater than 30 % give better values for the above statistics.

5.3. *Effect of error variance of states on forecast*

The standard deviation of the states used earlier in Table 1 was scaled and EnKF was run. The RMSE and the coefficient of efficiency are presented in Fig. 4. The performance of the EnKF does not appear to deteriorate with scaling factor up to a value of 5.

In the above experiment, the variance of all the states was scaled by the same scale factor. To see the effect of each store, the EnKF was run with only one store changed at a time and the results are presented in Fig. 5 for the standard deviation of store S_1. These results show that the RMSE

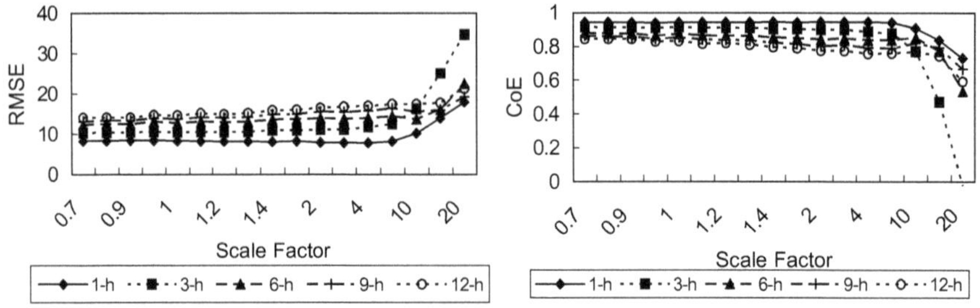

Fig. 4. RMSE and coefficient of efficiency as a function of the scale factor.

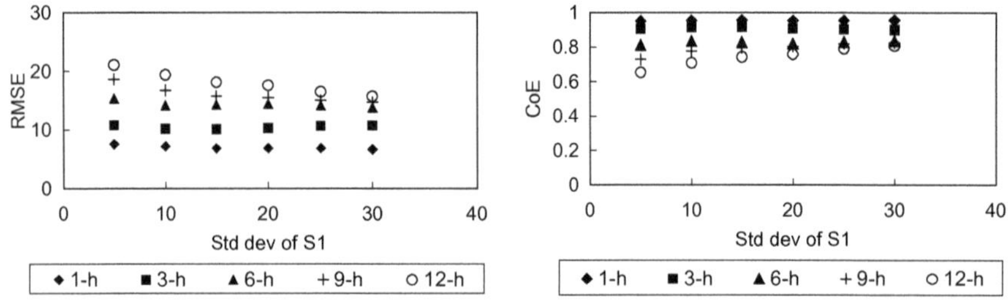

Fig. 5. RMSE and coefficient of efficiency as a function of the standard deviation of S_1.

is relatively insensitive to choice of process noise standard deviation. This suggests that the Kalman gain in Eq. (3) is large with the EnKF tracking the observed discharges and largely ignoring the state equation.

5.4. *Probability-probability plots*

Even when the forecast prediction limits largely contain the observed discharges, it is difficult to make any stronger statement about the quality of the forecast uncertainty. For instance, if the observed discharge was always below the expected discharge, this would be unsatisfactory. There is a simple graphical plot called the probability-probability (PP) plot (Thyer *et al.*, 2008) which can be used to better assess the appropriateness of the forecast uncertainty. It is a plot of the probability of the observed discharge within the discharge ensemble against a uniform distribution. If the forecast uncertainty is correct, then the exceedance probability of the observed discharge evaluated using the forecast ensemble should be uniformly distributed. Under these circumstances, the PP plot will be a 1:1 line. From the PP plots for the flood events in Gudgenby River with the previously chosen error variances, a value of 20 for the standard

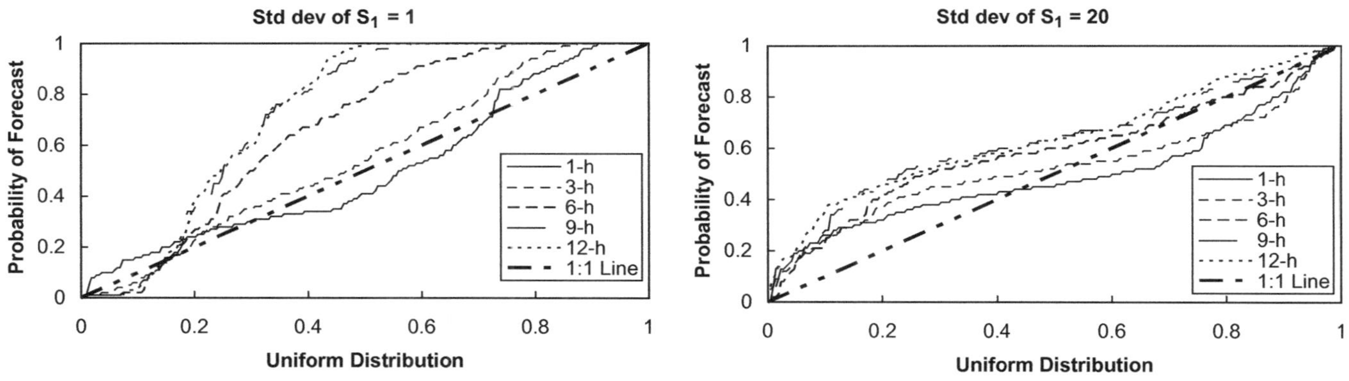

Fig. 6. PP plots for Gudgenby River.

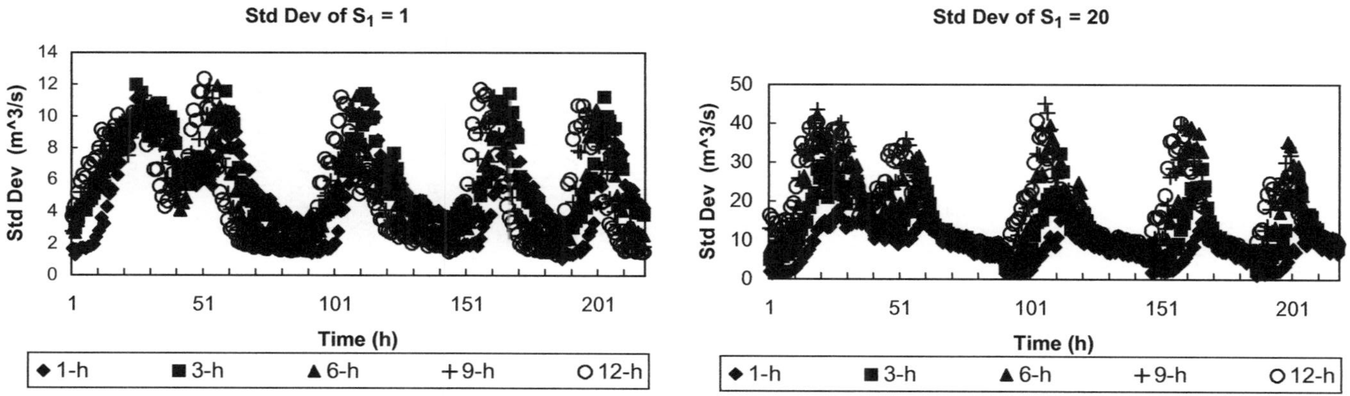

Fig. 7. Standard deviation of lead time forecasts.

deviation of S_1 appears to give reasonable results — see Fig. 6. Indeed, for the longer lead times, when the standard deviation of S_1 is 1, over half of the observed discharges lie above the 99% prediction limits. However the standard deviation of the forecast discharge becomes large (Fig. 7) compared to that with a value of 1 for the standard deviation of S_1.

6. Conclusions

In EnKF the error variances need to be specified *a priori*. There is no known objective way of specifying these variances. In the literature, these are often assumed or selected by trial and error procedures. In this report, the effect of error variances on the performance of EnKF and the uncertainty of the forecasts is investigated for state updating. The error variance of rainfall, discharge and states was varied one at a time and the effect monitored. The error variance of rainfall had a small effect on the forecast performance of the EnKF. However, the performance of EnKF marginally improved for long lead times with increasing error variance of discharge measurement. Also, the uncertainty increased with the increasing error variance of discharge measurements. Of the four stores, only S_1 had a large effect on the results. Longer lead time forecasts deteriorated with increasing standard deviation of S_1. Probability-probability plots were used to assess the forecast uncertainty. A value of 20 for the standard deviation of S_1 appears to give reasonable results. On the down side, it increases the standard deviation of the forecast discharge considerably.

The rainfall uncertainty was found to have little effect on forecast uncertainty. However, this may be an artefact of assuming perfect knowledge of future observed rainfall and catchment response time. In reality, forecasted rainfall ensembles will drive the discharge forecasts. The uncertainty of the discharge needs to be quantified and one possible way is to use the stream gauging data used for deriving the rating curve to achieve this. Apart from the PP plots, there appears to be no other objective way of identifying the process noise standard deviation of the stores.

References

1. G. Burgers, P.J. van Leeuwen, G. Evensen, Analysis scheme in the ensemble Kalman filter, *Monthly Weather Review* (126) (1998) 1719–1724.
2. K. P. Georgakakos, A generalized stochastic hydrometeorological model for flood and flash flood forecasting: 2 Case studies, *Water Resources Research* **22**(13) (1986) 2096–2106.

3. T. Koster, G. El Sefary, A.W. Heemink, H.F.P. van den Boogaard and A.M. Mynett. Input correction in rainfall-runoff models using Ensemble Kalman Filtering (EnKF), eds. J. H. W. Lee, & K. M. Lam, *Proc. 4th Int. Symp. On Environmental Hydraulics*, Hong Kong, 15–18 December 2004, Leiden Proc, Balkema, 2004, p. 1991–1996.
4. R. J. Moore, Probability-distributed principle and runoff prediction at point and basin scales, *Hydrological Sciences Journal* **30**(2) (1985) 273–297.
5. R. J. Moore, and V.A. Bell. Incorporation of groundwater losses and well level data in rainfall-runoff models illustrated using the PDM, *Hydrology and Earth System Sciences* **6**(1) (2002) 25–38.
6. R. J. Moore, The PDM rainfall-runoff model, *Hydrology and Earth System Sciences* **11**(1) (2007) 483–499.
7. J. E. Nash, and J. V. Sutcliffe. River flow forecasting through conceptual models Part 1 — A discussion of principles, Journal of Hydrology **10**(3) (1970) 282–290.
8. R. Srikanthan, G. E. Amirthanathan and G. Kuczera, Application of ensemble kalman filter to real-time flood forecasting, *2nd International Conference of GIS/RS in Hydrology, Water Resources and Environment (ICGRHWE' 07)*, Guangzhou, China, 2007.
9. R. Srikanthan, G. Amirthananthan and G. Kuczera, Application of ensemble kalman filter for flood forecasting in Australian rivers, *31st Hydrology and Water Resources Symposium*, Engineers Australia, Adelaide, 2008.
10. M. Thyer, B. Renard, D. Kavetski, G. Kuczera, S. Franks, and R. Srikanthan, Critical evaluation of parameter consistency and predictive uncertainty in Hydrological Modelling: A case study using Bayesian total error analysis, *Water Resources Research*, accepted 9/5/2008 subject to minor revisions, 2008.
11. A. H. Weerts and G. El Serafy, Particle filtering and ensemble Kalman filtering for State updating with hydrological conceptual Rainfall Runoff Models, *Water Resources Research* **42**, W09403, doi,10.1029/2005 WR004093, 2006.
12. P. Willems, Stochastic description of the rainfall input errors in lumped hydrological, *Stochastic Environmental Research and Risk Assessment* **15** (2001) 132–152.

Advances in Geosciences
Vol. 17: Hydrological Science (2008)
Eds. Namsik Park *et al.*

A STUDY ON THE ASSESSMENT OF MULTI-FACTORS AFFECTING URBAN FLOODS USING SATELLITE IMAGE; A CASE STUDY IN NAKDONG BASIN, S. KOREA

YOUNGJOO KWAK* and AKIHIKO KONDOH†
Center for Environmental Remote Sensing,
Chiba University, 1-33 Yayoi-cho Inage-ku,
Chiba 263-8522, Japan
**kwak_yj@graduate.chiba-u.jp*
†kondoh@faculty.chiba-u.jp

Floods are also related to the changes in social economic conditions and land use. Recently, floods increased due to rapid urbanization and human activity in the lowland. Therefore, integrated management of total basin system is necessary to get the secure society. Typhoon 'Rusa' swept through eastern and southern parts of South Korea in the 2002. This pity experience gave us valuable knowledge that could be used to mitigate the future flood hazards. The purpose of this study is to construct the digital maps of the multi-factors related to urban flood concerning geomorphologic characteristics, land cover, and surface wetness. Parameters particularly consider geomorphologic functional unit, geomorphologic parameters derived from DEM (digital elevation model), and land use. The research area is Nakdong River Basin in S. Korea. As a result of preliminary analysis for Pusan area, the vulnerability map and the flood-prone areas can be extracted by applying spatial analysis on GIS (geographic information system).

1. Introduction

1.1. *Background*

Natural disasters have been increasing every year due to various factors such as urbanization, population growth, destruction of natural environment and climate change. Furthermore, the Asian region accounts for nearly 90% of the total affected people in the world.[2] FEMA reports that floods are one of the most common disasters in the United States.[3] Flood effects can be local, impacting a neighborhood or community, or very large area, affecting entire river basins and multiple states. However, not all floods are alike.[a] In Japan, the importance of Flood Hazard Maps has undoubtedly been recognized in recent years, and now the preparation of the maps is progressing in a positive direction.[14] Korea's natural disasters are characterized by floods in rainy season. Particularly, typhoon 'Rusa' swept through eastern and southern parts of South Korea in the year 2002 and did serious damage over the large area.[11] This experience gave us valuable knowledge to mitigate future flood hazards. Governments and civilians demand readjustment of embankment and river control in basin.[15] The approach in current practice is two main strands (Flood Estimation Handbook) about rainfall and river flood frequency estimation in the UK.[1] The first of these is the so-called 'statistical method' based on frequency analysis of observed peak flow data. The second major strand is the 'rainfall-runoff method' which derives a unit hydrograph model.[9] Some more recent flood risk assessment (FRA) approaches took more aspects such as economically and explicitly integrating a wide variety of permanent and emergency flood control measures for floodplain management.[7] In this study, the advantages of FRA are to extract multi-factors applying statistic method and decide ranking of stream in Nakdong basin using an alternative approaches, integrated flood management. Next step, we improve the accuracy of a vulnerability area in local basin. Considering the regional parameters of flood factors, we improve and solve a disparity of risk assessment on downstream flooding. In order to determine the geomorphologic effectiveness of the flood, the authors use the flooding data, GIS data and satellite images in 2002. The authors suggest a methodology for mitigating floods using spatial analysis and image analysis.

1.2. *Objective*

Flood is serious natural disaster, so the mitigating of flood hazard is important especially for heavy rainfall and typhoon countries. The objective of this study is to extract the multi-factors related to flood hazard concerning geomorphologic characteristics, land-use, Normalized Difference Water Index (NDWI) and runoff régime. Parameters are considered geomorphologic functional unit, land use, wetness index of soil water.

1.3. *Study area*

The Nakdong basin is the second largest basin in S. Korea and is surrounded by mountain ranges. Its length is 521.5 km and its area is 23,817 km^2. It

[a] You can obtain these files from WWW pages at:
http://www.fema.gov/hazard/flood/index.shtm

is centered in Gyeongsang Province and is located between north latitude 127°29′ to 129°18′ and east longitude 35°03′ to 37°13′.[10] Geologically the Nakdong Basin represents part of the Archeozoic era, which is comprised of granite, sedimentary rocks and plutonic/hypabyssal rocks. The flows get wider and slower as the gentle slope of discharge increases.[15] The authors selected and produced the thematic map in Pusan watershed, one of the most urbanized area whole Nakdong basin. Figure 1 illustrates study area and flooding area during August 2002.

2. Method

2.1. *Study flow*

The authors produced the layers of a thematic map, which is a raw data acquisition from geomorphology, topography, satellite image part, and transfer to appropriate GIS (geographic information system) layers. GIS environments consist of ArcGIS9, ER Mapper and establish a GIS data table of flood information. Figure 2 illustrates study flow that it should be possible to define the following multi-factors in this study.

2.2. *Extraction to multi-factors*

This study extracts multi-factors such as history of flood, topography, geomorphology, land-use and NDWI (Normalized Difference Water Index) from Satellite Image by applying of GIS. The authors integrate these factors and produce the flood vulnerability map of basin (Table 1). At first, we are evaluating floods from previous occurrences and natural disasters in order to prevent future occurrences. Second, FA (flow accumulation), so-called 'an area of basin', is calculated to every cell using DEM (Digital Elevation Model). A large value of FA is more vulnerable than any other cell in mainstream. Generally, we also inspect that the inundated area was at the confluence point of two streams. Surface has intrinsic curvature, independent of an embedding. Third, Relative Digital Elevation Models (REMs) are an improved method that is developed to delineate DEM and flooding water level in basin. The surface information for calculating the slope and aspect can be expressed as raster of DEM.

$$H_{\_ij\ \mathrm{REM}} = H_{\_ij\ \mathrm{DEM}} - H_{\_ij\ (\mathrm{main\ stream})} \tag{1}$$

Where, H: height of surface, i: number of pixel, j: number of line.

Forth, Gao[4] developed the NDWI for determination of vegetation water content based on physical principles. The NDWI is an improved method

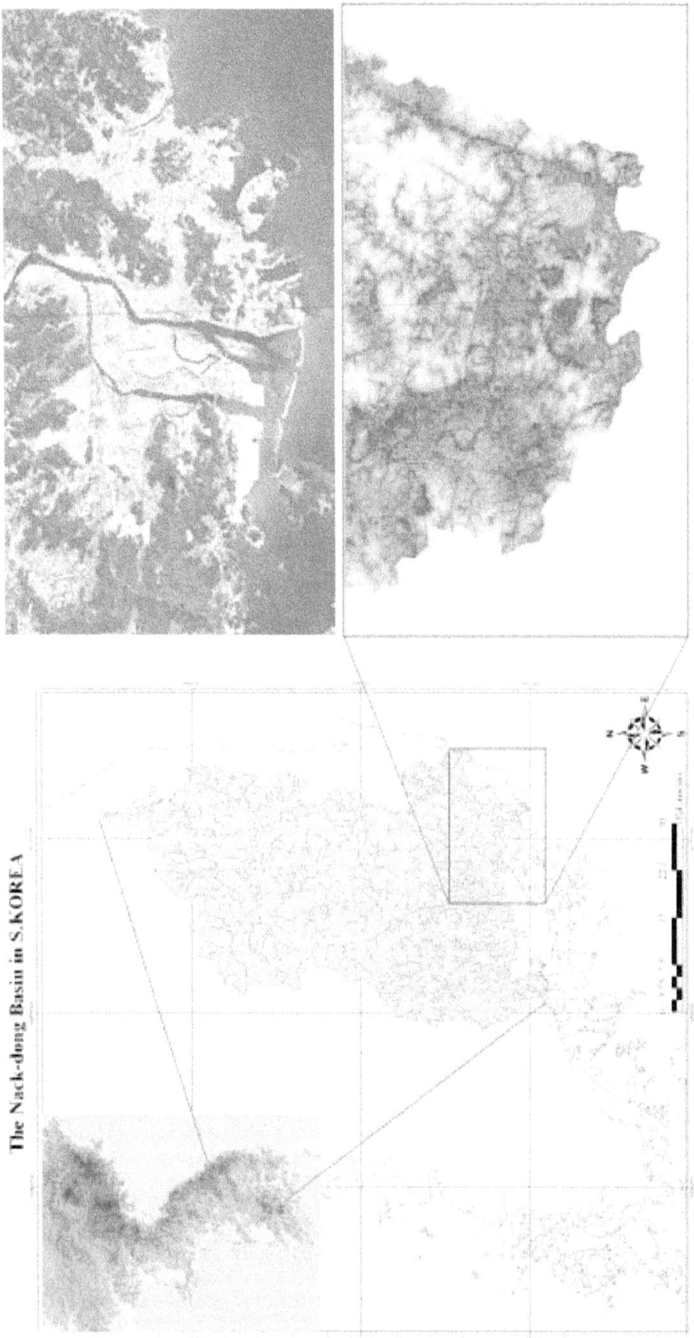

Fig. 1. Study area, Pusan watershed, is one of the most urbanized area in Nakdong basin.

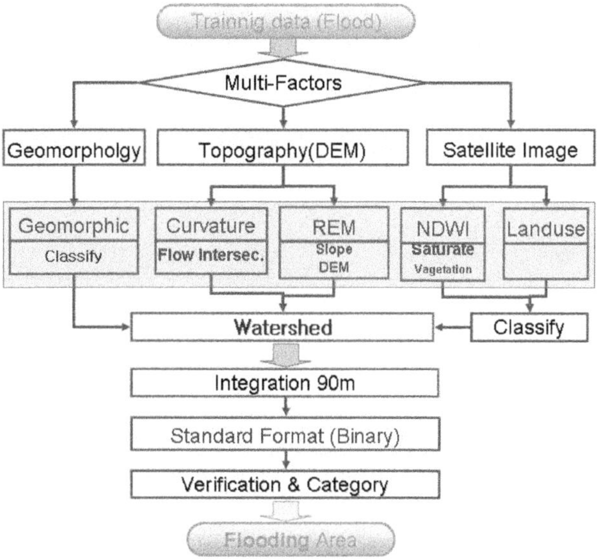

Fig. 2. Study flow.

Table 1. List of multi-factors.

Multi-factors	Source data	Shape	GIS Layer data
Criteria	Training data	Point	Flooding site survey
Step I	SRTM (DEM)	Polyline	REM, FA, Curvature
Step II	Landsat TM	Polygon	Land-use and NDWI: Soil water
Step III	Geomorphic map	Polygon	Distribution of geomorphology

that has been developed to delineate wetness and enhance its presence in remotely-sensed digital imagery. These indices are of the general form;

$$\mathrm{NDWI} = (\mathrm{NIR} - \mathrm{SWIR})/(\mathrm{NIR} + \mathrm{SWIR}) \tag{2}$$

where, NIR: near infrared radiometer, 0.6∼0.7 (band 3 of Landsat TM)

SWIR: shortwave infrared radiometer, 1.55∼1.75mm (band 5 of Landsat TM)

The NDWI may also provide researchers with turbidity estimations of water bodies using remotely-sensed digital data.[13] And we consider land-use as human activity and human-related factors(urbanization, etc) caused by the flood disaster. Finally, we apply geomorphic processes and classify characteristics and localities caused by a flood hazard in the lowland.

Table 2. The field data of a representative flooding site in Nakdong basin.

Site	Flood area [ha]	rainfall [mm/day]	Height of of F.A. [max. m]	Slope [°]	Water-level warning [m]	Height of banks [EL.m]	Curvature	Cause 30 Aug. 2002
A	100	814.5	H < 20		7	–	1.1	
B	100	814.5	H < 12		9.5	–	0.8	Collapse
C	320	760.1	H < 10	SL	10.5	15.86	2.3	of
D	720	1,114	H < 6	< 10	9	13.6	1.5	banks
E	520	695	H < 5		5	10.94	1.8	Flooding

3. Data

3.1. *Training data of previous flooding occurrence*

In Nakdong basin, late in the summer of 2002, remarkable flooding which occurred twice that August, brought by RUSA killed about 262 people and caused more than $60 billion in property damage. In order to determine the geomorphologic effectiveness of the floods, the authors use the present flooding data such as Table 2 which is illustrated the result of floods in 2002.[11]

3.2. *Topography*

Hydrologic processes are fundamentally different on hill slopes and in channels. In channels flow is concentrated. Certainly one of the most obvious factors controlling stream flow is the gradient, or slope, of the stream channel.[8] It is available for calculating at a catchment using the SRTM; DEMs extracted by Shuttle Radar Topography Mission (SRTM) imagery (90 meters) Data.

3.3. *Geomorphology*

The classification was to define naturally occurring clusters of geomorphic characteristics that would be indicative of discrete sets of geomorphic processes, with the intent that such a classification would be useful in river management.[16] The statistical classification was based on geomorphic characteristics of the river collected from geomorphology sheets spanning 1945 through 1970.

3.4. *Satellite image*

Normalized Difference Water Index (NDWI) and land use change are necessary to analyze the flood disaster. The NDWI and land use data were extracted from Landsat TM in April 2002 (Pusan city[12]). The NDWI has also recently been used to monitor moisture conditions of vegetation canopies over large areas.[13] An important parameter is the Manning roughness coefficient which comprises a flow resistant factor and a function of land cover in flooded areas.[6]

4. Results

4.1. *FA (flow accumulation) factor*

The result of flow accumulation is a raster of accumulated flow to each cell, as determined by accumulating the weight for all cells that flow into each down slope cell. The value of FA ranges from 0 to 24616.8. According to the characteristics of an accumulation and the Nakdong basin GIS information data from the flooding field, a large value of FA is more danger than any other cell. We determinate the location of most risky area among the 23 confluence point in Nakdong basin (Table 3) and approach using the field data (yangsan stream).

4.2. *Relative DEMs*

We propose an improved model that use relative digital elevation models (REMs) and heights of main stream in hydrology. Our results are based on the use of flow-accumulation height, which implies water level ($h = 0$ meter) acquired from a main stream, such as Eq. (1). The range of heights values in meter is 0 to 1410 meters height-cells which are established from DEM; with primary topographic attributes by 90m. The vulnerable areas of a flood in the study were evaluated in near river and lower urban areas (Fig. 3). Table 4 illustrates a category score of REM.

Table 3. Flow accumulation of confluence point in Nakdong basin.

Confluence point		Main stream	Sub stream	Difference	Area of basin (km^2)	Rank
Up	Banbyeng stream	447625	283533	164092	3601.71	23
...	...	...	...	...	...	...
Middle	Wi stream	959338	203515	755823	7974.85	15
...	...	...	...	...	...	...
	Kangryu stream	2079427	25853	2053574	18264.78	3
	Han stream	2137993	19277	2118716	20570.56	2
Down	Milyang river	1945686	211584	1734102	21351.11	6
	Yangsan stream	2407946	35604	2372342	23030.64	1

Area of basin is measured by K-water 2005[11]

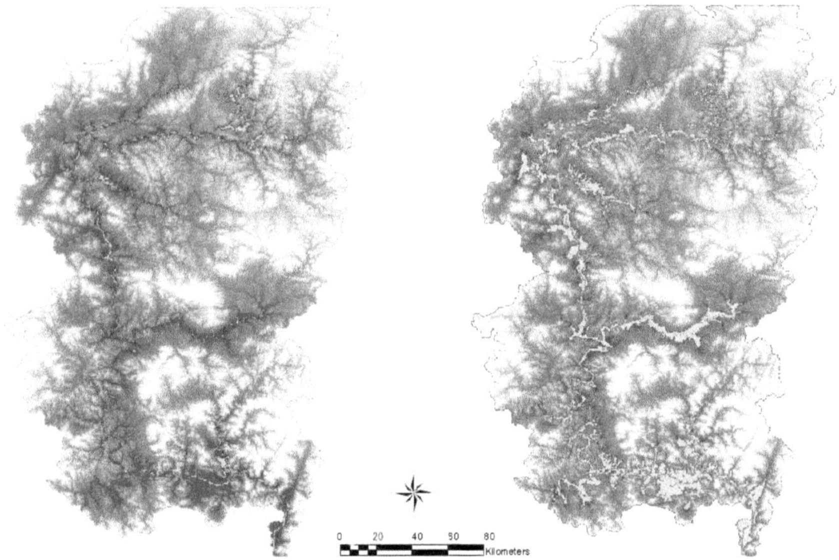

Fig. 3. The vulnerable areas of a flood estimated using REM method (left 0, right 10 meter).

Table 4. Category score of REM.

Range of REM [m]	0	1	2	3	4	5	6	7	8	9	10
Category	1	2	3	4	5	6	7	8	9	10	11
Category score	10	9	8	7	6	5	4	3	2	1	0

4.3. *Curvature of surface*

The value of curvature was determined the relationship between a flow and a curvature in Nakdong basin. The inundated area was at the joint area of two streams, thus the flood water overflowed bounding discharge capacity of the stream. Maximum value of curvature and the previous flooding site do accord well together. It is known that sub-stream meets a main stream and flows backward against high speed energy and curvature of main stream.[12]

4.4. *Geomorphic classification*

Distribution of geomorphology is classified in 6 classes based on micro-geomorphology in 1973; marshy, dry river bed, valley plain, innings, delta, natural levee. As a result, the authors approach distributing flood

Table 5. Category score of geomorphic factor.

	Marshy	Dry river bed	Valley plain	Innings	Delta	Natural levee
Category	1	2	3	4	5	6
Category score	5	4	3	2	1	0

hazard from a geomorphologic point of view, rather than a regional one. Geomorphic factors are classified category obtained from the literature to assess the damage for the 2002 flood event in Nakdong basin. Table 5 illustrates that the category has the value ranges from 0 to 5.

4.5. *Land use*

Supervisor classification was used in order to examine multiple possible criteria data extraction. To focus on a weak point of flooding, Fig. 4 illustrates classes derived from assessing values of Manning roughness coefficient.[5] The authors more importantly considered a paddy field, urban, and waterside than other classes, because flood risk assessment caused by human activities. The spatial distribution of all classes was extracted from Landsat TM in April 2002. Table 6 illustrates the category score; the value ranges from 0 to 10.

4.6. *Wetness index*

NDWI is an effective technique for discriminating water responses to flooding. In case of the flood disaster analysis, raster data of NDWI is one of the important factors to analyze floods and the most effective technique for discriminating wetland responses to flooding. NDWI is a nonlinear function that varies between -1 and $+1$. For the extraction of inundated areas by flood, conditions of NDWI were set as follows; NDWI ranges from 1 to 0, we can make an estimate of a flood potential area that NDWI values were higher under flooding conditions (NDWI$>$0) than under drought conditions (NDWI$<$0) in April 2002.

5. A Case Study of Vulnerability Map

A case study shows simple models and local actions often present the starting point among the diversity of circumstances in Nakdong basin. The authors produced a vulnerability map from the hazard-related multi-factors

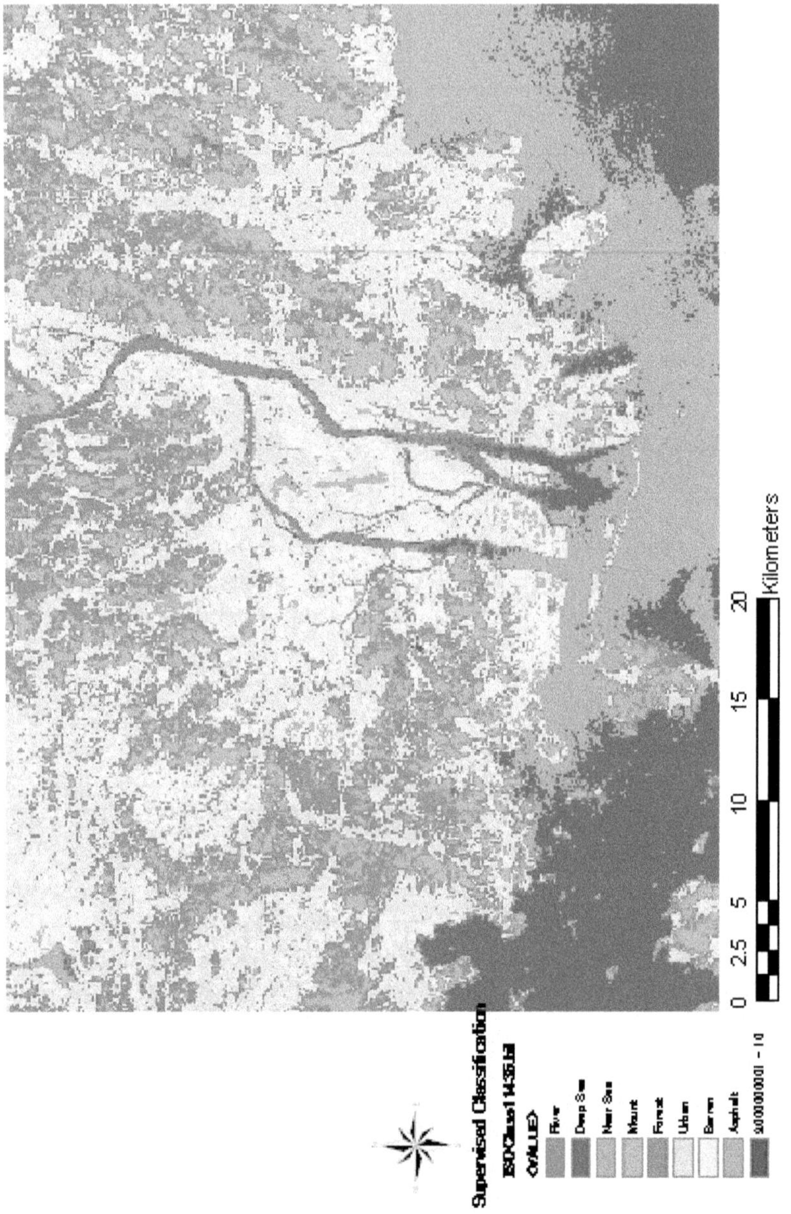

Fig. 4. Supervised classification in Pusan watershed.

Table 6 Category score of land use.

Classes	Paddy field	Urban	Water side	Soil	Barren	Glass	Water	Loose forest	Density forest	Road
Category	1	2	3	4	5	6	7	8	9	10
Category score	10	9	8	7	6	5	4	3	2	1

and calculated the following: existing history of flooding, the layer value of REM ranges from 0 to 10, the value of FA (flow accumulation) ranges from 0 to 24616.8, calculate and modify using an adjust variable. The value of GF (geomorphic factor) ranges from 0 to 5 (curvature of stream less than 2), the value of LU (land use) ranges from 0 to 10, for instance urban and lower land has high value. The value of NDWI ranges from 1 to 0. The probability, P (Eq. (3)) of raster calculator indicates a flooding index in the rate of occurrence in Nakdong river such as Table 7. The method of $W_i$setting is carried out manually, try and error method. At first of all, we verify W_i based on the field data, inundation area of Yangsan stream, using calculation of the Eq. (3) and we adapt to weight of regional factor (REM = 0.3, GF = 0.25, LU = 0.2, NDWI = 0.15, FA = 0.1) in Pusan watershed.

$$P = W_i \times REM_{jk} + W_i \times FA_{jk} + W_i \times GF_{jk} + W_i \times LU_{jk} + W_i \times NDWI_{jk} \quad (3)$$

Where, $\Sigma W_i = 1$; weight of parameters. j = number of line, k = number of pixel.

As a result, Fig. 5 illustrates a probability of floods in which the dark blue area shows the most dangerous inundation area (hazard area). The orange area (risk area) illustrates more danger than any other place (light blue) in Pusan watershed. Table 8 illustrates a classified area which is calculated by multi factors.

Table 7. The assessment of a hazard multi factors.

Factor	Range	Adjusted a coefficient	Max score	Weight	Risk value
REM	0∼10	1	10	0.30	3.0
GF	0∼5	2	10	0.25	2.5
LU	1∼10	1	10	0.20	2.0
NDWI	1∼0	10	10	0.15	1.5
FA	0∼24616	0.004	10	0.10	1.0
Total	–	–	50	1	10

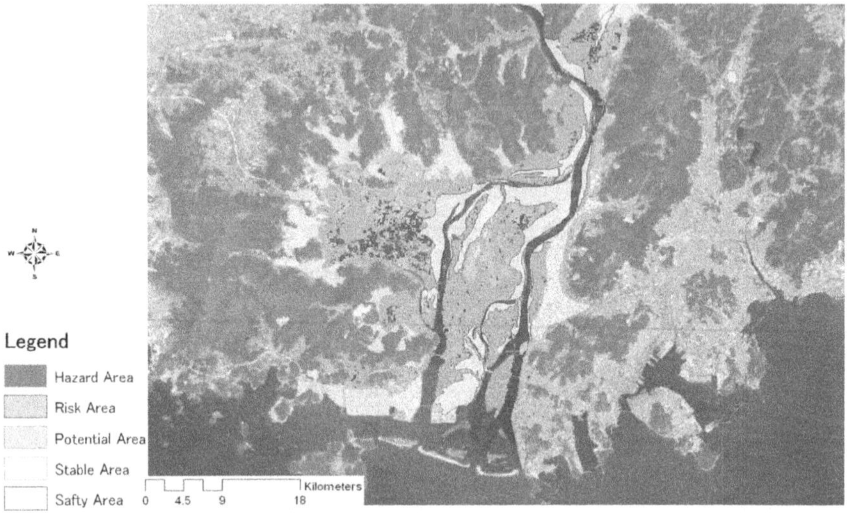

Fig. 5. The vulnerability map of Pusan city watershed.

Table 8. The result of a risk assessment.

Classify	Probability value
Hazard area	10∼8
Risk area	8∼7
Potential area	7∼6
Stable area	6∼5
Safty area	5∼0

6. Conclusions

The 2002 floods in the Nakdong basin provided important lessons about the flood monitoring and training data. The authors suggest a basic function and produce the vulnerability map applying spatial analysis on GIS. This paper focuses on the improvement of the extraction of the multi factors to assess flood hazard in future event. What factors should be considered for improving the flood risk management. Third kinds of the hazard-related multi-factors are investigated: First, flow accumulation making the channel network using DEM. Second, land-use representing urbanization, which can be used to analyze the urban area developed over an alluvial plain by comparing past geomorphologic map. Particularly, we conclude that the location of utmost danger is the marshy district in geomorphic classification.

Third, wetness index of soil water assesses the susceptible relationship between the basin saturation states. Finally, river flood risk assessment conduct a large area through integrated remote sensing and GIS techniques. Besides precipitation the rapid urbanization and human activity in the lowland also increase the disaster damage. Although this method includes some assumption and experience value to calculate multi factors, flood risk multi factors indicate the current capability available for strategic flood risk management and urban planning.

References

1. A. Calver, E. Stewart and G. Goodsell, Comparative analysis of statistical and catchment modeling approaches to river flood frequency estimation, *J Flood Risk Management* (2009) 24–31.
2. ADRC, *Data Book on Asian Natural Disasters in the 20th century, Asian Disasters Reduction Center* (2002) 2–30.
3. FEMA, *Federal Emergency Management Agency*, http://www.fema.gov
4. Gao, bo-cai,NDWI a normalized difference water index for remote Sensing of Vegetation Liquid water from space, *Remote Sens. Environ.* **58** (1996) 257–266.
5. C. J. van der Sande, S. M. de Jong and APJ. de Roo, A segmentation and classification approach of IKONOS-2 imagery for land cover mapping to assist flood risk and flood damage assessment, *International Journal of Applied Earth Observation and Geoinformation* **4** (2003) 217–229.
6. A. P. J. De Roo, LISFLOOD: A rainfall-runoff model for large river basins to assess the influence of land use changes on flood risk, in Balabanis, *Management and Flood Mitigation*, EUR 18287 EN, 1999, pp. 349–357.
7. J. R. Lund, Flood plain planning with risk-based optimization, *J. Water Resources Planning and Management* **127**(3) (2002).
8. J. McCoy and K. Johnston, Using AcrGIS Spatial Analyst, ESRI, http://www.esri.com/ (2005).
9. T. R. Kjeldsen, The revitalised FSR/FEH rainfall-runoff method, Flood Estimation, *Centre for Ecology and Hydrology*, 2007.
10. K-water, *Korea Water Resources Cooperation*, http://www.wamis.go.kr.
11. K-water, National Water Resources Plan 2006, http://waterplan.kwater.or.kr/ (2005, Accessed 26 August 2008).
12. Y. Kwak, and A. Kondoh, A study on the Extraction of multi-factor influencing floods from RS Image and GIS data, *21th Congress of ISPRS* **37** (2008).
13. S. K. McFeeters, The use of the Normalized Difference Water Index (NDWI) in the delineation of open water features, *International Journal of Remote Sensings* **17** (1996) 1425–1432.
14. MLIT, Flood hazard map manual for technology transfer, *Ministry of Land, Infrastructure and Transport*, Japan, March 2003.

15. NRFCO, *Nakdong River Flood Control Office*, http://www.nakdongriver.go.
16. R. B. Jacobson and C. M. Elliott, Geomorphic Classification and Assessment of Channel Dynamics in the Missouri National Recreational River, South Dakota and Nebraska, *US. Geological Survey*, 2006.

Advances in Geosciences
Vol. 17: Hydrological Science (2008)
Eds. Namsik Park et al.

A DISTRIBUTED-LUMPED MODEL FOR FLOOD AND SEDIMENT YIELD PREDICTIONS AT THE CATCHMENT SCALE

APIP
Graduate Student, Dept. of Urban and Environmental Eng., Kyoto University, Kyoto 615-8540, Japan
apip@flood.dpri.kyoto-u.ac.jp

TAKAHIRO SAYAMA, KAORU TAKARA* and YASUTO TACHIKAWA[†,‡]
**Disaster Prevention Research Institute, Kyoto University, Uji 611-0011, Japan*
[†]Dept. of Urban and Environmental Eng., Kyoto University, Kyoto 615-8540, Japan
[‡]tachikawa@mbox.kudpc.kyoto-u.ac.jp

This study proposes a lumping method for a physically-based distributed sediment runoff model. A distributed-lumped model obtained by the lumping method enables to predict flood and sediment yield on a real-time basis by considering hydrological and transport processes. First, model lumping of distributed kinematic wave rainfall runoff and erosion sediment transport equations is developed, which is the integration processes of spatially distributed equations assigned to each grid-cell over landscape. Then, lumped discrete relations between the variables, namely a stream discharge and a maximum sediment storage capacity is produced as a function of stored water volume in the catchment. Stream flow and sediment yield are predicted by using these discrete time-invariant relations and lumped continuity equations. The performance of distributed-lumped model was confirmed from the case study the Lesti River catchment, Indonesia.

1. Introduction

The Brantas River, 320 km length with 39 tributaries and a catchment area about 11,800 km^2, is the second largest river located in East Java. The Lesti River catchment (351.3 km^2), a tributary catchment in the Upper Brantas River basin is selected as study area. At the confluence point of the Lesti River and the Brantas main reach, the Sengguruh dam was constructed in 1988 for water resources and hydropower generation. Its original gross storage in 1989 was 22.4 million m^3 and reduced to 5.5 million m^3 in 1993. Unexpectedly, most of the gross storage has been already filled with the large amount of sedimentation from the Lesti River.[1] In order to protect the capacity of the reservoir and manage water resources from sedimentation caused by flood events, it is necessary to predict water and sediment inflows to the reservoir.

Many complex distributed models have been developed to understand the spatial dynamic sediment yield and transportation at the catchment scale. A physically-based distributed sediment runoff model considers multiple sources of sediment transport, namely soil detachment by raindrops and soil detachment by overland flow developed by the authors.[2] However, the direct applications of the model to the real-time flood and sediment yield predictions are still limited because of the high computational requirement especially when stochastic algorithms are included. Thus, the model originated from a physically meaningful model structure and parameter with less computational requirement has a great potential to be used to the real-time flood and sediment predictions on coping with predictive uncertainty. It due to the most of proposed uncertainty assessment approaches conduct Monte Carlo type simulations with long time-consuming process.

One of the potential ways to step forward and overcome the high computational burden is by structural lumping of physically-based distributed models as pointed out by various researchers.[3,4] The lumping kinematic wave equation of a physically-based distributed sediment runoff model is addressed in the study. This paper focuses on (1) proposing a new method to lump a physically-based distributed sediment runoff model at the catchment scale and (2) subsequently constructing a distributed-lumped model as a new type of lumped sediment runoff model which preserves the physical meaning of the hydrologic and sediment transport processes. The paper describes the overall model structures and the lumping method, as well the reliable application of the model at the Lesti River catchment, Indonesia. Mathematical formulation of the physically-based distributed sediment runoff model and its distributed-lumped version are provided following section.

2. Physically-Based Distributed Sediment Runoff Model

The hydrological model[5] considers three principal water flux pathways within a catchment: subsurface flow through unsaturated flow (capillary pore), subsurface flow through saturated flow (non-capillary pore), and surface overland flow. Based on stage-discharge relationship,[6] after the water depth is greater than the surface soil layer, the net rainfall will accumulate as surface water and begin to flow as overland flow. Subsurface and surface flows in both land surface and river channel networks are computed as one-dimensional kinematic wave equation. In a given rainfall event, rainfall is directly added to subsurface or surface flow according to the water depth, depending on the rainfall occurring in a grid-cell.

The stage-discharge, $q - h$, relation for both surface and subsurface flows:

$$q = \begin{cases} v_m d_m (h/d_m)^{\beta}, & 0 \leq h \leq d_m \\ v_m d_m + v_a(h - d_m), & d_m \leq h \leq d_a \\ v_m d_m + v_a(h - d_m) + \alpha(h - d_a)^m, & d_a \leq h \end{cases} \tag{1}$$

$$\nu_m = k_m i, \quad \nu_a = k_a i, \quad k_m = k_a/\beta, \quad \alpha = \sqrt{i}/n$$

where q is discharge per unit width, h is water depth, i is the slope gradient, k_m is the saturated hydraulic conductivity of the capillary soil layer, k_a is the hydraulic conductivity of the non-capillary soil layer, d_m is the depth of the capillary soil layer, d_a represents the depths of the capillary and non-capillary soil layers, β is the exponent constant of unsaturated flow, v_m and v_a are the flow velocities of unsaturated and saturated subsurface flows, respectively, and n is the roughness coefficient based on the land cover classes.

A sediment transport algorithm was newly added to the hydrological model. Soil erosion and deposition are described in terms of processes occurring on hillslopes. The sediment transport algorithm includes multiple sources of sediment transport, namely soil detachment by raindrops (d_r) and soil detachment or deposition driven by overland flow (d_f).

Soil detachment by raindrops is given by an empirical equation in which the rate is proportional to the kinetic energy of effective rainfall. From observation of rainfall characteristics in the study area,[7] d_r is given as

$$d_r = \mu\ k\ k_e = \mu\ k\ 56.48\ r \tag{2}$$

where μ is the soil erodibility, k is a tuning parameter, k_e is the total kinetic energy of the rainfall, and r is the rainfall intensity.

The concept of sediment transport capacity was used to determine soil detachment or deposition by overland flow. Sediment transport capacity of overland flow is defined as the maximum value of sediment concentration to transport; d_f represents the sediment yield by surface flow shear stress[8]:

$$d_f = \alpha(T_c - c)h_s \tag{3}$$

where α is the detachment/deposition efficiency factor, h_s is the overland flow depth, and T_c is the maximum sediment concentration transport capacity. Following the T_c value, if the current sediment concentration c is lower than this capacity, erosion occurs; otherwise an excess of soil deposition exists.

The transportation capacity was calculated based on the unit stream power theory that can be applied for sediment transport in open channels and surface land erosion.[9] Hence T_c is estimated as

$$T_c = \log c_t = I + J \log((vi - v_{critical}\ i)/\omega) \tag{4}$$

where $v * i$ is the unit stream power (where v is the flow velocity and$i$ is the slope gradient), $c_{critical*}i$ is the critical unit stream power ($v_{critical}$ is the critical flow velocity), ω is the sediment fall velocity calculated by Rubey's equation, as well I and J are the dimensionless parameters.

The catchment is divided into an orthogonal matrix of square grid-cells (i.e. 250m $\times$ 250m), assumed to represent homogeneous conditions according to the digital elevation model (DEM). This allows the use of DEM to derive flow direction map to define the interaction between the objects which simulate sediment runoff at each grid-cell. Runoff generation, soil erosion or deposition are computed for each grid-cell and are routed between grid-cells using the kinematic wave model following the water flow direction, which defines the routine order for the water flow and sediment transport propagation.

The physically-based distributed sediment runoff model is a suitable tool allowing for the simulation and investigation of hydrological process, sources of erosion and deposition within internal location of the catchment (Fig. 1).

3. Distributed-Lumped Sediment Runoff Model Derived from the Lumping Method

A method to lump a physically-based distributed rainfall runoff model[4] was used and extended for erosion and sediment transport model. The lumping method has been developed to produce a distributed-lumped model as a new type of lumped sediment runoff model version. The original distributed model is used to derive a lumped relation between water storage volume, discharge, and maximum sediment storage of the catchment by considering spatial distribution of topographic variables, water content, and sediment

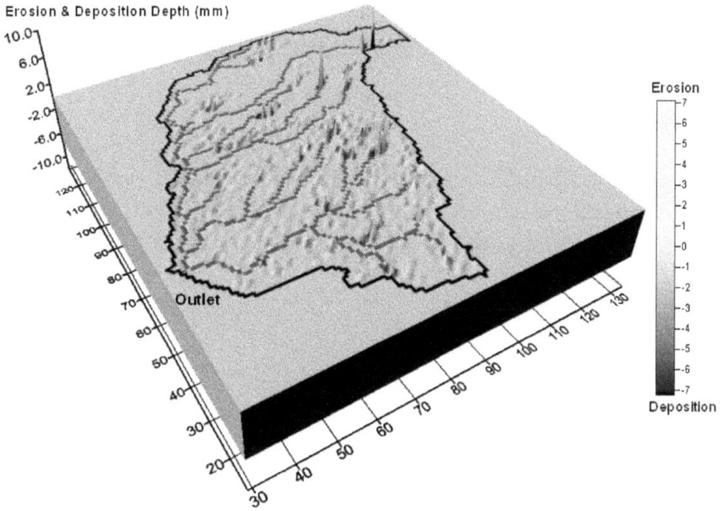

Fig. 1. Spatial distribution of erosion and deposition sources for internal location of the catchment after rainfall event on September 13–15, 2004 at the study site.

concentration transport capacity. The method proves that lumped version is derived from the distributed version. The objective of the development is not only to improve the computational efficiency but more importantly to construct a lumped formulation of the hydrologic and sediment transport equations at catchment scale by keeping the physical meaning of the distributed model. All the represented dominant processes are preserved in the lumping procedures.

The fundamental assumption of the lumping method is that the rainfall runoff process of the catchment system reaches the steady state condition with spatially uniform rainfall input. From this assumption, discharge flux can be expressed as the product of rainfall intensity and the upslope contributing areas. Then, the relation between total stored water and the outflow discharge in each grid cell is theoretically derived by integration processes of spatially distributed equations assigned to each grid cell over the landscape:

$$q(x) = q(0)/w + r \int_0^x (x)\, dx = r\, U/w + rx \tag{5}$$

where U is the upslope contributing area, x is the horizontal distance from the upstream end of a grid cell, and w is the width of the grid cell.

The stored water volume of a grid cell, s, is given as

$$s = w \int_{0}^{L} h(x)\, dx \tag{6}$$

where L is the horizontal distance at the downstream end of a grid cell. By substituting the variable of integration from x to q using the relation given by Eq. (5), Eq. (6) becomes

$$s = \frac{w}{r} \int_{q(0)}^{q(L)} f(q)\, dq = \frac{w}{r}\left[F(q(L)) - F(q(0))\right] \tag{7}$$

It is assumed that q is a function of h and can be analytically integrated with h. If the value of q is known from Eq. (5), and h is numerically obtained using Eq. (1), then $F(q(x))$ can be calculated.[10]

The stored water at the catchment scale, S_w, can be calculated by summing the s from each grid cell as follows:

$$S_w = \sum_{i=1}^{N} s_i \tag{8}$$

where N is the total number of grid cells for a whole catchment. Finally, q at the outlet is linked to S_w as a function of the topographic and physical characteristics of each grid cell, as well as effective rainfall intensity.

The sediment transport process is affected by the dynamic spatial distribution of overland flow. For lumped representation of the catchment, soil detachment or redeposition depends on the balance between the current sediment concentration and maximum sediment storage capacity. The maximum sediment storage variable is produced from lumping distributed erosion-sediment transport equations as

$$S_s^{\max} = \frac{\sum_{i=1}^{N} (T_c s_{ws})_i}{S_w} \tag{9}$$

where s_{ws} is the surface water stored in each grid cell if $h > d_a$. Variable T_c is then calculated using Eq. (6). According to the relation between the current sediment storage, S_s, and S_w for each time-step calculation, the value of c from the hillslope area can be solved by

$$c = \frac{S_s}{S_{ws}} \tag{10}$$

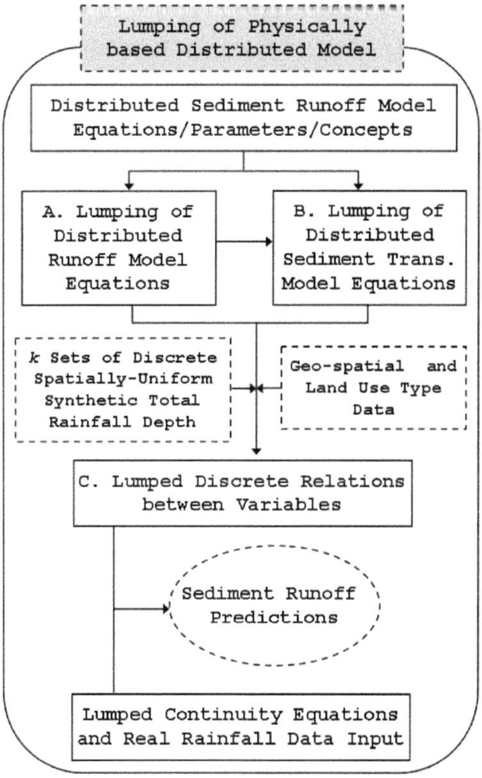

Fig. 2. Schematic diagram of the procedures to develop a distributed-lumped sediment runoff model based on the lumping method.

where S_{ws} is the total stored water surface of the catchment. For each time-step calculation, c is assumed to be uniform over the catchment and is the variable of lumped continuity.

In brief, the procedures to develop the distributed-lumped sediment runoff model version (see Fig. 2) may be summarized as follows:

1. Based on the original distributed equations and their physically meaningful parameterization, lumping process of the distributed model is done by spatially integrating kinematic wave runoff and sediment transport equations for each grid-cell over the entire catchment system.
2. Obtain the lumped discrete relations between the variables, namely the stream discharge, the total water stored amount, and the maximum sediment storage capacity. It may be conducted by spatially distributed

information on land use, soil properties, geomorphology and topology, as well for k sets of different synthetic rainfall intensity (i.e. from 1, 2,... to 500 mm/hr). Finally, it prepares the lumped relations in the "look-up" table form, which is time-invariant values and considered as parameters of the lumped model. Thus the innovative aspect of this method is these lumped model parameters (Fig. 3) can be derived directly from the distributed model without adding any calibration or parameterization. This feature can solves the major disadvantage of traditional lumped models, which lacks physical background of the parameters.

3. Streamflow and sediment concentration are predicted and updated using lumped continuity equations and real rainfall intensity. They related to each of discrete relations which summarized in the "look-up" table. Lumping of original distributed model leads to transforming the water flow and suspended sediment transport routing processes into a cascade of non-linear reservoirs represents the cacthment as a whole of the form as shown in Fig. 4. It transforming kinematic wave routing process into storage routing process as lumped continuity equations.

4. Model Evaluation and Discussions

The sediment runoff models both the distributed and lumped have been applied to the Lesti River catchment shown in Fig. 5. The parameter values of the original distributed model version: roughness coefficient n (0.001–0.3 $m^{-1/3}s$); depth of capillary and non-capillary soil layer d_a (0.08–1.2 m); depth of the capillary soil layer d_m (0.04–0.8 m); hydraulic conductivity of the non-capillary soil layer K_a (0.015 m/sec); exponent constant of unsaturated flow β (8); median of grain size D_{50} (0.062 mm); soil detachability k (0.004 kg/J); and detachment/deposition efficiency factor α (0.98).

Figures 6 shows a comparison between the values of simulated streamflow and sediment concentration at the outlet (Tawangrejeni station) for two scenarios of soil thickness with the same rainfall intensity and total depth calculated through the distributed and the lumped models. The evaluation results suggest that in the case when soil thickness is shallow (Case 1), the discrepancy of estimated values for both the streamflow and sediment concentration by the lumped and distributed models is generally less than that of the case the soil is thicker (Case 2).

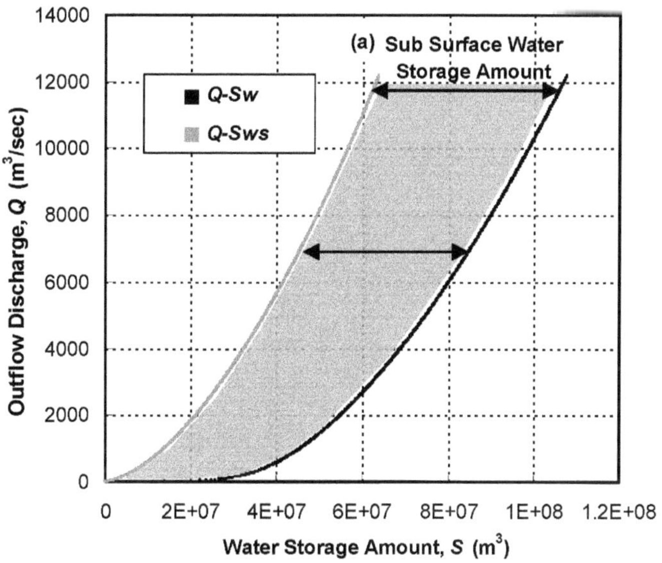

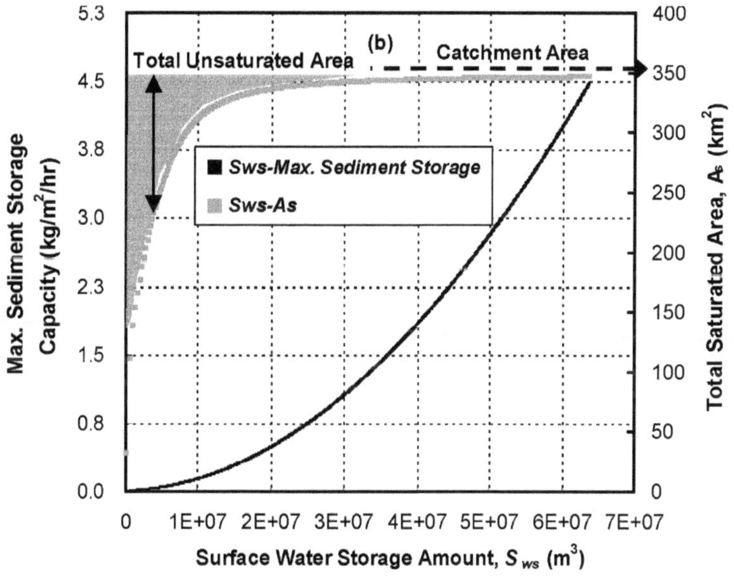

Fig. 3. (a) Plots of lumped discrete relations between variables of $Q - S_w$ and $S_{ws} - S_w$; and (b) discrete relations between $S_s^{max} - S_{ws}$ and $A_s - S_{ws}$ for 1150 different sets of rainfall intensity.

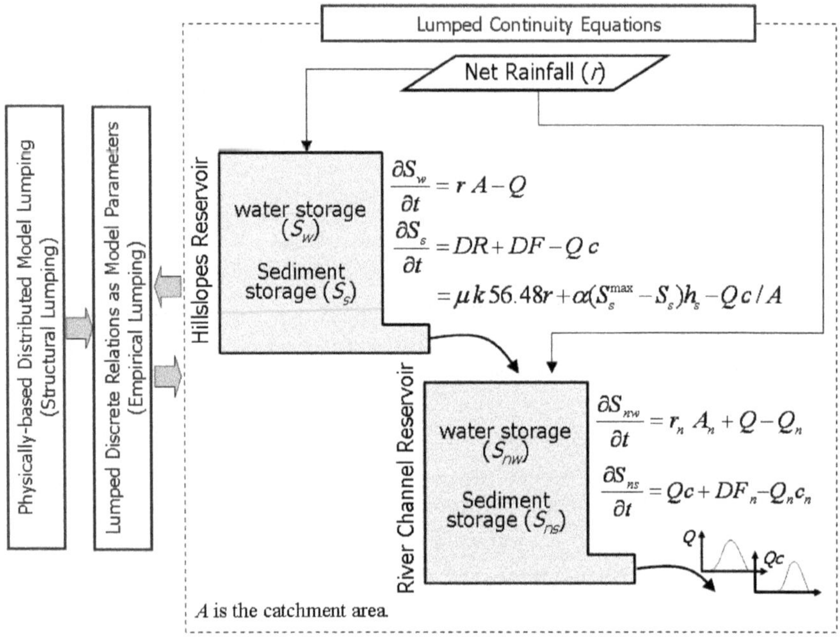

Fig. 4. Conceptual scheme of the new lumped sediment runoff model structure.

The difference increases with the soil thickness, where the lumped model results generally have a tendency to underestimate streamflow and sediment concentration at the rising limb, and over estimate at the falling limb of hydrograph and sedimentgraph. This difference is due to the assumed steady state condition in deriving the lumped model. In Case 2, at the beginning time of event, the catchment system does not reach a steady state yet, rainfall stored in the soil layer flows is rather slow, and therefore streamflow is insensitive to rainfall intensity. Afterwards, the soil layers are easily saturated and once the water depth at the most area of catchment exceeds the capacity of the capillary soil layer, streamflow rises up suddenly. This implies it was not correct for the event to assume the steady state condition at the beginning of the simulation in Case 2. Thus the second discrete relation of $Q - S_w$ for the lumped model shows some difference from the distributed one. Another possible reason is that in the distributed model, the setting of initial conditions causing some parts of the catchment are already saturated (i.e. river channel area) and overland flow happens thus the rising limb of the hydrograph is faster than the lumped

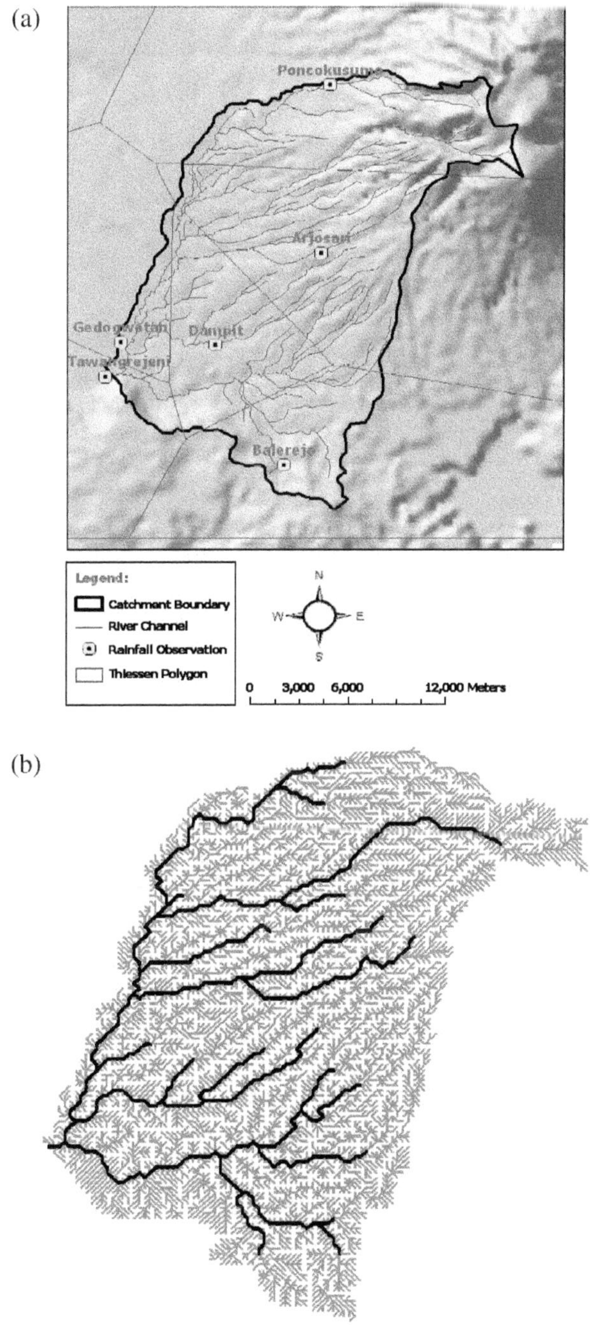

Fig. 5. (a) Lesti River catchment: river system, catchment boundary, and rainfall observation sites; and (b) its flow direction map.

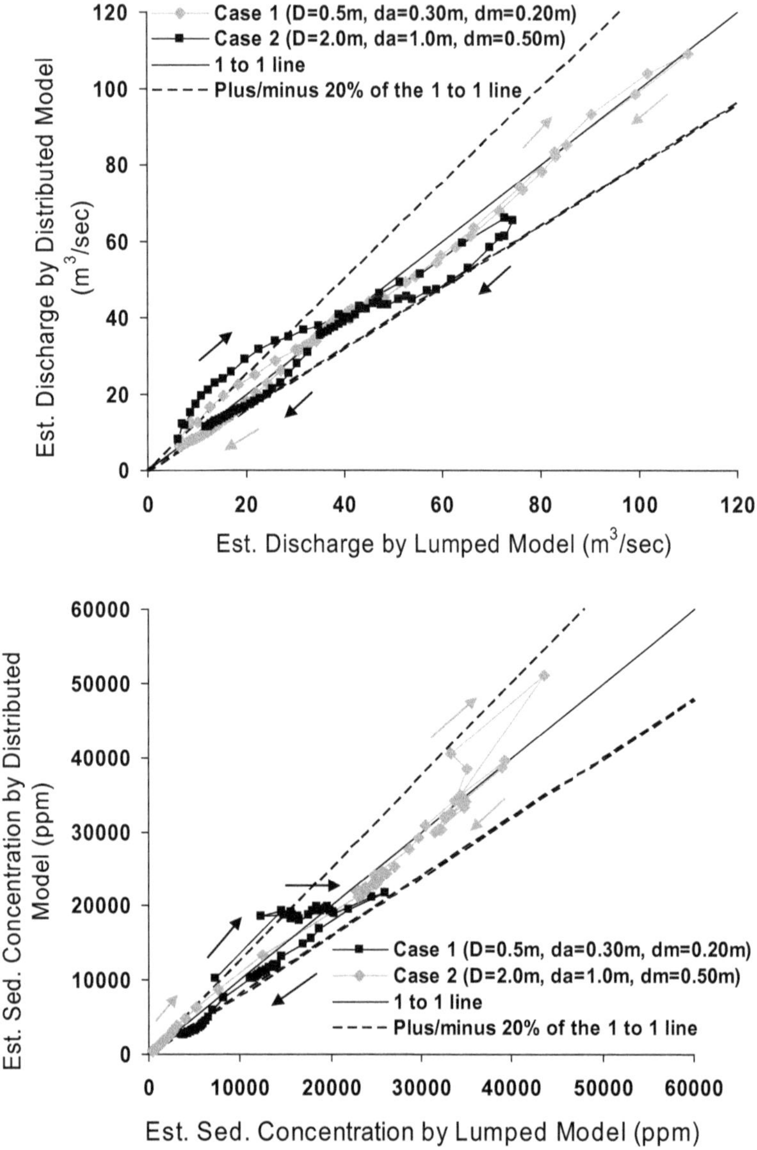

Fig. 6. Comparisons of hysteresis loops of outflow discharge and sediment concentration for various cases of soil thickness (D) through original distributed model and its lumped model version.

Table 1. Simulated cumulative streamflow and sediment concentration for two simulation cases.

Variables	Cases	Distributed model	Lumped model	Difference (%)
Runoff (m^3)	Case 1	2724.20	2776.63	1.56
	Case 2	2192.48	2293.90	4.63
Sediment yield (kg)	Case 1	1517.05	1548.86	2.10
	Case 2	856.45	909.13	6.15

one. On the contrary, the capacity of non-capillary layer $d_a - d_m$ in Case 1 is smaller as compared to Case 2. The catchment system rises up to dynamic steady state condition, and lumped model results show better performance belonging to the original distributed model than in Case 2.

The above experiments showed that the lumped model version, structured according to the proposed lumping method, produces acceptable results. Within the range of possible soil depths, the difference is small and does not cause severe problems for the application. In addition, the lumped model successfully reduced the computational burden. Simulation time of the lumped model was about 1/37 of that of the original distributed model. Table 1 summarizes the differences in cumulative total runoff and sediment yield calculated by lumped and distributed models, which are all lower than 7%.

Figure 7 plots computed streamflow discharges and sediment concentrations by the lumped model compared to the observed data. The simulation results of the lumped model shows good agreement with the observed one.

5. Conclusions

The Lumping method of a physically-based distributed sediment runoff model has been developed. The advantage of the method is that the lumped model preserves hydrological and sediment transport processes, as well the model parameters derived from the original distributed model does not require additional calibration. From the case study in Indonesian river catchment, we confirmed both distributed model and the newly developed lumped model can indicate the reliable application compared to the observed streamflow discharge and sediment concentration. The discrepancy between the two models increases as the soil layer becomes

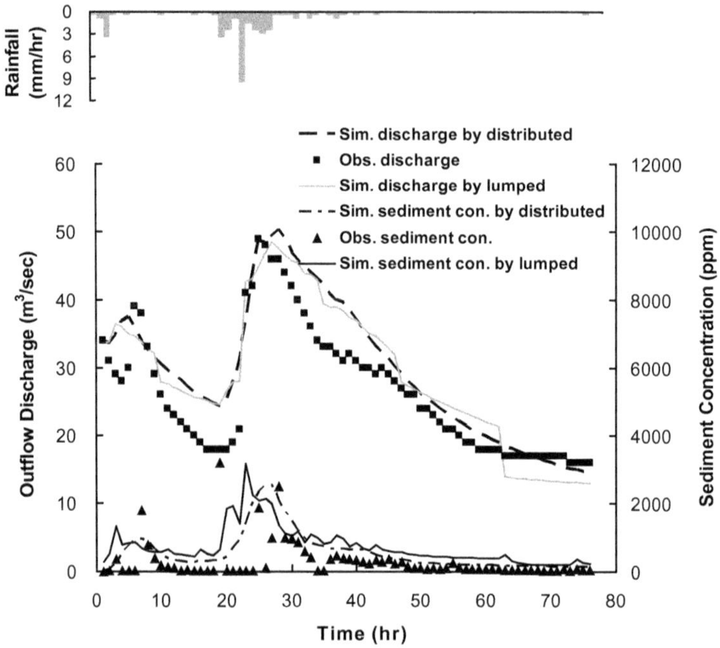

Fig. 7. Event sediment runoff observation and simulation at the catchment outlet in 2003.

thicker, rainfall amount decreases, and the spatial-temporal variability of input rainfall intensity become more significant. These are mainly due to the steady state assumption that lies under the lumping method. The proposed distributed-lumped sediment runoff model is an effective way, particularly for the practical applications on the real-time basis in combination with stochastic analyses.

Acknowledgments

This study was supported by the MEXT Coordination Fund for Promotion of Science and Technology, Japan Science and Technology Agency (PI: Prof. Kaoru Takara, DPRI, Kyoto University).

References

1. K. Takara, T. Yamashita, S. Egashira, P. R. Dyah, S. Irwan, A. R. Syamsudin and Anton, Application of remote sensing and GIS to research on disasters

caused by floods and sedimentation. In *Proceedings of Workshop on Disasters Caused by Floods and Geomorphological Changes and Their Mitigations (WDFGM-1996)* (1996) 62–77.

2. T. Apip, Y. Sayama, Tachikawa and K. Takara, The spatio-temporal predictions of rainfall-sediment-runoff based on lumping of a physically-based distributed model. *Annuals of Disaster Prevention Research Institute*, Kyoto University **50B** (2007) 79–94.
3. E. Todini. Rainfall-runoff modeling-past, present and future, *Journal of Hydrology* **100** (1988) 341–352.
4. Y. Ichikawa, T. Oguro, Y. Tachikawa, M. Shiiba and K. Takara. Lumping general kinematic wave equation of slope runoff system, *Annual Journal of Hydraulic Engineering, JSCE* **44** (2000) 145–150 (in Japanese with English summary).
5. T. Kojima and K. Takara. A grid-cell based distributed flood-runoff model and its performance, weather radar information and distributed hydrological modeling. *IAHS Publ.* **282** (2003) 234–240.
6. Y. Tachikawa, G. Nagatani and K. Takara. Development of stage-discharge relationship equation incorporating saturated-unsaturated flow mechanism. *Annual Journal of Hydraulic Engineering, JSCE.* **48** (2004) 7–12.
7. S. Oishi, T. Sayama, H. Nakagawa, Y. Satofuka, Y. Muto, D. Sisinggih, and K. Sunada, Development of estimation method for impact energy of raindrop considering raindrop size distribution and the relationship between the impact energy and local sediment yield, *Annual Journal of Hydraulic Engineering, JSCE* **49** (2005) 1087–1092.
8. G. Govers, and G. Rauws, Transporting capacity of overland flow on plane and on irregular beds. *Earth Surface Processes and Landforms* **11** (1986) 515–524.
9. C. T. Yang, *Sediment transport theory and practice*, The McGraw-Hill Companies, Inc., 1996.
10. T. Apip, Y. Sayama, K. Tachikawa and K. Takara. Lumping of a physically-based distributed model for sediment runoff prediction in a catchment scale, *Annual Journal of Hydraulic Engineering, JSCE* **52** (2008) 43–48.

Advances in Geosciences
Vol. 17: Hydrological Science (2008)
Eds. Namsik Park et al.

INFLUENCE OF TYPHOON CHARACTERISTICS ON FLOOD FORECASTING

GWO-FONG LIN*, GUO-RONG CHEN†, PEI-YU HUANG‡
Department of Civil Engineering, National Taiwan University,
Taipei 10617, Taiwan
**gflin@ntu.edu.tw*
†grchen@ntu.edu.tw
‡d90521010@ntu.edu.tw

The development of flood forecasting models with typhoon characteristics is presented in this paper. To investigate the influence of typhoon characteristics on flood forecasting, four different types of model inputs are designed to yield one- to six-hour ahead forecasts of runoff. A performance comparison of models with and without typhoon characteristics is conducted. With typhoon characteristics as input to the model, the forecasting performance is significantly improved for both medium and long lead-time forecastings, especially when the rainfall data are not available. In conclusion, the typhoon characteristics should be used as input to the flood forecasting models.

1. Introduction

Flood forecasting plays an essential role in a flood warning system. Improving flood forecasting, especially for long lead-time, is always desired. In the Taiwan area, rainfall brought by typhoons is an important water resource, but floods caused by typhoons frequently lead to serious disasters. The island of Taiwan is situated in one of the main paths of North Western Pacific typhoons. On average, three to four typhoons attack the island each year. To improve the typhoon-flood forecasting is always an important task in Taiwan. The highly non-linear and extremely complex process between typhoon rainfall and runoff leads to a lot of difficulties in constructing a physically-based mathematical model. Artificial neural networks (ANNs) are used to construct flood forecasting models herein because of their great ability to model nonlinear systems without the need to make any assumptions.

In this paper, typhoon characteristics, which are seldom regarded as key input in literature, are added to the flood forecasting models to further improve the long lead-time forecasting. In literature only the antecedent runoff and rainfall are frequently used as inputs to typhoon-flood forecasting models.[1,2] However, the forecasting performance of models, which only use the antecedent runoff and rainfall as input, usually decreases rapidly when the forecasting lead-time exceeds the time of concentration. Lin and Chen[3] indicate that typhoon characteristics are capable of showing the trend of rainfall when a typhoon is nearby. Typhoon characteristics seem to be capable of providing valuable information for longer lead-time forecasting. Such a reasonable speculation has prompted an investigation into the influence of typhoon characteristics on flood forecasting, especially for long lead-time forecasting.

The objective of this paper is to assess the use of typhoon characteristics to improve the long lead-time flood forecasting. For this purpose, four different types of model inputs are designed to yield one- to six-hour ahead forecasts of runoff. A total of 24 ANN-based forecasting models are constructed and the coefficient of efficiency is selected to evaluate the overall performance of forecasting models. In addition, considering the rainfall data may not be available in some real cases. For example, some rain gauges may be out of function during typhoon periods and rainfall data records may be missing or too short to calibrate forecasting models. Hence, according to the availability of rainfall data, four input designs are classified into two categories herein for discussions. In the first category the rainfall data are available, whereas in the second category the rainfall data are not available. Finally, an application is given to show the influence of typhoon characteristics on flood forecasting.

2. The Study Area and Data

The study area is located upstream of the Jen-Shou Bridge on the Kee-Lung River, a tributary of the Tan-Shui River in northern Taiwan. The watershed has an area of 94.75 km^2 and an average river slope of 1/45. The flood forecasting of a watershed with such a small area and a steep slope is always a challenging task. Furthermore, the study area is abundant in rainfall. The rain gauge (Huo-Shao-Liao) in the study area often has the highest yearly rainfall in Taiwan. The average yearly rainfall from 1955 to 2007 is 4,660 mm. Another important matter is that Taipei (the capital of Taiwan) is only 28 km downstream of the outlet of the study area. Therefore, a well-performed and efficient flood forecasting model is needed for the study area.

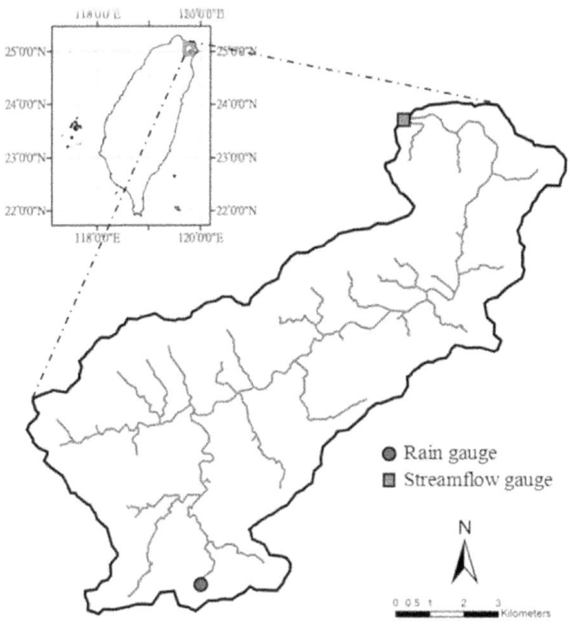

Fig. 1. The study area.

Figure 1 shows the study area and the locations of rain gauge and streamflow gauging station. The rainfall and the runoff data are obtained from the Water Resources Agency and the typhoon characteristics are collected from the Central Weather Bureau. A total of 15 typhoon events are used to develop the forecasting models. Typhoon characteristics include the latitude and longitude (degree) of the typhoon centre, the maximum wind speed (m/s) near the center, the atmospheric pressure (hPa) of the center, the radius (km) of winds over 15 m/s, and the speed (km/h) of the typhoon movement. To develop the forecasting models, of these 15 typhoon events, six events are taken as the training set, six events are selected as the validation set, and the remaining 3 events are used as a testing set for testing the forecasting performance.

3. Model Development

3.1. *ANN-based models*

The flood caused by a typhoon is an extremely complex, nonlinear and fragmented physical process. Hence, it is very difficult to develop a physically-based model. ANNs, which have great ability to model nonlinear systems, are used to construct flood forecasting models herein. In decades, ANNs have found a growing number of applications in hydrology and comprehensive reviews of the applications of ANNs in hydrology have been done by ASCE Task Committee.[4,5] As to ANN-based flood forecasting models, there are many studies have been conducted in recent years.[6–8] We have applied radial basis function networks (RBFNs) to 1- to 3-hour lead-time flood forecasting[9] and also have proposed a systematic approach to determining the inputs of NN-based rainfall-runoff models.[10] Generally speaking, ANNs are information processing systems. ANNs are composed of a number of interconnected processing elements. These elements (neurons) are joined together with weighted connections. The parameters associated with each of these connections are called weights. An ANN has to be trained so as to produce the desired outputs when the associated inputs are received. The training process is an iterative process for determining appropriate weights that minimize the error. In this paper, back-propagation networks (BPNs), which are trained by the back-propagation algorithm, are used to construct flood forecasting models.

3.2. *Input design*

In this paper, four different types of model inputs are designed: (a) only antecedent runoff for ANN1, (b) antecedent runoff and rainfall for ANN2, (c) antecedent runoff, rainfall and typhoon characteristics for ANN3, and (d) antecedent runoff and typhoon characteristics for ANN4. One- to six-hour ahead forecasts of runoff are desired. Hence, a total of 24 forecasting models are constructed. The ANN1 has a general form as

$$Q_{t+\Delta t} = f\left(Q_t, Q_{t-1}, \ldots, Q_{t-(n_1-1)}\right) \tag{1}$$

where t is the current time, Q_t is the runoff at time t, Δt is the lead-time period ($\Delta t = 1$ represents one-hour-ahead forecast, $\Delta t = 2$ represents two-hour-ahead forecast and so on.), and n_1 is the lag length of runoff.

With antecedent runoff and rainfall as inputs, the ANN2 is written as

$$Q_{t+\Delta t} = f\left(Q_t, Q_{t-1}, \ldots, Q_{t-(n_1-1)},\ R_t,\ R_{t-1},\ \ldots,\ R_{t-(n_2-1)}\right) \tag{2}$$

where R_t is the rainfall at time, and n_2 is the lag length of rainfall.

Based on the ANN2, typhoon characteristics are added to develop the ANN3. The form of the ANN3 is

$$Q_{t+\Delta t} = f\left(Q_t,\ \ldots,\ Q_{t-(n_1-1)},\ R_t, \ldots,\ R_{t-(n_2-1)},\ Ty_t, \ldots,\ Ty_{t-(n_3-1)}\right) \tag{3}$$

where Ty_t represents the six typhoon characteristics which include the latitude and longitude of the typhoon centre, the maximum wind speed near the center, the atmospheric pressure of the center, the radius of winds over 15 m/s, and the speed of the typhoon movement and n_3 is the lag length of typhoon characteristics.

Finally, to consider the situation that the rainfall data are not available for a watershed, ANN4 is proposed as

$$Q_{t+\Delta t} = f\left(Q_t,\ Q_{t-1},\ \ldots,\ Q_{t-(n_1-1)},\ Ty_t,\ Ty_{t-1},\ \ldots,\ Ty_{t-(n_3-1)}\right) \tag{4}$$

in which only antecedent runoff and typhoon characteristics are used as input. The model development is schematized in Fig. 2.

3.3. *Determination of the most appropriate input lengths*

The criterion for selecting the lag lengths is the relative percentage error (RPE):

$$\mathrm{RPE} = \frac{E_{\text{valid}}(n_i) - E_{\text{valid}}(n_i + 1)}{E_{\text{valid}}(n_i)} \times 100 \tag{5}$$

where $E_{\text{valid}}(n_i)$ and $E_{\text{valid}}(n_i + 1)$ are the RMSEs for the validation set of ANN models with n_i and $n_i + 1$ lag lengths, respectively. In general, the RMSE decreases with increasing lag term. When the RPE is less than 5%, the increase of lag lengths is stopped. According to this determination process of lag lengths, finally the values of n_1, n_2 and n_3 are confirmed.

3.4. *Performance measures*

In this paper, the coefficient of efficiency (CE) is selected to evaluate the model performance. For a single typhoon event, the CE is defined as

$$\mathrm{CE} = 1 - \sum_{i=1}^{T} (Q_i - \hat{Q}_i)^2 \Big/ \sum_{i=1}^{T} (Q_i - \bar{Q})^2 \tag{6}$$

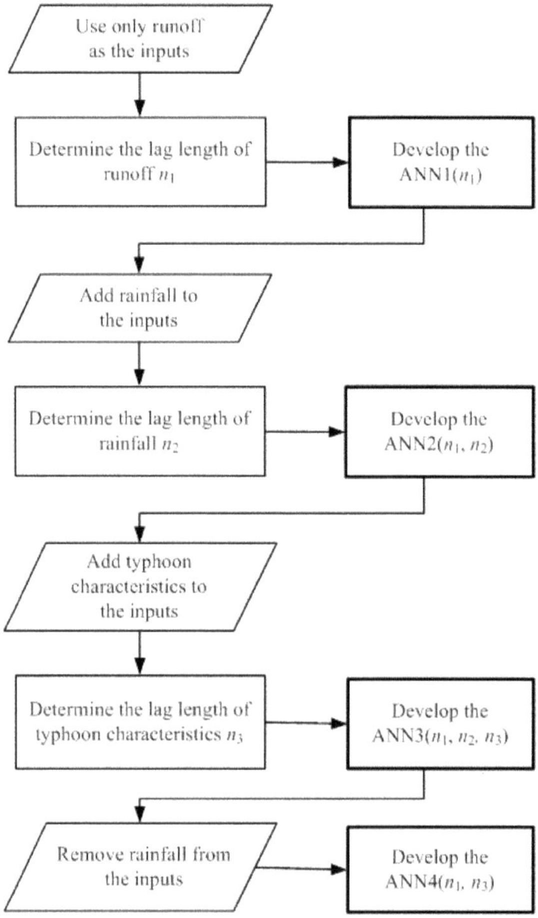

Fig. 2. Flowchart of the model development.

where Q_i and $\hat{Q}_i$ denote the observed and forecasted runoff at time i, respectively, $\overline{Q}$ is the average of the observed runoff, and T is the number of the time steps. If the CE value is equal to one, the forecasts are perfect. The mean CE of N typhoon events is written as

$$\mathrm{MCE} = \frac{1}{N}\sum_{j=1}^{N}\mathrm{CE}_j \tag{7}$$

where N is the number of typhoon events, and CE_j is the coefficient of efficiency (CE) for the jth typhoon event.

4. Influence of Typhoon Characteristics

The influence of typhoon characteristics on forecasting performance is presented in this section. According to the availability of rainfall data, four input designs are classified into two categories herein for discussion. In the first category the rainfall data are available, whereas in the second category the rainfall data are not available. In the first category, the comparison between ANN2 (which uses antecedent runoff and rainfall as input) and ANN3 (which uses antecedent runoff, rainfall and typhoon characteristics as input) are made. In the second category, the comparison of ANN1 (which uses only antecedent runoff as input) and ANN4 (which uses antecedent runoff and typhoon characteristics are as input) is conducted.

4.1. *Category I: Rainfall data are available*

The MCE values for three models (ANN1, ANN2 and ANN3) are shown in Fig. 3. The performance of ANN1 is obviously worse than that of the other two models (ANN2 and ANN3), especially for both medium and long lead-time forecastings. Only using antecedent runoff as input cannot yield effective forecasts, when the forecast lead-time is greater than 2 hours.

To highlight the influence of typhoon characteristics on flood forecasting, we focus on the comparison between the performance of ANN2 and that of ANN3. In Fig. 3, the MCE values decrease with increasing forecast lead-time and the MCE values of ANN2 decrease more rapidly than those of ANN3. The results show that ANN3 forecasts runoff more accurately than ANN2, especially for long lead-time forecasting. The MCE values of ANN2 and ANN3 are 0.43 and 0.54, respectively. For long lead-time forecasting, ANN3 yields an increase of MCE value of 26%. Figure 4 shows the observed runoff with the 6-hour-ahead forecasts using ANN2

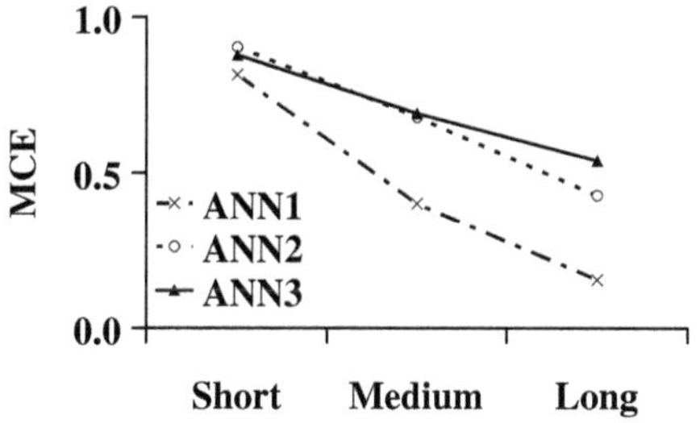

Fig. 3. MCE of ANN1, ANN2 and ANN3 (Category I).

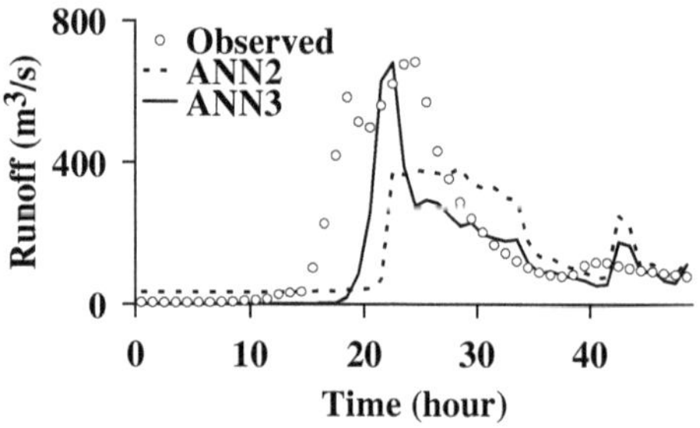

Fig. 4. Comparison of the observed runoff with the 6-hour-ahead forecasts using ANN2 and ANN3 for Typhoon Herb (Category I).

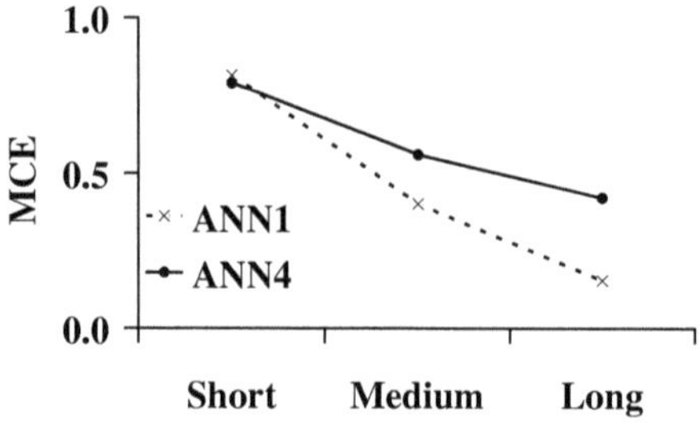

Fig. 5. MCE of ANN1 and ANN4 (Category II).

and ANN3 for typhoon Herb. It is clearly shows that ANN3 is better than ANN2 for long lead-time, especially for forecasting peak runoff (magnitude and time). The results indicate that typhoon characteristics can provide valuable information to decision makers of a flood warning system.

4.2. *Category II: Rainfall data are not available*

Figure 5 presents the MCE values of ANN1 and ANN4. Similar to Category I, both the MCE values of ANN1 and ANN4 decrease with increasing forecast lead-time, and those of ANN1 decrease more rapidly than ANN4. In addition, the long lead-time forecasting has the most significant improvement of the overall performance as compared to short and medium lead-time forecastings. As compared to ANN1, ANN4 increases MCE values of 39% and 171% for medium and long lead-time forecastings, respectively. Figure 6 shows the observed runoff with the 6-hour-ahead forecasts using ANN1 and ANN4 for typhoon Herb. It is clearly shown that

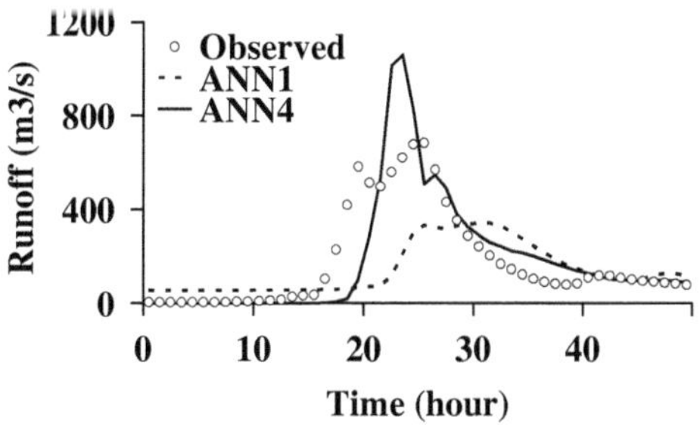

Fig. 6. Comparison of the observed runoff with the 6-hour-ahead forecasts using ANN1 and ANN4 for Typhoon Herb (Category II).

ANN4 provides better performance than ANN1. Typhoon characteristics effectively improve the overall performance, not only for long lead-time forecasting but also for medium lead-time forecasting.

4.3. *Comparison of Categories I and II*

Table 1 lists the MCE values of four flood forecasting models. Typhoon characteristics provide significant improvements mainly for medium and long lead-time forecastings. The models with typhoon characteristics (ANN3 and ANN4) perform better than those without typhoon characteristics (ANN1 and ANN2). Among the four models, ANN3 (which uses antecedent runoff, rainfall and typhoon characteristics as input) yields the best results. If the rainfall data are not available, ANN4 (which uses antecedent runoff and typhoon characteristics as input) is recommended. In Category I, ANN3 increases MCE values of 2% and 39% respectively for medium and long lead-time forecastings as compared to ANN2. In Category II, ANN4 increases MCE values of 39% and 171% respectively for medium and long lead-time forecastings as compared to

Table 1. MCE of four flood forecasting models.

	MCE			
	Category I		Category II	
Lead time	ANN2	ANN3	ANN1	ANN4
Short	0.90	0.88	0.82	0.79
Medium	0.68	0.69	0.40	0.56
Long	0.43	0.54	0.16	0.42

Table 2. The influence of typhoon characteristics on forecasting performance.

	MCE	
Lead time	Category I	Category II
Short	No	No
Medium	No	Yes
Long	Yes	Yes

Yes: There is a significant improvement.
No: There is no improvement.

ANN1. Based on the results, one can find that typhoon characteristics do improve the performance of forecasting models and the improvement increases with increasing lead-time, especially when the rainfall data are not available. Table 2 summarizes whether the models with typhoon characteristics can provide a significant improvement in MCE (forecasting performance). It clearly shows that the typhoon characteristics are more necessary for improving the accuracy of forecasts if the rainfall data are not available.

5. Conclusions

This paper describes the development of flood forecasting models to improve the long lead-time flood forecasting. In addition to the antecedent inflow and rainfall, which are frequently used as input to flood forecasting models, typhoon characteristics are added to the proposed model to further improve the long lead-time forecasting. To investigate the influence of typhoon characteristics on flood forecasting, four different types of model inputs are designed to yield one- to six-hour ahead forecasts of runoff. A total of 24 forecasting models are constructed. The overall performance of all models decreases with increasing forecast lead-time, and the models without typhoon characteristics decrease more rapidly than those with typhoon characteristics. Typhoon characteristics can improve the overall forecasting performance, and the improvement increases with increasing lead-time especially when the rainfall data are not available. In conclusion, the typhoon characteristics should be used as input to the flood forecasting models. The proposed modeling technique is expected to be useful to improve the medium and long lead-time flood forecasting.

References

1. E. Toth, A. Brath and A. Montanari, *Journal of Hydrology* **239**(1–4) (2000) 132–147.
2. T. Y. Pan and R. Y. Wang, *Journal of Hydrology* **297**(1–4) (2004) 34–50.
3. G. F. Lin, L. H. Chen, *Hydrological Processes* **19**(9) (2005) 1825–1837.
4. ASCE Task Committee on Application of Artificial Neural Networks in Hydrology, *Journal of Hydrologic Engineering* **5**(2) (2000) 115–123.
5. ASCE Task Committee on Application of Artificial Neural Networks in Hydrology, *Journal of Hydrologic Engineering* **5**(2) (2000) 124–137.
6. G. Kim and A. P. Barros, *Journal of Hydrology* **246**(1–4) (2001) 45–62.
7. H. Moradkhani, K. L. Hsu, H. V. Gupta and S. Sorooshian, *Journal of Hydrology* **295**(1–4) (2001) 246–262.
8. W. R. Huang, B. Xu and A. Chan-Hilton, *Hydrological Processes* **18**(13) (2004) 2545–2564.
9. G. F. Lin and L. H. Chen, *Journal of Hydrology* **289**(1–4) (2004) 1–8.
10. G. F. Lin and G. R. Chen, *Hydrological Processes* **22**(14) (2008) 2524–2530.

Advances in Geosciences
Vol. 17: Hydrological Science (2008)
Eds. Namsik Park et al.

ESTIMATION OF UNCERTAINTY IN PREDICTION OF A FLOOD FORECASTING MODEL

HADI KARDHANA* and AKIRA MANO†
Graduate School of Engineering, Tohoku University,
6-6-11-1110 Aoba Aramaki Aoba-ku,
Sendai 980-8579, Japan
**kardhana150@ppk.itb.ac.id*
†mano@civil.tohoku.ac.jp

Quantitative precipitation forecast (QPF) by Japan Meteorological Agency (JMA) is main source of rainfall-runoff model based flood prediction. An uncertainty that appears on the forecast will affects accuracy of flood prediction. The clear estimation of uncertainty may lead to wider forecast application. Present research reproduces total seven flood events on Shichikashuku Dam in Japan during year of 2002 to 2007. The QPF based on short range forecast of Japan operational numerical weather prediction. The rainfall-runoff model is using conceptual-distributed modified tank model. Uncertainty is estimated by relative error analysis of multi events of the a) QPF and b) discharge prediction and reflected by level of confidence.

The results suggest that uncertainty of flood forecast mainly rises as a response of QPF'. It is also shown the correlation between forecast lead time and uncertainty. Correspond with the magnitude prediction error of discharge and precipitation, uncertainty is higher on longer lead time.

1. Introduction

Application of continuous advances in Numerical Weather Prediction (NWP) has the potential of allowing delivery of flood warning to extend well beyond current research. Present research conducted a short term forecast for a small catchment respectively. The term short forecasts was adopted from NWP which lead-times from a few hours to 2 or 3 days are termed short-term forecasts.

The term is distinct from a long-term forecast which initiated by Ad P.J. De Roo *et al.*,[1] and demonstrates by Werner[2] in developing European flood forecasting system (EFFS). The forecast system is based on global

NWP to provide ensemble prediction up to 10 days ahead. However the mesh size of global model of present time can not express the nature of cyclone, convective, or mesoscale system, Weisman[3]. The system may drive heavy rainfall in such small catchment. In other condition the prediction time may be too short to issue warning by only relying on monitoring from observation system or nowcasting rainfall. This type of catchment is mostly found in archipelagic region such as Japan. These lead to the needs of utilization of short range forecast that enables expressing small weather system while it can be provided sufficient time delay between prediction and people at risk response.

In order to follow up the issuance, the time needed for people at risk must be estimated. US Army Corp. of Engineer proposed a standard of response time based on storm event that shows a time range of people responds to the warning is about 6-12 hours. This figure has been agreed by study conducted Lindell[4].

Pre-2006, JMA runs short term forecast based on regional spectral model (RSM) twice a day (51-hour forecasts) and mesoscale model (MSM) four times a day (18-hour forecasts) which are obtained from the global model by nesting fine meshes. RSM horizontal resolution is 20 km and MSM's 10 km. In the end of 2007, RSM had been discontinued due to fining the resolution of Global Spectral Model (GSM).

Considerable uncertainty in the future prediction should be treated stochastically, means that resulting forecasts should be provided with estimation of uncertainty. As rainfall-runoff model, it has been shown that flow simulation uncertainty is strongly dependent on catchments scale of prescribed parametric and radar-rainfall input uncertainty (Carpenter and Georgakakos[5]). Present research made an effort to expand the uncertainty estimation of flood prediction up to 12 hours prediction.

Two main assets of this paper are runoff model and QPF as a product of NWP run by JMA. The runoff model is modified tank model which has been well exercised for continuous event and single events (Kato and Mano[6], Tatesawa[7], Kardhana[8]).

Case study was done for Shichikashuku Dam basin, upstream of Shiroishi River, Miyagi Prefecture, Japan. Model domain was selected from one outlet point to upstream area of about 180 km^2. The outlet point discharge data is hourly and estimated based on rating curve on elevation observation.

2. Flood Forecast Model

2.1. *Runoff model*

The runoff model used in this research is conceptual, semi-distributed, tank model attributes with some physical-based features (Kato and Mano[6], Tatesawa[7], Kardhana[8]). Topography and channel for the model expressed by 250 m × 250 m DEM. The basin consists of sub basins, each attached to one channel's grid. Sub-basin storage consists of three linear cascade tanks that represent soil layers. Precipitation observation is from JMA's operational analysis precipitation called RADAR AMeDAS. The resolution was 2.5 km in space and 1 hour in time. We used precipitation prediction from MSM, with temporal resolution 1 hour.

In each tank, main hydrological process is taken into consideration. Interception estimation use Crown Method[9]. Througfall is calculated from net precipitation after interception. Infiltration process is distributed based on saturated hydraulic conductivity, k_h^* and water content on the top tank. The infiltration is capable of expressing three conditions of ponding. Direct runoff is lumped into the channel with assumption of no time lag during its travel in the surface of sub-basin and contributes to surface runoff. Flux from each layer was expressed by Darcy's Law. The distribution between interflow, and percolation in each layer is driven by correspond k_h^* and water content, λ_i in lower tank. Interflow from three tanks contributes to the subsurface runoff. The k_h^* is approximated using Jabro's Method[10] from soil fraction data. Groundwater runoff was estimated from observation data and added as the initial base flow from the tank.

$$\lambda_i = \frac{H_i}{H_{i\,\max}} \tag{1}$$

$$q_i = ck_{h_i}^* I\lambda_i \tag{2}$$

Where c is deviation coefficient from the Darcy's Law; H_i is water depth in tank; and I is slope of sub basin. Subscript i denotes layer of soil; H_i follows by subscript *max* denotes tank depth. Tank depth represents soil layer thickness. Soil fraction from each layer and its properties information are extracted from Japan soil map.

Channel routing use one dimension kinematics wave approximation (KWA) with manning equation for motion equation and rectangular river section assumption. KWA is solved by 1D finite difference.

2.2. *Updating procedure*

The procedure was conducted by updating forecasted discharge with real discharge. Physical information in an outlet point represent process in its basin, initial updating conducted in this particular point. Absolute error can be obtained by reducing calculated discharge, Q_{cal} with observed discharge, Q_{obs} on the outlet within a time interval. This represents volume error in the basin.

The updating performed based on updating correction factor, Collischonn *et al.*[11] denotes by *FCA* which is ratio of Q_{obs} and Q_{cal} in outlet.

$$FCA \equiv \frac{Q_{\mathrm{obs}}}{Q_{\mathrm{cal}}} \quad (3)$$

Forecasted discharge is updated by real discharge in the outlet river mesh. Updating discharge in other river mesh used *FCA*. In order to maintain water balance after discharge updated, water depth in channel, h must also be updated. The replacement to the new-observed discharge and new-predicted water depth prompt distribution the volume error into sub-basins. This amount changes the condition of subsurface water.

This leads to proposed water depth updating factor (*FCB*). *FCB* is calculated by using Manning's Equation with following assumption; spatially constant manning coefficient n, slope I, and river width b, and insignificant h/b. River width distribution was estimated based on linear regression of catchment area of each river mesh. Thus the correlation between *FCA* and *FCB* is obtained as

$$FCA = FCB^{5/3} \quad (4)$$

First step to determine volume error is by revising discharge and water depth using *FCA* and *FCB*. Summation in unit volume determines total error on the basin. This error is distributed to each sub basin. For small catchment, fraction of error for each sub basin is approximated from rational method.

Afterward, subsurface water updating were done by distribute each fraction of error to each tank in a sub basin by following conceptual of runoff model. For cold start, the procedure is conducted on 6 hours window, from t_o-6 until from t_o. The time interval of discharge and water depth updating is one hour.

2.3. *Forecasting procedure*

Discharge was calculated based and rainfall prediction MSM as input to the distributed runoff model. The procedure is conducted on 18 hours window, from t_o+1 until t_o+18. Time step of the calculation is 60 seconds. Computation time takes about 1.5 minutes on desktop computer (P4 2.8 GHz HT processor, 2 GB RAM).

3. Reproduction of Flood Events

Reproduction of flood events was conducted for total seven selected events in the catchment. The runoff model was calibrated for two characteristics of events of single or continuous event. This step has been taken based on the evident that both have different range of error prediction. The difference between the observed and calculated results made a different treatment of model setup in attempt to express initial condition of water content. Its response to the subsurface runoff has been matched with base flow data. The performance coefficient used Nash Sutchliffe Coefficient, *NS* and relative error, η.

$$NS = 1 - \frac{\sum(Q_{\text{obs}} - Q_{\text{cal}})}{\sum(Q_{\text{obs}} - Q_{\text{cal}})} \tag{5}$$

$$\eta = \frac{|Q_{\text{cal}} - Q_{\text{obs}}|}{|Q_{\text{obs}}|} \tag{6}$$

Discharge prediction based on rainfall-runoff process of runoff model gives results as depicted on Table 1. Input rainfall uses analysis precipitation.

Table 1. Resume of flood reproduction.

Event	Event period	Q_{peak} (m^3/s)	*NS*		Category	Purpose
A1	9–12 July, 2002	600	0.92	13.33	Isolated	Calibration
A2	30 Sept.–2 Oct., 2002	275	0.96	1.82	Isolated	Calibration
A3	24–26 August, 2005	230	0.69	17.39	Continuous	Calibration
A4	14–16 July, 2007	185	0.82	24.32	Isolated	Validation
A5	8–10 Sept., 2007	240	0.87	20.83	Isolated	Validation
A6	19–22 May, 2004	75	0.74	8.00	Continuous	Calibration
A7	8–10 Oct., 2004	95	0.65	31.58	Continuous	Calibration

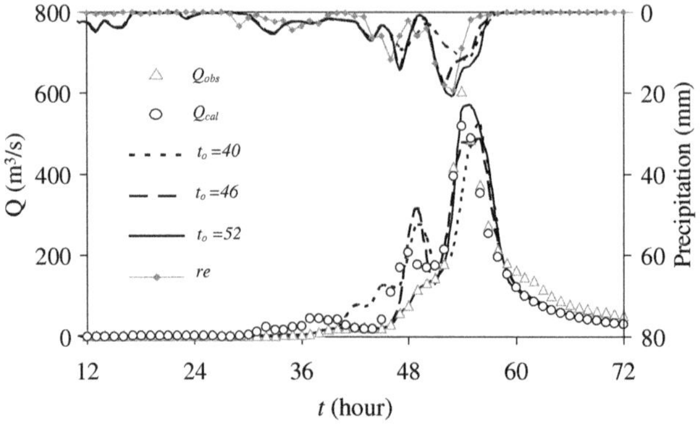

Fig. 1. Flood forecast on Event A1 based on MSM, origin $t = 8$ July 2002, 15.00 UTC.

Table 2. Contingency table for given threshold.

Scenario	Status
$\eta < th$	Hit
$\eta > th, Q_{\text{cal}} > Q_{\text{obs}}$	False
$\eta > th, Q_{\text{cal}} < Q_{\text{obs}}$	Miss

Discharge prediction based on rainfall-runoff process of runoff model gives results as samples depicted on Fig. 1. Input rainfall uses precipitation prediction of MSM.

Where Q_{obs} is observed discharge, Q_{cal} is calculated discharge, t_o is issuing time, *re* is observed catchments precipitation. *UTC* refers to Coordinated Universal Time. Primary y-axis (lower graph) is discharge; Secondary y-axis (upper graph) is catchment precipitation.

4. Forecasting Skill

The forecast performance indicates by common forecasting skill used by World Meteorological Agency, such as *BIAS* score on Eq. (7) and threat score or Critical Success Index/*CSI*, Eq. (8). Bias skill has range from 0 to infinity, which 0 for no skill and 1 for perfect score. *CSI* skill has range 0 to 1, which 0 for no skill and 1 for perfect score.

$$BIAS = \frac{\text{Hit} + \text{False}}{\text{Hit} + \text{Miss}} \tag{7}$$

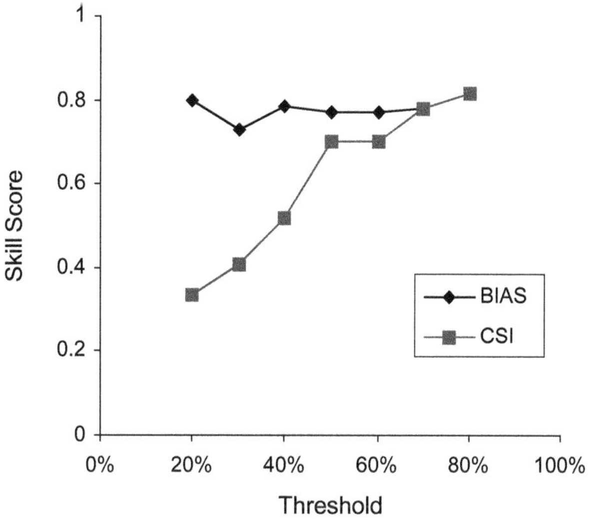

Fig. 2. Skill scores for given threshold.

$$CSI = \frac{\text{Hit}}{\text{Hit} + \text{Miss} + \text{False}} \tag{8}$$

A threshold is needed to determine such skill. M. Verbunt *et al.*[12] used a flood discharge as given threshold and a contingency table in the case of runoff prediction. Since on this paper, flood events are pre-selected, a threshold (th) is used as assumed acceptable η of peak discharge (η_{peak}).

Analysis of events prediction gives average $BIAS$ about 0.78, shows slight under-forecasting of flood occurrence. This skill seems independent from value of the threshold. CSI depends on given threshold, for example the assumed of acceptable relative error 40% (th = 40%,) will leave CSI = 0.52. It has meaning that slightly more than half of flood event magnitude was correctly forecast. Clearly that setting the threshold to relative error more 50% will give good skill. The figure has been shown a definitive degree of prediction uncertainty; however this figure has disregarded time parameter of prediction.

5. Estimating Uncertainty

In order to include time parameter, present research take main sources of uncertainty as a component and treat it separately, Ewen *et al.*[13]. The precipitation prediction as a main source is predicted continuously and each has time prediction, also as discharge. The term lead time, T_L in present research refers to time from issuing forecast until an event time point, i.e. precipitation point or discharge point. This paper used standard deviation, s_x to express uncertainty. As assets of the forecast model, precipitation forecast and discharge forecast are set to be the main components.

To avoid a very extreme value of error, a non dimensional parameter that reflected the significance of predicted point is introduced. The error of each component is modified by including predetermined extreme precipitation and discharge.

$$\alpha = \frac{\theta_{\text{obs}}}{\theta_{\text{max}}} \tag{9}$$

Thus the relative error is modified to express the non dimensional components.

$$\eta^* = \frac{|\theta_{cal} - \theta_{\text{obs}}|}{|\theta_{\text{obs}}|}\alpha \tag{10}$$

This paper used most extreme value of precipitation or discharge that occurred during events for θ_{max}.

The error is calculated on point basis, it has time factor and calculated every point on event and lead time. Figure 3 shows the averaged η*of precipitation and discharge prediction for one hour interval. The uncertainty at discharge prediction is total uncertainty. Figure 4 shows the uncertainty discharge prediction and its comparison with precipitation.

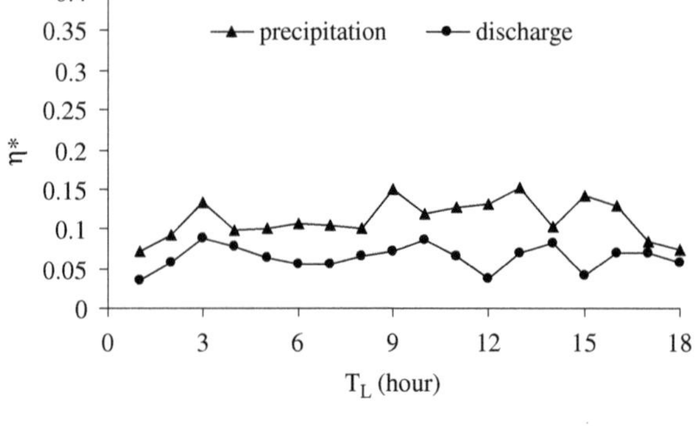

Fig. 3. Error of forecast based on MSM.

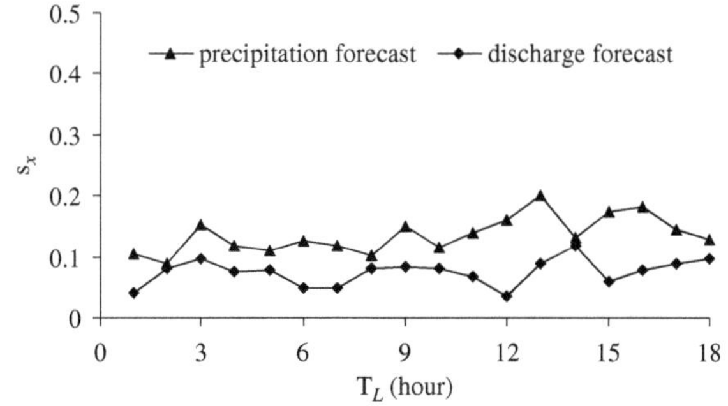

Fig. 4. Total uncertainty based on MSM.

It is understandable that the error of discharge forecast is smaller than precipitation forecast since the precipitation dynamic tends to change more dramatically then discharge. Discharge forecast also carried momentum from previous flow and corrected by the updating process.

The prediction is aiming peak discharge which occurs in the mid range of forecast, thus the beginning and the end lead time has lesser N. The results of time variation error and uncertainty had been suggested that filtering is necessary. Therefore filtering procedure is applied to the figures up to 12 hours lead time as needed. The method used quintiles method. The result of filtering showed on Fig. 5.

$$x_{0.25} + c_b(x_{0.25} - x_{0.75}) \leq \mu_X \leq x_{0.75} + c_b(x_{0.75} - x_{0.25}) \tag{11}$$

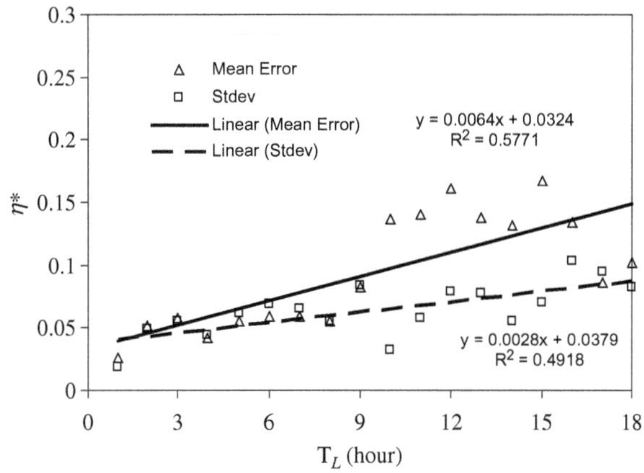

Fig. 5. Discharge mean forecast error and standard deviation.

Where:
$\mu_X$is expected value, $x_{0.25}$ is 25 % quantile, $x_{0.75}$ is 75 % quantile, and c_b is constant.

Pursuing stochastically approach, T distribution was used because it is good approximation for both Gaussian and skewed distribution in the condition of number of sample, N more then 30, Barrett and Goldsmith[14]. It is likely that N will be reached in operational scale. On present research N are about 5–20 samples depending on lead time.

$$\Pr\left\{m_X - \frac{t_{N-1,1-\alpha/2}\cdot s_x}{N^{0.5}} \leq \mu_X \leq m_X + \frac{t_{N-1,1-\alpha/2}\cdot s_x}{N^{0.5}}\right\} = 1-\alpha \quad (12)$$

Where:

$1-\alpha$ is confidence level (1 perfect confident, α = un-confidence value), m_X is mean, and s_x = standard deviation.

Based on Eq. (12), Fig. 6 shows confidence level of past event prediction for 90 % confidence. Two peak events were selected to reproduce reliability of the confidence figure.

Validation of the application in real event is conducted on event A4, A5. Most of the prediction is well validated, however a few out of its confidence level. This abnormality occurs on event A5 caused by miss-forecast of MSM precipitation. The lead time provided is 4 and 10 hours. The predicted discharge of 10 hour lead time far below threshold, it would not raise a warning. However after 6 hours the warning will be raised because the

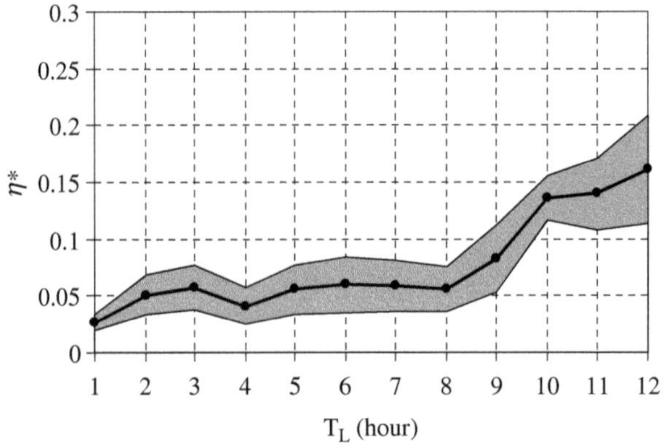

Fig. 6. Discharge prediction in 90 percent confidence level.

Table 3. Events Validation.

Event	Lead Time (hr)	Peak Discharge (m^3/s)	Discharge Prediction (m^3/s)	η^*
A4	4	185	160	0.04
	10	185	115	0.12
A5	4	240	200	0.07
	10	240	80	0.27

prediction is above threshold. This result had emphasized the importance of rapid issuance of forecast.

6. Conclusion

The error and uncertainty of discharge prediction is about half of the precipitation's emphasizes amount advantage of this particular model.

The result suggests that uncertainty of flood forecast mainly as a response of precipitation prediction. Filtered result shows the correlation between forecast lead time and uncertainty. Correspond with the magnitude prediction error of discharge and precipitation, uncertainty is higher on longer lead time.

An optimal lead time of the specific case enable to be estimated by constructed level of confidence. The figure may feed reliable warning; however greater exercise on events must be continuous to reach stronger confidence level.

References

1. A. De Roo, *et al.*, Development of European flood forecasting system, *International Journal of River Basin Management* **1** (2003) 49–59.
2. W. Micha, *et al.*, Flood forecasting and warning at the River Basin and at the European scale, *Natural Hazards* **36** (2005) 25–42.
3. Weisman, Skamarock, and Klemp, The resolution dependence of explicitly modeled convective systems, *Monthly Weather Review* **125** (1997) 527–548.
4. M. K. Lindell, C. S. Prater, R. W. Perry and J. Y. Wu, EMBLEM: An empirically-based large scale evacuation time estimate model, Texas A&M Univ. Hazard Reduction and Recovery Center, College Station, 2002.
5. T. M. Carpenter, K. P. Georgakakos, Impacts of parametric and radar rainfall uncertainty on the ensemble streamflow simulations of a distributed hydrologic model, *Journal of Hydrology* **298** (2004) 202–221.

6. H. Kato and A. Mano, Flood runoff model on one kilometer mesh for the Upper Chang Jiang River, *Proceeding of GIS & RS in Hydrology, Water Resources, and Environment*, 2003.
7. H. Tatesawa, Study on the transformation of flood hazard risk by the countermeasure facilities, Dissertation, Tohoku University, 2007.
8. H. Kardhana, H. Tatesawa and A. Mano, Flood forecasting based on distributed runoff model and numerical model prediction, *River Basin Management*, 2007, pp. 201–212.
9. S. Hattori, H. Chikaarashi and N. Takeuchi, Measurement of the rainfall interception and its micrometeorological analysis in Hinoki Stand, *Bulletin of the Forestry and Forest Products Research Institute* **318** (1982) 79–102.
10. J. D. Jabro, Estimation of saturated hydraulic conductivity of soils from particle size distribution and bulk density data, *ASAE* **35**(2) (1992) 557–560,
11. W. Collischonn, R. Haas, I. Andreolli and C. E. M. Tucci: Forecasting River Uruguay flow using rainfall forecasts from a regional weather-prediction model, *Journal of Hydrology* **305** (2005) 87–98.
12. M. Verbunt, M. Zappa, J. Gurtz and P. Kaufmann, Verification of a coupled hydrometeorological modelling approach for alpine tributaries in the Rhine Basin. *Journal of Hydrology* **324** (2006) 224–238.
13. J. Ewen, G. O'Donnell, A. Burton and E. O'Connell, Errors and uncertainty in physically-based rainfall-runoff modelling of catchment change effects, *Journal of Hydrology* **330** (2006) 641–650.
14. J. P. Barret and L. Goldsmith. When is N large enough? *The American Statiscian* **30** (1976) 600–71.

Advances in Geosciences
Vol. 17: Hydrological Science (2008)
Eds. Namsik Park et al.

MALAYSIA'S EXPERIENCE IN THE MONITORING OF INVESTMENT AND RESULTS OF WATER RESOURCES MANAGEMENT

SALMAH BINTI ZAKARIA
Director-General, National Hydraulics Research Institute of Malaysia (NAHRIM)
drsalmahz@gmail.com

LEE JIN
Vice-Chairman 2, Malaysian Water Partnership
drleejin@gmail.com

The current status of IWRM implementation in Malaysia, overview of development planning process and financial allocations together with monitoring of IWRM in Malaysia are first presented. This is followed by a case study review of Malaysia's 9th 5-year Development Plan (2005–2010) so as to provide a deeper understanding and appreciation of the current approach adopted by the Malaysian government in the formulation of development policies, project planning, budget allocation and macro monitoring of projects. Arising from the review it was highlighted that the water-related, sectoral developmental objectives are found in all the 5 developmental thrusts in the 9th Malaysia Plan. This is because water is an essential natural resource and is also an economic good. Thus, it was concluded that in order for the monitoring of investments in IWRM and results to be effective and useful to support policy formulation to achieve the goal of IWRM there is a need for a collation of all information reported under the different, water-related, sectoral developmental objectives in the 9th Malaysia Plan to be organised within an IWRM framework.

It was also concluded that a major IWRM challenge for Malaysia is how to make the transition from the current, narrow, sector-specific, financial and budgeting paradigm of the sectorial agencies to the "total cost" paradigm that involves taking into consideration and integrating the costs in other sectors affected by any projects proposed within a sector. The current, sectorial-focus approach, adopted in the 9th Malaysia Plan, logically measures the effectiveness of sectorial agencies by their ability to achieve their immediate sectorial goals and targets, most often with emphasis on infrastructure projects since the impacts of such projects are usually immediate and visible. However, the use of such approach alone, without taking into account the costs that are borne by other interfacing sectors, and also within the sector over the long term, is unsustainable.

1. Status of IWRM Implementation in Malaysia

The IWRM concept was first introduced in Malaysia in various forms in the early 1990s by water-related agencies, such as the Department of Irrigation and Drainage and is also promoted by the Malaysian Water Partnership (MyWP), an NGO formed in 1997, with a mission to promote the acceptance and implementation of IWRM in Malaysia. Thus, Malaysia has since accepted and adopted the concept of IWRM in its overall economic development. This is embodied in the country's Vision 2020 (1990–2020) when it was stated that "....Malaysia will be a developed nation by 2020, with the environment remaining pristine...." It is also verbalized in an overarching policy in the medium term plan, the Outline Perspective Plan 3 (OP3), 2001–2010, and the short term 5-year development plan, starting with the 8th Malaysia Plan, 2001–2005 and in the current Ninth Malaysia Plan (2006–2010).

In 2004, the government reorganized its ministries and set up 2 new key ministries dealing with water; namely, the Ministry of Natural Resources and Environment (MNRE) to look after resource management of water; and the Ministry of Energy, Water and Communication (MEWC) to undertake the services requirements including domestic and industrial water supply, sewerage and hydro power. Service requirement for agriculture remains with the Ministry of Agriculture (MOA) while MNRE is expected to look at the requirements for the environment as a whole. At the same time, approval of development project on-site remains with the state government authorities, such as the Local Governments. The Department of Local Government and the Department of Town and Country Planning (which prepares the National Physical Plan) are within the Ministry of Housing and Local Governments (MHLG). The biggest public developer is the Public Works Department (PWD) within the Ministry of Public Works (MPW).

Even though the governing machineries for water resources management and utilisation has been reorganised to facilitate the effective implementation of IWRM, the overall policy decisions regarding water development and financial matters still remains with the generic managers at both the federal and state levels. Thus, there is a need to sensitise top level decision makers on the IWRM concepts and approaches, and the need to have a streamlined National IWRM Strategy and Implementation road map.

2. Development Planning and Financial Allocation in Malaysia

The role for national development planning, financial allocations and monitoring in Malaysia are centralised under two special units in the Prime Minister's Department. They are the Economic Planning Unit (EPU) of Malaysia (http://www.epu.gov.my/) and the Implementation and Co-ordination Unit (ICU) of Malaysia (http://www.icu.gov.my/). The EPU is responsible for development planning and financial allocations to development projects, whereas the ICU is responsible for the detail monitoring of the implementation of the development projects. In view of the above, the monitoring of investment in IWRM and results arising from it lies within the purview of both the EPU and ICU.

2.1. *Development planning in Malaysia*

Planning in Malaysia is a two-way interactive process between the EPU, on the one hand, and the line ministries and agencies, on the other, as shown in Fig. 1 below. This top-down and bottom-up processes ensure that national policies and strategies are realized and development concerns at

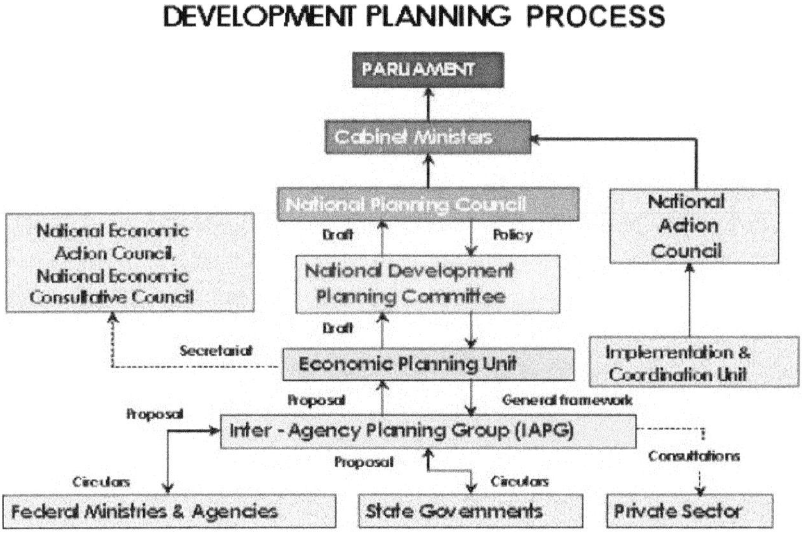

Fig. 1. Development planning process in Malaysia.

sub-regional level are fully integrated into the overall national development thrusts. Planning from the top which is confined to setting macro level parameters is determined in the context of the Inter-Agencies Planning Groups (IAPGs). The EPU is the secretariat for each of the IAPGs, whose work precedes the formulation of any development plan. Planning from the bottom, on the other hand, essentially involves the line ministries, agencies and the state governments which translate the sectoral master plans into specific programmes and projects. The EPU plays the key role in matching the micro-level programmes and projects with the macro-level plans for each economic sectors.

2.2. *Development financial allocation in Malaysia*

Development planning in Malaysia has a three-tiered cascading planning horizon, covering the long (10 and 20 years plan), medium (5 years plan) and short-term (annual) planning horizons. The *Outline Perspective Plan* sets the broad thrusts and strategies in the development agenda for the nation over a long-term. It sets the macroeconomic framework and long-term targets through which the vision of the society is to be achieved. The *medium-term plan* is for a five-year period. The plan is formulated in the context of and within the framework set by the Outline Perspective Plan. This, in a sense, is the key working document for the implementation of the Government's development programme. The five year development plans set out the macroeconomic growth targets as well as the size and allocation of the public sector development programme. In addition, it states the indicative role envisaged for the private sector.

The *Mid-Term Review* (*MTR*) of the five-year plan is carried out in the middle of the five-year cycles. The MTR is not only a stock taking exercise to determine whether the plan is being implemented in accordance with the stated targets and development schedules, but it also reviews macroeconomic and sectoral policies and strategies and makes adjustments if needed.

The third tier is the *annual budget*. This annual planning exercise is undertaken by the Ministry of Finance in conjunction with their annual budget preparation. Apart from taking into consideration the views of the private sector through budget dialogues, it also consults other central agencies through the IAPGs. The EPU plays the key role in determining the details of the annual development budget.

The Federal Government development allocation for every five year development plan will be provided for projects that will generate economic growth to achieve balanced development among states and regions. Development budget allocation is based on sectoral needs and current economic situations. The total budget allocation for five year development plan is derived from macro economic framework. The total allocated development budget will be distributed among four major sectors namely economic, social, security and general administration sectors. The economic and social sectors are provided the major portion of the total development allocation, followed by security and general administrative sectors.

3. Development Monitoring in Malaysia

3.1. *Monitoring of government investments in IWRM*

The ICU is the key government agency that monitors the implementation of the national development programmes. The main functions of the ICU are:

- Act as Secretariat to the National Development Council and National Development Committee
- Streamlines, manage, monitor and evaluate the basis and implementation of development projects
- Develop, manage, implement, monitor and evaluate the information system for national development projects

The mechanism that the ICU uses to monitor the implementation of the development projects defined in the national development plans is an integrated, computer monitoring system known as the SPP II. The SPP II was operationalised during the 8th Malaysia Plan. The SPP II is a computer system that links all federal ministries, departments and government agencies and requires the entry of project monitoring information at source. The system has increased the efficiency of project monitoring and ensures timely reporting of project progress implementation by removing the need for multiple information entry and compilation at different levels of the federal government bureaucracy.

3.2. *Monitoring of private sectors investments in IWRM*

There are no specific mechanisms to monitor the investments and results of private sectors and civil society's investments in IWRM in Malaysia.

4. Case Study — Review of 9th Malaysia Plan

4.1. *Case study review*

In order to have a deeper understanding and appreciation of the current approach adopted by the Malaysian government in the formulation of development policies, project planning, budget allocation and macro monitoring of projects a review of the 9th Malaysia Plan (2005–2010) was carried out. In order to set the stage for the definition of the development thrusts in the 9th Malaysia Plan there is also a section that provides an overview of the progress made by Malaysia in 5 key development areas over a period of 3 Malaysia Plan (6th, 7th and 8th) from 1991–2005. The 5 key areas are:

- The Economy
- Human Capital: Education, Innovation and Values
- Socio-Economic Equity
- Quality of Life
- Institutional Strength

The review on the progress made in the 5 key areas sets the stage for the definition of the 5 key development thrusts of the Plan. They are:

- First: To move the economy up the value chain
- Second: To raise the capacity for knowledge and innovation and nurture 'first class mentality'
- Third: To address persistent socio-economic inequalities constructively and productively
- Fourth: To improve the standard and sustainability of quality of life
- Fifth: To strengthen the institutional and implementation capacity

For each of the 5 development thrusts, the progress made in the 8th Malaysia Plan (2000–2005) for each sectoral developmental objectives under the thrusts were reported, before the prospects that is to be achieved during the period of the 9th Plan from 2006-2010 were described together with the institutional support and financial allocations.

From the review of the 9th Malaysia Plan, the water-related, sectoral development objectives under the 5 development thrusts of the Plan have been identified and are listed below:

Thrust 1 Development Objectives

- Strengthening agriculture and the agro-based industry
- Realising tourism potential

Thrust 2 Development Objectives

- Enhancing human capital
- Harnessing science, technology and innovation
- Women and development
- Empowering youth for the future
- Fostering family and community development

Thrust 3 Development Objectives

- Achieving growth with distribution
- Attaining balanced regional development

Thrust 4 Development Objectives

- Improving infrastructure, utilities and urban transportation
- Sustainable energy development
- Achieving better health
- Providing quality housing and urban services
- Promoting environmental stewardship

Thrust 5 Development Objective

- Efficient public service delivery system

It can be seen from the above lists that the water-related, sectoral developmental objectives cut across all the 5 developmental thrusts in the 9th Malaysia Plan. This is because water is an essential natural resource and is also an economic good.

4.2. *A possible framework for monitoring IWRM investments*

In view of the above, in order for the monitoring of investments in IWRM and results to be effective and useful to support policy

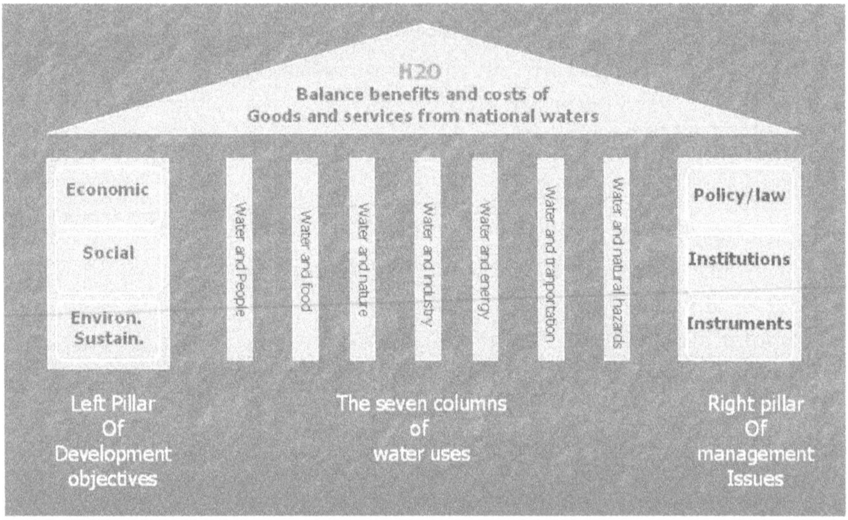

Fig. 2. Integrated Water Resources Management required for Management of the various facets of water and its uses.
Source: Mogens Dyhr Nielsen, 2006 — Personal Communications.

formulation to achieve the goal of IWRM there is a need for a collation of all information reported under the different, water-related, sectoral developmental objectives in the 9th Malaysia Plan to be organised within an IWRM framework.

A possible IWRM framework (Dyhr-Nielsen, 2006) is illustrated in Fig. 2, where the various facets of water uses have to be managed, through changes and fine tuning of the policy/law, institutions, and use of management instruments, to achieve an optimum balance of benefits between the 3 development objectives of economic progress, social equity and environmental sustainability. In this way, it is then possible to monitor the implementation of all the relevant, water-related projects under the respective sectoral developmental objectives to see how they are aligned to the IWRM Framework.

5. Conclusions and Recommendations

The challenge Malaysia faces in the water financing area is how to make the transition from the current, narrow, sector-specific, financial and budgeting paradigm of the sectorial agencies to the "total cost" paradigm that involves taking into consideration and integrating the costs in other sectors affected by any projects proposed within a sector. The current, sectorial-focus approach, adopted in the 9th Malaysia Plan, logically measures the effectiveness of sectorial agencies by their ability to achieve their immediate sectorial goals and targets, most often with emphasis on infrastructure projects since the impacts of such projects are usually immediate and visible. However, the use of such approach alone, without taking into account the costs that are borne by other interfacing sectors, and also within the sector over the long term, is unsustainable.

The achievement of the immediate goals and targets in one sector may contribute to increasing the negative impacts in other sectors, or "sew the seed" of a bigger problem within the sector over the longer term. This will logically lead to increased budgetary requests for infrastructure projects from the other affected sectors, to implement immediate remedial measures to mitigate the negative impacts, or for bigger budgets within the sector in the future to address the same, recurring problem in the longer term, respectively. Good examples of this can be seen in the relationship between the land use and development sector, with the sectors responsible for addressing the issue of increasing frequency and severity of urban floods, and increased pollution of river water quality and the health concerns of the public over the potable water supply derived from such river sources.

In view of the preceding discussions on the current status of IWRM implementation in Malaysia and the current approach in development planning, budget allocation and project monitoring, it is recommended that the following be implemented to move the IWRM agenda forward in Malaysia.

(a) Form a joint National IWRM co-ordinating committee among key IWRM sectorial agencies to provide the leadership to streamline and optimise the respective sectorial agencies' goals and to address the implementation and enforcement gaps so that the desired transition to the "total cost" paradigm can be made.
(b) Develop the details of the National IWRM Strategic Plan, based on the National Water Vision and National Framework for Action that was developed in 2000.
(c) Develop the IWRM Monitoring Framework for collating all information reported under the different, water-related, sectoral developmental objectives in the 9th Malaysia Plan.

In terms of regional cooperation and support needs, the following is recommended:

(a) Support a regional pilot project to develop a computerised IWRM Monitoring Framework system to collate information reported under the different, water-related, sectoral developmental objectives. This is because the sectoral approach to development planning is the normal approach adopted by almost all countries and is effective in monitoring and meeting sectorial objectives. What is needed is a computerised IWRM Monitoring Framework system that can complement the sectorial monitoring approach using the same information reported in the individual sectors.
(b) Support the development of a network for the exchange of experiences in monitoring IWRM investments among IWRM practitioners in the region, through support of capacity building and awareness activities on the use and contribution of IWRM monitoring case studies to the GWP-SEA IWRM Toolbox website (gwpsea.multicentric.com) and the AguaJaring (SEA) IWRM expertise database website (www.aguajaring.org).

Advances in Geosciences
Vol. 17: Hydrological Science (2008)
Eds. Namsik Park *et al.*

WASTE LOAD ALLOCATION MODELING BASED ON A NEW EQUITY MEASURE

S. MOHAN and K. PAVAN KUMAR*
Professor and Research Scholar, Environmental and Water Resources Division,
Department of Civil Engineering,
Indian Institute of Technology, Madras 600036
**pavankumar1978@gmail.com*

In this paper, a novel approach to measure equity among the polluters through a waste load allocation model is presented. The concept behind arriving at the proposed new equity measure is that a polluter may be penalized according to the amount of DO deficit that the polluter's waste load causes in a river within the reach, before other polluter's load joins. The waste load allocation model is developed as a multiobjective model with two objectives of treatment cost minimization and equity maximization. The developed model was applied to Thambraparni river in Tamil Nadu, India. Since the flow in the river varies from month to month, the monthly tradeoff curves between treatment levels versus equity have been presented.

1. Introduction

Water quality management is part of an all encompassing water resources management. Waste load allocation (WLA) model refers to the problem where each polluter is allocated a certain maximum amount of waste load (measured in terms of kg BOD_5/day) that he can discharge into the river so that the water quality (measured in terms of mg/l of DO available) within the reach before other polluter's load joins, and meets the water quality standards as prescribed by a Pollution Control Agency (PCA). The polluter can limit the waste load discharge to an allocated waste load by providing certain level of treatment. This involves treatment cost on part of the polluter. The more is the required treatment level, higher is the treatment cost. Hence, it is obvious that a polluter will always try to minimize his treatment cost by providing a minimum treatment level to his waste load (which means discharging more waste load into the river). However, this approach conflicts with the interest of PCA which will prefer the polluter to minimize the total effluent waste load (i.e. provide higher treatment to his influent) so that higher levels of water quality standards can be met within the reach. The conflict between PCA and a polluter can be resolved partially by developing tradeoff curves between treatment levels and desired DO levels within the reach. The tradeoff curves can be obtained by solving a multiobjective waste load allocation model. However, in a waste load allocation model, the conflict is not just between a polluter and PCA, but there is conflict among the polluters also. The conflict among the polluters occurs regarding the question of what should be the treatment cost share of a polluter within the overall treatment cost of maintaining the desired water quality in the river. The burden of resolving these conflicts among the polluters should be one of the primary objectives of PCA, apart from the objectives of treatment cost minimization and water quality improvement. One way to resolve the conflict among the polluters is by introducing the concept of equity into the waste load allocation model formulation.

The importance of equity in water quality management was recognized as early as 1960's when Johnson[8] postulated two criteria to be followed for an equitable distribution of water quality maintenance cost among the polluters. The first criterion was that "equals should be treated equally" and the second criterion was "non-equals should be treated differently". The earliest way of addressing the issue of equity among the polluters was by suggesting a uniform treatment for all the polluters. This was however criticized by Johnson[8] and Brill *et al.*[2] as being way too costlier and also for being too much burdening on smaller polluters. Even though equity was recognized as an important aspect to be considered in water quality management problems, its applications in waste load allocation models have been limited. This can be seen from the fact that from late 1960's to until now there have been very few works which have equity as one of the objectives. Even when equity has been used in waste load allocation model, the mathematical formulation for equity measure has been adopted from some other domain and applied to waste load allocation problem. Hence it is felt that there is a need to develop an equity measure from the context of waste load allocation taking into consideration of various complexities that are specific to the problem.

In this paper, a new equity measure based on the principle that a polluter should be held responsible only for the damage his waste load causes within his reach rather than the amount of waste load he is discharging into the stream has been proposed and applied. The damage in this context is taken as the reduction in available dissolved oxygen (DO) at the end of the reach. It is assumed that the marginal cost of effluent treatment is linearly proportional to the marginal cot of improving DO levels within the reach. The detailed mathematical formulation for the equity measure is given in subsequent sections.

2. Equity Measures in Waste Load Allocation Model

Before discussing the theory behind proposed equity measure, it is necessary to discuss about various equity measures that have been used in WLA problems. This will help the reader in understanding the difference between the existing equity measures and the proposed one.

One of the earliest ways of addressing the issue of equity among the polluters was to adopt a uniform treatment program (UTP). In a uniform treatment program, all the polluters are required to treat their effluent to a uniform treatment level.[8] The mathematical formulation for a uniform treatment program can be written as:

$$\min s \tag{1}$$

$$x_i - s \leq 0 \tag{2}$$

However, this formulation resulted in high treatment cost and it also does not differentiate between polluters with different waste load contribution. Owing to the aforementioned limitation with UTP, Johnson[8] proposed two criteria to be considered for achieving equity among the polluters. They were: (i) a program must treat equals in a like manner. Equality here is defined as individuals being the same on the basis of the criterion that marginal damages caused by a unit waste input from any individual are equal or approximately equal and (ii) polluters who are different should receive different treatment i.e. a waste polluter whose input causes no damage should not be assessed at the same costs as a polluter whose waste cause damage. Though it was recognized that equity is an important issue in water quality management, it was generally consigned to a constraint but not as a separate objective. On the other hand, Brill *et al.*[2] suggested that equity should be treated as an objective rather than as a mere constraint in water quality management model. They proposed three equity measures, as listed below:

a) A Summed Deviation Equity Measure (SDEM)

$$\min |u - v| \tag{3}$$

$$u + x_i - v - x_{avg} = 0 \tag{4}$$

$u =$ deviation above x_i; $v =$ deviation below x_i
$x_{avg} =$ average treatment level
$u \geq 0$; $v \geq 0$;

b) A Range Equity Measure (REM)

$$\min\ (x_{\max} - x_{\min}) \tag{5}$$

$$x_{\max} = \max[x_i] \quad \forall i \in n \tag{6}$$

$$x_{\min} = \min[x_i] \quad \forall i \in n \tag{7}$$

n = number of polluters

c) Maximum Efficiency Equity Measure (MEEM)

$$\min\ x_{\max} \tag{8}$$

$$x_{\max} = \max[x_i] \quad \forall i \in n \tag{9}$$

n = number of polluters

All the above objective functions were subjected to the constraint on treatment cost. The principle behind all the three equity measure was same i.e. to distribute the treatment cost among the polluters in such a way that the treatment cost for each polluter is as near as possible to an overall average treatment cost. These equity measures would be more suitable when there is not much difference among the polluters in terms of the waste load magnitude. However, when the difference among the polluters is large (for example when the stream receives waste load from mainly large and small scale industries) the use of SDEM or REM equity measure will be disadvantageous to small polluters. In some cases when the overall treatment cost of the system is largely dependent on a single large polluter, MEEM would be better than other two equity measures. Other than the aforementioned equity measures, Brill *et al.*[2] also suggested zonal effluent treatment as an alternative treatment program for achieving equity among the polluters, provided the polluters are willing to cooperate among themselves to form suitable zones (either geographical or based on waste load magnitude). But in practice it is very difficult to achieve any cooperation among polluters to demarcate the zone. Also zoning may also lead to some socio-political complexities like when polluters who as individuals may not have much socio-political influence can become stronger when formed as a group. However zoning can be helpful for small scale polluters since more than as individuals they will be better served when as a group. Arikol and Basak[1] proposed an equity measure based on the principle that the treatment cost for a polluter should be proportional to his waste load contribution. This is somewhat different from penalizing a polluter according to the damage (i.e. depletion of DO) that he causes within his reach. Mathematically it can be expressed as:

$$\min{(x_i - x_t)^2} \tag{10}$$

$$x_t = \frac{\left(\sum_{i=1}^{n} f_i x_i\right) L_i}{f_i \left(\sum_{i=1}^{n} L_i\right)} \tag{11}$$

x_i = treatment level for polluter i
x_t = target treatment level for a polluter
f_i = cost associated with unit waste removal rate
L_i = BOD level of the i-th discharger when no waste removal is provided

Hathhorn and Tung[7] developed a multiobjective waste load allocation model, in which one of the objectives was equity maximization. They proposed two equity measures,

a) Minimization of difference between treatment levels among any two polluters (Equal Percent Removal, EPR)

$$\min{|x_i - x_j|} \tag{12}$$

b) Minimization of difference between effluent waste loads among any two polluters (Equal Effluent Concentration Removal, EECR)

$$\min{|B_i - B_j|} \tag{13}$$

B_i, B_j are the waste loads for polluters i and j after treatment

Burn and Yulianti[3] proposed an equity measure such that increased fairness occurs when the ratio of the removal fraction at a source to the average removal level is as close as possible to the ratio of the pollutant input at a source to the average pollutant input. Mathematically it is expressed as:

$$\min{EQ} \tag{14}$$

$$EQ = \sum_{i=1}^{n} \left| \frac{x_i}{x_{avg}} - \frac{W_i}{W_{avg}} \right| \tag{15}$$

From the brief review of existing equity measures in waste load allocation problems, they can be classified as:

1. Equity measures in which a target level is fixed (mostly an average treatment level) to which all the treatment levels are aspired to reach. SDEM, UT program, Zonal effluent scheme, and equity measure proposed by Arikol and Basak belong to such a category. This type of equity measure can be referred to as 'reference level' or 'target level' equity measures.
2. There are those equity measures where rather than trying to achieve some idealized target level the difference among the polluters is minimized. REM, MEEM, EPR, and EECR belong to this category. They can be termed as 'equalizing' equity measures.
3. Then there are equity measures which are based on the concept that a polluter should be penalized according to the share of his waste load. They can be called as 'proportionate' equity measures.
4. In all the above described equity measures, the polluter is treated as external to the river system i.e. the waste load distribution among the polluters is seen as independent to the effect on water quality caused by the effluent of an individual polluter.
5. An equity measure in the context of waste load allocation problem will be more realistic if we define it in such a way that the polluter is penalized according to the damage his waste load causes within his reach. There is no equity measure that have been reported considering this fact. Hence in the present work a new equity measure is being proposed which is based on the above stated principle.

3. Mathematical Formulation for the Proposed Equity Measure

The proposed equity measure is based on the principle that a polluter should be penalized according to the damage (decrease in DO level) caused by the waste from industry within a given reach. There are however the following assumptions are made in developing the new equity measure.

a) It is very difficult in general to assess the economic loss caused due to reduced water quality within the river. However if it is assumed that dissolved oxygen (DO) is the major water quality parameter being considered, then the damage to water quality within a reach can be related to the decrease in DO levels.
b) The marginal treatment cost for a polluter is linearly proportional to the cost involved in improving DO level within a river reach by a unit DO level.

c) Each polluter is responsible only for his reach i.e. he is not responsible for the deterioration of water quality within his reach due to the effluent from polluters on the upstream side, and also the pollution caused in the downstream of the reach.

Streeter-Phelps equation is used as water quality model. The standard one dimensional, steady state Streeter-Phelps equation is given by:

$$D_{def} = \frac{k_d}{k_a - k_d} L_m \left(e^{-k_d t} - e^{-k_a t}\right) + d_i \left(e^{-k_a t}\right) \tag{16}$$

where the above terms are:

D_{def} is the DO deficit at any time t in the river in mg/l.

k_d and k_a are the de-oxygenation and re-aeration coefficients respectively in d^{-1}.

L_m is the BOD of the river at the point of discharge and is given by simple mass balance equation in mg/l

$$L_m = \frac{Q_r L_r + Q_e L_e}{Q_r + Q_e} \tag{17}$$

Q_r and Q_e are the river discharge and effluent discharge in m^3/sec.

L_r and L_e are the initial BOD of the river and BOD of the effluent entering the river, respectively.

d_i is the initial DO deficit in the river in mg/l and $L_e = L_p(1 - x_i)$, where L_p is the BOD of the influent and x_i is the treatment level provided by the polluter i.

t is the time of travel from the point of discharge to any given downstream point.

From the above equation DO deficit due to a polluter's waste load can be given by:

$$DO_d = \frac{Q_e L_e}{Q_r + Q_e} \left(\frac{k_d}{k_a - k_d}\right) \left(e^{-k_d t} - e^{-k_a t}\right) \tag{18}$$

$$= f A L_e$$

$$f = \left(\frac{k_d}{k_a - k_d}\right) \left(e^{-k_d t} - e^{-k_a t}\right) \tag{19}$$

$$A = \frac{Q_e}{Q_r + Q_e} \tag{20}$$

The DO profile for a polluter due to his effluent BOD is shown in Fig. 1. If C (\$/(mg/l)) is the cost of improving DO level in the river by one unit, then

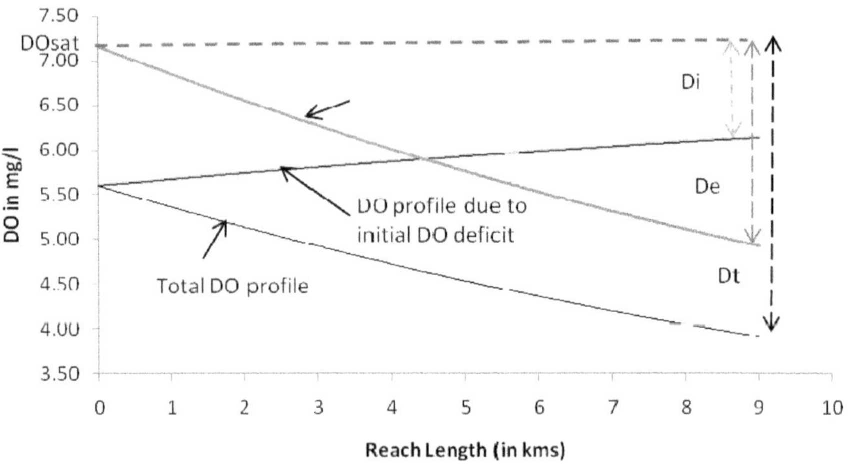

Fig. 1. Dissolved oxygen profile for a polluter due to his waste load.

the total cost of bringing DO level at the end of the reach to the original DO level for a polluter i (assuming originally the DO was at saturation level) can be given by:

$$M_i = C_i(DO_{sat} - D_e)_i \tag{21}$$

$$M_i = C_i(DO_d)_i \tag{22}$$

The total cost of improving DO level can be given by:

$$M_t = \sum_{i=1}^{n} C(DO_d^i) \tag{23}$$

$$n = \text{number of polluters}$$

DO_d^i is the DO deficit at the endof reach due to waste load of polluter i

Assuming that the cost of improving the DO level within the reach for a polluter is linearly proportional to the treatment cost, one can then define the equity measure as:

$$EM = 1 - \left| x_i - \frac{M_i}{M_t}\left(\sum_{i=1}^{n} x_i\right)\right| \tag{24}$$

Once the formulation for equity has been developed, the waste load allocation model can be modeled as a multiobjective optimization model with treatment cost minimization and equity maximization as objective functions. The multiobjective model formulation and solution are given in the next section.

4. Multiobjective Waste Load Allocation Model

The minimization of treatment cost and maximization of equity among the polluters are conflicting objectives. Maximizing equity results in increased treatment cost, which may be undesirable from economic point of view whereas minimizing treatment cost leads to reduced equity among the polluters, which may not also be desirable for PCA. Hence, it would be appropriate to solve the waste load allocation problem as a multiobjective model so that tradeoff curve can be obtained between treatment cost and equity. From this tradeoff curve, a suitable solution point that satisfies both polluters and PCA could be selected. Although there are many traditional and advanced methods available to solve multiobjective models, genetic algorithm (GA) has been selected as a solution method to solve the problem due to some of its inherent advantages over other multiobjective solution methods.[5,6] GA has garnered considerable attention in solving multiobjective waste load allocation models. One of the earliest applications of GA to waste load allocation was done by Chen and Chang (1996, 1998). Since then there have been a gradual increase in the application of GA in multiobjective waste load allocation models. Some of the notable works of GA application in waste load allocation are by Vasquez *et al.*,[12] Burn and Yulianti,[3] Cho *et al.* (2004), Murty *et al.*,[9] Tufail and Ormsbee.[11] In all these works, it has been established that genetic algorithm is an effective optimization tool for solving a multiobjective waste load allocation model. Thus GA has been used in the present work for solving the multiobjective waste load allocation model.

The mathematical formulation for the multiobjective waste load allocation model can be written as:

$$Z = [z_1(x), z_2(x)] \tag{25}$$

$$z_1(x) = \min \sum_{i=1}^{n} x_i \tag{26}$$

$$z_2(x) = \max (E_i) \tag{27}$$

The constraints for the model are:

$$c_i - a_i x_i \leq DO_{sat} - DO_{des} \tag{28}$$

$$x_{\min} \le x_i \le x_{\max} \tag{29}$$

$$x_i \ge 0 \; for \; i = 1, 2, \ldots\ldots n \tag{30}$$

x_i is the treatment level for a polluter i

$n$is the number of polluters discharging effluent into the stream

$z_1(x)$ and $z_2(x)$ are the objective functions defining treatment cost minimization and maximization of equity respectively.

DO_{sat} and DO_{des} are the saturation DO level and desirable DO at the end of the reach (in mg/l).

x_{min} and x_{max} are the minimum and maximum treatment levels.

c_i and a_i are the transfer coefficient parameters, which are obtained from Streeter-Phelps water quality model. They are function of river discharge, re-aeration coefficient, de-oxygenation coefficients, temperature, and travel time in the reach.

Since GA optimizes a fitness function of an objective function, rather than the objective function itself, it is necessary to define a proper fitness function. The fitness function used in our model is given by:

$$F = \left(\frac{\sum_{i=1}^{n} x_i}{X} \right) + \left(\frac{\sum_{i=1}^{n} E_i}{n} \right) \tag{31}$$

F is the fitness function

x_i is the treatment level for an individual polluter

n is the number of polluters

$$X = \sum_{i=1}^{n} x_i$$

The details of the GA parameters at which the model showed convergence to a constant solution, is given in Table 1.

Table 1. Values of GA parameters adopted.

GA Parameter	Value
Population size	20 times the number of decision variables
Number of generations	320
Crossover probability	0.9
Mutation probability	0.1
Proportion of best Pareto-optimal set	0.25

5. Application of Model to the Study Area

The developed model was applied to a case study of Tambraparni river in Tamil Nadu. The river for its major part flows through the district of Tirunelveli and a small part of its catchment also lies in the district of Tuticorin. The total length of the river from its origin at Western Ghats to its fall in Bay of Bengal is about 120 km. Tambraparani River Basin lies between geographic co-ordinates lat. 8° 26' 45" N to 9° 12' 00" N and long 77° 09' 00" E to 78° 08'30" E. Along its course it is joined by numerous major and minor tributaries, the most notable among them being Servalar, Manimutthar, Gadana, Pachaiyar, and Chittar. The slope of the river is towards southeast, with an average slope of about 4.5 m/km.

The major sources of pollution in the river are, industrial effluent, domestic sewage and agricultural runoff. There are only a few major industries along the sides of the river, the major ones being Madhura Coats at V. S. Puram, and Sun Paper Mills at Cheranmahadevi. The major share of pollution in the river is due to uncontrolled disposal of domestic sewage from nearby cities or towns and non point source pollution from agricultural runoff. There is no treatment plant existing for any of the major or minor towns along the river.

This makes the effect of domestic sewage on the river water quality even more pronounced. Most of the small towns and villages in the basin area have no sewerage disposal facilities. The DO in the river varies from a lowest value of 1.7 mg/l during the month of March to a highest value of 6.8 mg/l during the monsoon season of August and September whereas the BOD varies from 0.13 mg/l to 10.2 mg/l. Figure 3 shows various towns located along the river which are discharging their domestic sewage directly into the river. In the development of models for water quality certain assumptions need to be made and they are:

1. The population details obtained from Tamil Nadu Census Board were available up to the year 2001. However the present model was applied to the conditions for a projected population as on year 2010 with an assumption of a linear growth rate of 2% for all towns (TWAD, 2003).
2. The daily per capita water consumption is taken as 90 liters per capita per day (TWAD, 2003) and the waste water volume that reaches the river is assumed to be 80% of 90 liters per capita per day (TWAD, 2003).
3. The BOD load per person is taken to be about 0.08 kg/day.

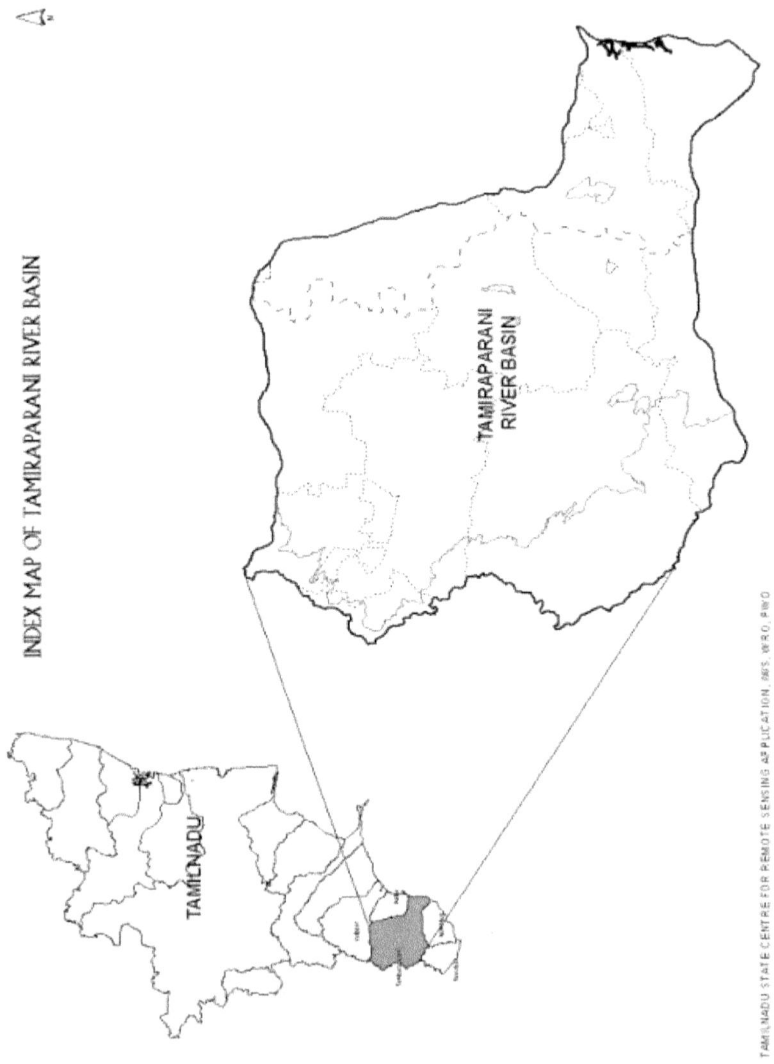

Fig. 2. Layout map of the Tamraparni river basin.

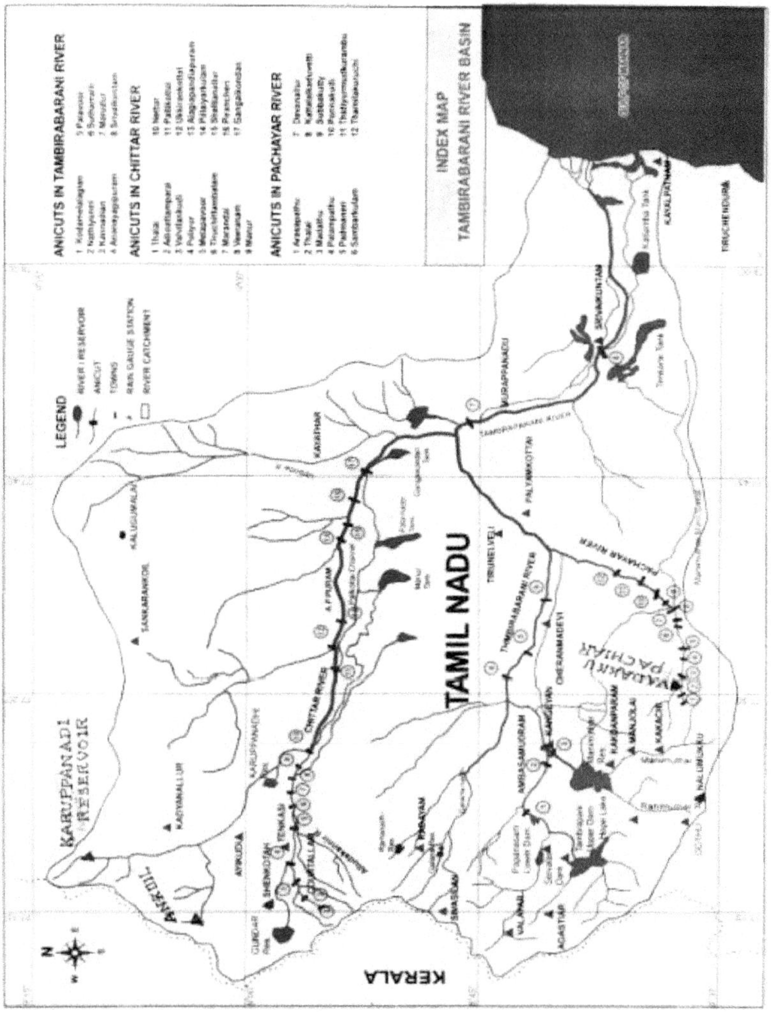

Fig. 3. Index map of Tamrapa-ni river basin with details of location of various towns and anicuts.

Table 2. Population detail, BOD load (kg/d), and effluent discharge for the towns discharging effluent directly into the river.

Polluter	Town	Location (km)	Population	BOD Load (kg/d)	Q_{avg} (m^3/sec)	Q_{max} (m^3/sec)
P1	V. S. Puram	13	97,258	7,781	0.0605	0.309
P2	Ambasamudram	22	49,376	3,950	0.041	0.223
P3	Cheranmahadevi	39	12,9813	10,385	0.0286	0.165
P4	Melapalayam	52	99,525	7,962	0.083	0.402
P5	Palayamkottai	55	14,2250	11,380	0.1185	0.541
P6	Thirunelveli	58	49,2370	39,389	0.41	1.522
P7	Thatchanallur	82	97,353	7,788	0.0811	0.394
P8	Srivaikundam	96	25,312	2,025	0.0211	0.128

6. Results and Discussion

The multiobjective waste load allocation model was solved with treatment cost minimization and equity maximization as objective functions. The treatment cost versus equity curves were obtained for different desirable DO levels. The Pareto-optimal curves are shown in Fig. 4. From the figure, it can be observed that as the desirable DO level decreases the Pareto-optimal front recedes further towards the ordinate. This means that as the desirable DO level decreases, the total treatment cost decreases for the same level of

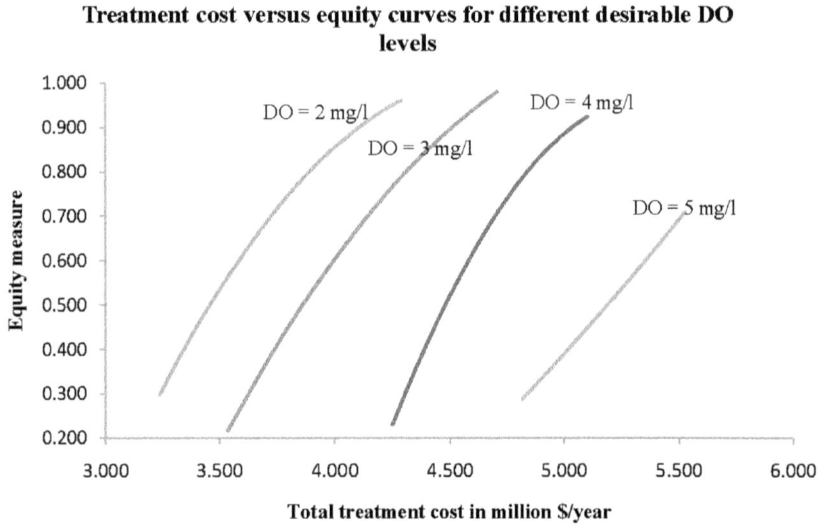

Fig. 4. Pareto-opitmal curves for total treatment cost versus equity for different desirable DO levels.

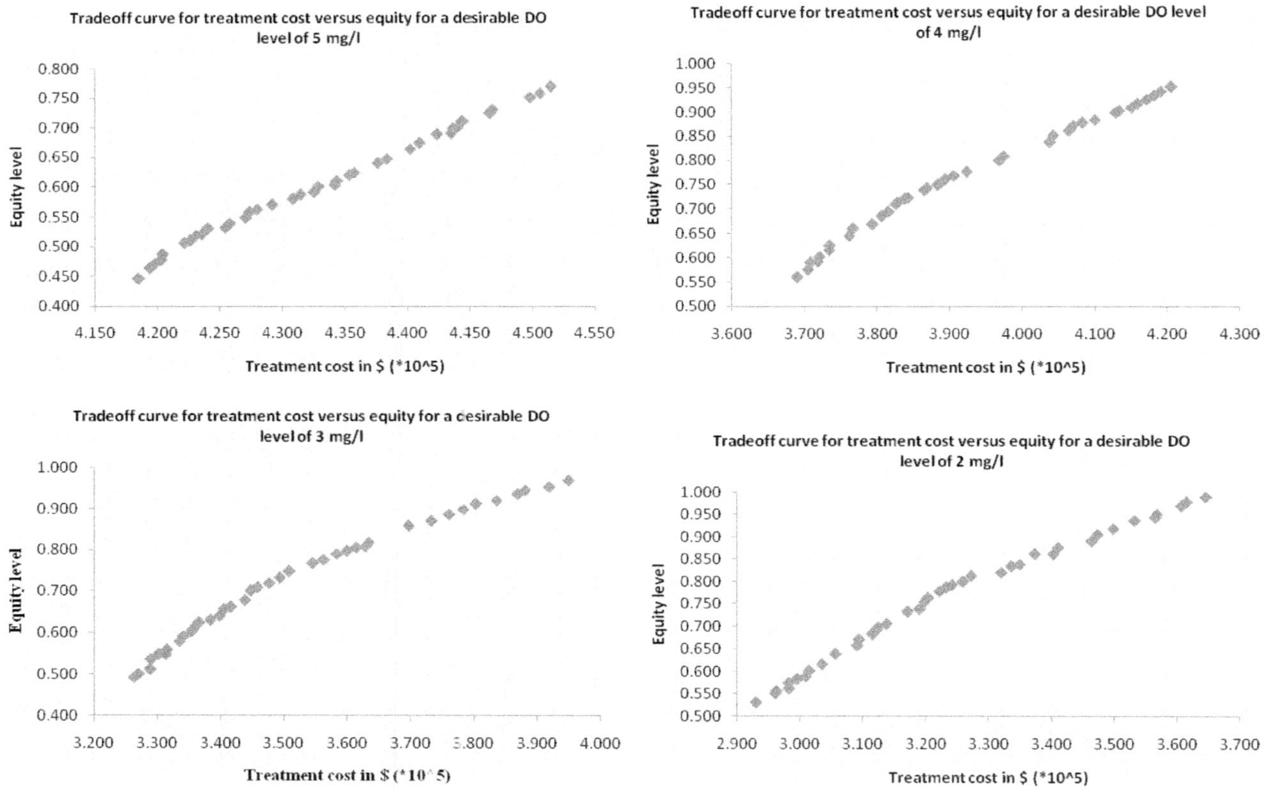

Fig. 5. Tradeoff curves equity versus treatment cost for different DO levels.

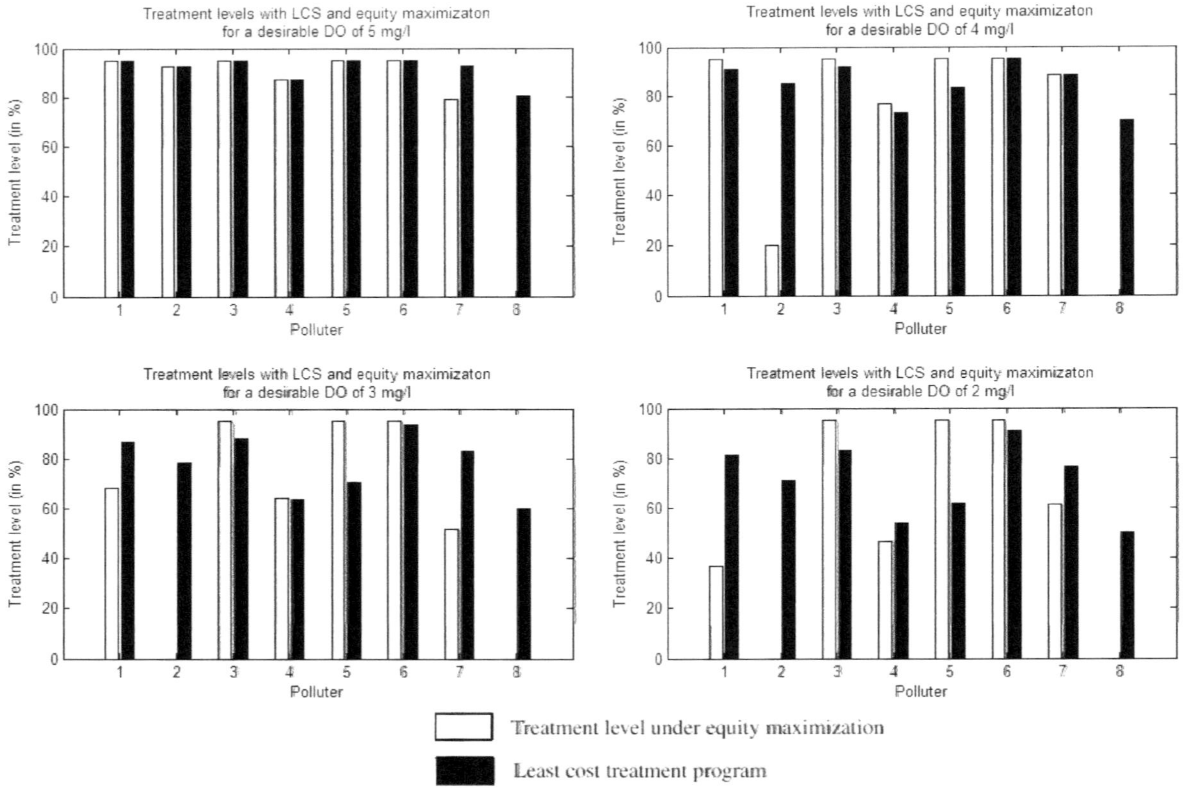

Fig. 6. DO levels at the end of a reach under cost minimization program and equity maximization program.

equity measure. For example in Fig. 4, for an equity measure of 0.6, the total treatment cost under 2 mg/l, 3 mg/l, 4 mg/l and 5 mg/l of desirable DO levels is 3.567, 3.978, 4.578 and 5.347 million $/year, respectively. The set of figures in Fig. 5 shows the tradeoff curve between treatment cost and equity level under different desirable DO levels. The decision maker can select an appropriate solution point from the curves depending on his desirable DO levels and equity satisfaction level. The set of figures in Fig. 6 gives the bar chart for DO levels at the end of the reach under cost minimization and equity maximization programs.

In order to show how the proposed equity measure is different from the existing equity measures, it would be appropriate to compare the proposed equity measure with an existing equity measure which has appeared recently in the waste load allocation literature and also which has been cited more frequently in the literature. Based on the aforementioned criterion, the equity measure as proposed by Burn and Yulianti was selected for the comparison and also to highlight some special features of the proposed new equity measure. Tables 3 and 4 lists the treatment levels for each polluter

Table 3. Treatment level for each polluter under various waste load allocation models and for different desirable DO levels.

	Desirable DO of 5 mg/l			Desirable DO of 4 mg/l		
Polluter	Least cost solution	Treatment level with equity measure as proposed by B&Y*	Treatment level under new proposed equity measure	Least cost solution	Treatment levels with equity measure as proposed by B&Y*	Treatment level under new proposed equity measure
P1	95.0	95.0	95	95	84.5	91
P2	92.7	93.1	92.6	20.1	73.4	85.1
P3	95.0	94.8	95	95	87.9	92
P4	87.1	93.3	87.1	76.5	84.8	72.8
P5	95.0	95.0	95	95	88.8	83
P6	95.0	95.0	95	95	92.8	95
P7	79.1	93.2	93	88.04	84.5	88.32
P8	0.0	78.0	80.2	0	58.7	69.78
Total treatment cost in million $/year	4.814	5.556	5.524	4.25	4.992	5.1
Equity measure	0.152	0.804	0.717	0.296	1	0.951

Table 4. Treatment level for each polluter under various waste load allocation models and for different desirable DO levels.

Polluter	Desirable DO of 3 mg/l			Desirable DO of 2 mg/l		
	Least cost solution	Treatment level with equity measure as proposed by B&Y*	Treatment level under new proposed equity measure	Least cost solution	Treatment levels with equity measure as proposed by B&Y*	Treatment level under new proposed equity measure
P1	68.3	77.4	86.74	36.8	70.3	81.4
P2	0	63.4	78.7	0	54.6	71.28
P3	95	82	88.2	95	76	83.3
P4	64	77.7	63.56	46.5	70.83	53.8
P5	95	83.3	70.81	95	77.6	618
P6	95	89.08	93.7	95	85	90.86
P7	51.7	77.38	83.07	61.3	70.4	76.6
P8	0	47.1	60	0	38.2	50.1
Total treatment cost in million $/year	3.534	4.502	4.71	3.24	4.09	4.28
Equity measure	0.317	1	1	0.344	1	1

for different desirable DO levels, when waste load allocation model is solved with proposed new equity measure (PNEM) and Burn and Yulianti's[3] equity measure (BYEM). From the results shown in Tables 3 and 4, it can be seen that except for polluters P4 and P5, for all other polluters the treatment level with PNEM is higher than the one with BYEM. For polluters P4 and P5, there is a considerable decrease in treatment levels under PNEM. It is also interesting to note that since polluters P1, P4 and P7 are discharging about same magnitude of waste load, hence using the BYEM they all are required to provide about same level of treatment. This however does not take into account the fact that though the waste load magnitude for all the polluters P1, P4 and P7 is nearly same but the reach conditions for the polluters is different. The waste load allocation with PNEM is able to consider this difference in reach conditions among the polluters. This can be seen from comparing the treatment levels for the polluters P1, P4 and P7. From comparison it can be seen that the treatment level for the polluters under PNEM scheme is not equal and it varies accordingly with reach parameters like river discharge, reach length, k_a

(re-aeration coefficient), k_d (de-oxygenation coefficient). Hence the PNEM is more dynamic in nature.

References

1. A. M. Arikol, and N. Basak, An equity approach to stream water quality management, *European Journal of Operational Research* **20** (1985) 182–189.
2. E. D. Brill, Jr., C. J. Liebman, and S. C. ReVelle, Equity measures for exploring water quality management alternatives, *Water Resour. Res.* **12**(5) (1976) 845–851.
3. D. H. Burn, and J. S. Yulianti, Waste-load allocation using genetic algorithms, *J. Water Resour. Plan. Manage.* **127**(2) (2001) 121–129.
4. Chief Engineer Micro-level study of the Tambraparni river basin, Indian Water Studies, Public Works Department, Government of Tamil Nadu, India, 2005.
5. K. Deb, *Multi-objective Optimization Using Evolutionary Algorithms* (Wiley Student Ed., 2005).
6. D. E. Goldberg, *Genetic Algorithms in Search, Optimization, and Machine Learning* (Addison-Wesley, Reading, Mass, 1989).
7. W. E. Hathhorn, and Y. K. Tung, Bi-objective analysis of waste load allocation using fuzzy linear programming, *Water Resour. Manage.* **3** (1989) 243–257.
8. E. L. Johnson, A study in the economics of water quality management, *Water Resour. Res.* **3**(2) (1967) 291–305.
9. Y. S. R. Murty, K. Srinivasan, and B. S. Murty, Multiobjective optimal waste load allocation models for rivers using nondominated sorting genetic algorithm-II, *J. Water Res. Plan. Manage.* **132**(3) (2006) 133–143.
10. H. Streeter, and E. Phelps, U.S. Public Health Service Bulletin, No. 146, 1925.
11. M. Tufail, and L. E. Ornsbee, Genetic algorithm for water quality management in an urban watershed, Genetic and Evolutionary Computation Conference, 2006.
12. J. A. Vasquez, H. R. Maier, B. J. Lence, B. A. Tolson, and R. O. Foschi, Achieving water quality system reliability using genetic algorithms, *J. Environ. Eng.* **126**(10) (2000) 954–962.

Advances in Geosciences
Vol. 17: Hydrological Science (2008)
Eds. Namsik Park *et al.*

THE IMPACT OF WEATHER VARIABLES ON SUMMER WATER USE IN THE CITY OF TORONTO

ROMILA VERMA
Research Associate, Canadian Institute for Environmental Law and Policy, Toronto, Ontario
romilav@rogers.com

The present study examines the impact of daily rainfall and temperature on the daily per capita municipal water use in the City of Toronto for the summer months (May–August). Water use peaks between the months of May to August and starts declining after September. Water use does not fall below 500 ML irrespective of rainfall or temperature i.e. a "base" component of water use is not related to weather variables. Daily water use is fairly low and steady when the temperature is below 15°C. However, it increases when the daily temperature is over 15°C and peaks between the temperatures of 20 to 30°C. In the next stage of model development, multiple regression is used. Rainfall explains three per cent of the variation. The regression coefficient of rainfall is – 3.57, which is statistically significant, and has the expected inverse relationship with water use. For every additional mm of rain, on average, 3.7 ML less of water will be used. Temperature however, explains 32% of the variation. For every additional degree in temperature, on average, an extra 13.79 ML of water will be used. The R Square for both temperature and rainfall is 0.36 and the models are significant at 99% confidence level. Water use modelling plays an important role in urban water resource planning and management. Climate change can affect three major aspects of water- its availability, demand and quality. To mitigate and adapt to the increasing temperature, the City of Toronto needs to develop and implement conservation policies and techniques.

1. Introduction

Studies have indicated that the earth's climate is undergoing a major change. The average global temperature is increasing gradually. This, in conjunction with rapid population growth, increased urbanization and industrialization, will very likely have a significant impact on municipal water use. It is, therefore, essential to address the issue of water supply and demand vis-à-vis climate change. Municipal water supply is a high priority in the City of Toronto.

Several socioeconomic and climatic variables affect water use levels in an urban area. Socioeconomic variables such as population, income, the price of water, cost of wastewater treatment and housing characteristics can impose long-term changes in urban water demand patterns (Grima, 1972); while weather variables such as precipitation and temperature may induce short-term changes (Miau, 1990). These weather-related factors gain new significance in the context of climate warming because they may provide a way to predict the changes in municipal water use with increasing global warming. The role of weather variables on municipal water use is the focus for this research.

2. Research Questions and Objectives

In order to analyze the impact of weather variables on water use, temperature and rainfall are considered as independent variables. The Following question is explored in this study:

What is the role of weather variables, i.e. temperature and rainfall in influencing municipal water use, particularly during the peak months?

The research objectives include:

- To examine the effects of weather variables on daily water use per capita in the City of Toronto during the growing season (May–August).
- To propose and test a statistical model of the impact of weather variables on water use in the City of Toronto during May through August.

The data on daily rainfall and temperature for eight years (1988–1995) was obtained from Environment Canada records. Daily total water use for the City of Toronto for eight years (1988–1995) was accessed from the Works and Emergency Service Department, Water Supply Division.

3. Study Area

The study area for this research is the City of Toronto, capital of Ontario. It comprises the municipalities of City of Toronto, North York, East York, York, Etobicoke, and Scarborough (Fig. 1). According to 2006 Census, the population of City of Toronto is 2.48 million people.

The climate of Toronto has some fairly unique features, including four well-marked seasons. Winter is normally cool to cold and summers are warm to sometimes hot. Monthly precipitation averages around 65 mm of rain and/or snow equivalent. The warmest weather normally occurs around mid to late July when daytime maximum temperatures reach 27°C on average.

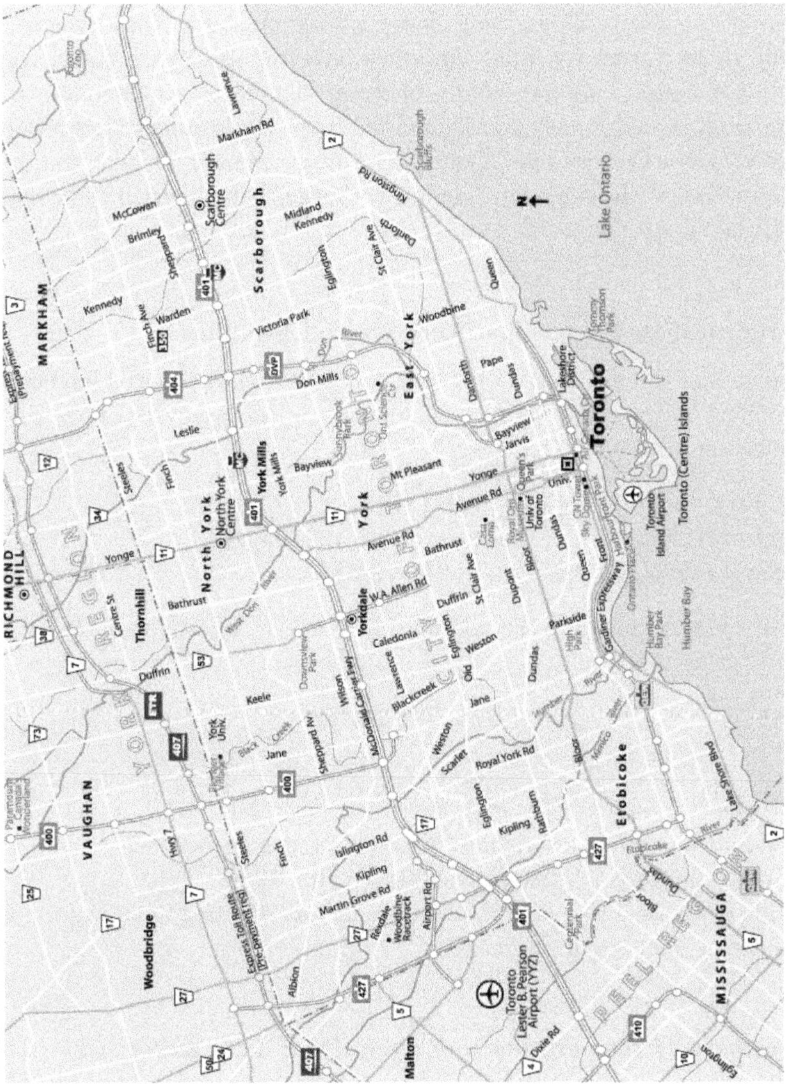

Fig. 1. The city of Toronto.

The coldest period occurs in early February when average minimum temperatures fall to −8°C. The winter months are noted for large temperature ranges and occasionally, for large day-to-day temperature changes. Because of the urban heat island effect, average temperature in the core of the city is higher than those reported outside of the city core.

Due to its proximity, Lake Ontario exerts the greatest influence on Toronto's climate. This large body of water significantly moderates the climate from the extremely continental one that would otherwise prevail. Winter temperatures average about 3°C warmer than those reported for other localities at the same latitude but without the influence of the Great lakes (Auld, *et al.* 1989).

4. Structural Relationships: Analysis and Findings

To understand the relationship between water and weather variables, eight years of data on water use, rainfall and temperature are analyzed to describe the following:

- General trend of water use (1988–1995);
- Effect of temperature of water use (1988–1995); and
- Effect of rainfall on water use (1988–1995).

A linear model expressing the relationship between climatic variables and daily water use per capita is expressed as follows:

$$Ct = a + b1Xit + b2X2t \ldots Ut \quad (1)$$

where,

t is the time index; Ct is daily water use per capita in ML in period t;
Xi are the predictor variables; bi is the regression coefficient;
a is the intercept; Ut is the residual or error term.

Equation (1) is used to model the weather factors affecting daily water use per capita in the City of Toronto.

4.1. *Impact of temperature and rainfall on daily water use*

The daily per capita water use shows an expected seasonal pattern of increase and decrease during a year (Fig. 2). In 1988, the increase in water use starts around the end of May. Water use peaks towards the end of June and starts to decline by the first week of September. The water demand

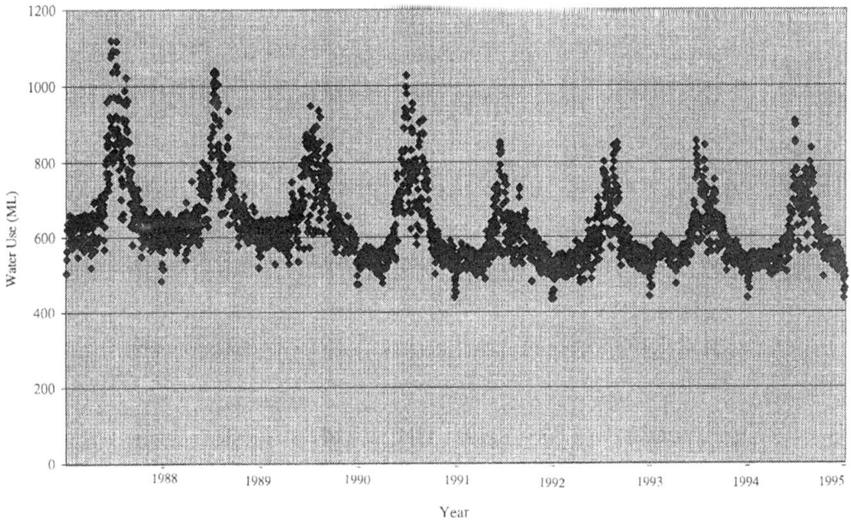

Fig. 2. Daily Per Capita water use in the city of Toronto, January–December, 1988–1995.

during the successive years shows the same trend. Similar patterns are reported by Asibey *et al.* (1993), for the city of Calgary. In the City of Toronto, during 1988–1991, the per capita water use is higher than in the period 1992–1995 partly because of the unusually wet summer between 1992–1995. The total rainfall for the summer months (May–August) in 1988–1991 was 1133.6 mm whereas the total rainfall during the summer months of 1992–1995 was 1162.0 mm.

Another reason for the low use may be Toronto's water conservation initiatives which resulted in an overall decrease in use of 8 percent since 1991. Additionally, estimated savings from retrofit of city buildings are more than $700,000 per year (Waller *et al.* 1994). In the past, the focus was on the water conservation efforts in the municipality but lately, the City of Toronto is also focusing on the issue of climate change and its impact on the future water demand.

4.2. *Impact of rainfall on water use*

The relationship between the daily water use per capita and rainfall during the growing season (May–August) is shown in Fig. 3. As expected, the scattergram shows that as the daily rainfall increases, it depressed the daily

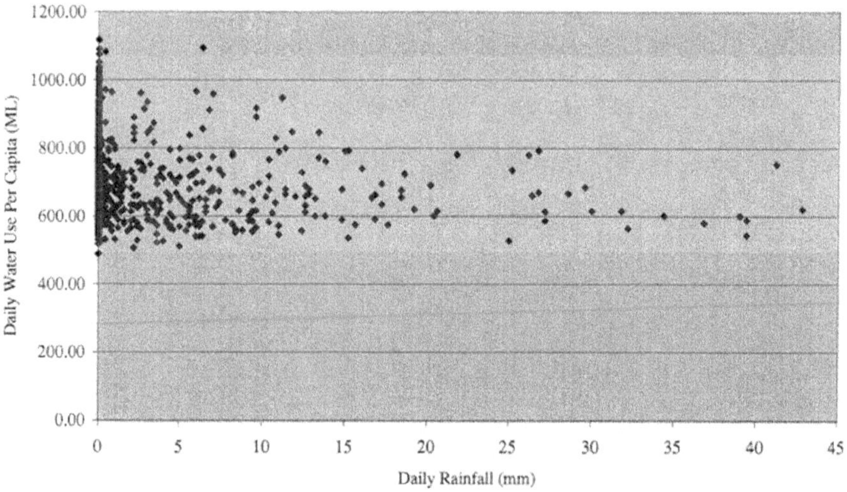

Fig. 3. Daily Per Capita water use and total daily rainfall for the city of Toronto, May–August, 1988–1995.

water use. During the growing season (May–August) rainfall events tend to reduce lawn watering and outdoor water-using activities. The scattergram also shows that water use does not fall below 500 ML irrespective of rainfall. This level reflects water uses not affected by weather conditions (e.g. water use in offices, industries, hospitals and indoors residential water use).

A similar pattern is reported by Aisbey *et al.* (1993): the water use in Calgary does not fall below 400 ML irrespective of the rainfall. The Toronto level is about seven times higher than in Calgary and this reflects the more intensive industrial, commercial and institutional land use patterns in the largest metropolis in Canada.

Using Eq. (1) to show the relationship between water use and rainfall (Table 1), the R square is 0.03, the regression coefficient is -3.57, the F-statistic is 31.6, and the t-statistic is -5.6. The regression coefficient for rain is significant statistically and has the expected sign.

Table 1. Regression analysis daily water use and rainfall in the city of Toronto, May–August, 1988–1995.

	Coefficients	Standard Error	t-Stat	P-value
Intercept	721.14	3.88	185.39	0
RAIN (mm)	−3.57	0.63	−5.62	2.46E-08

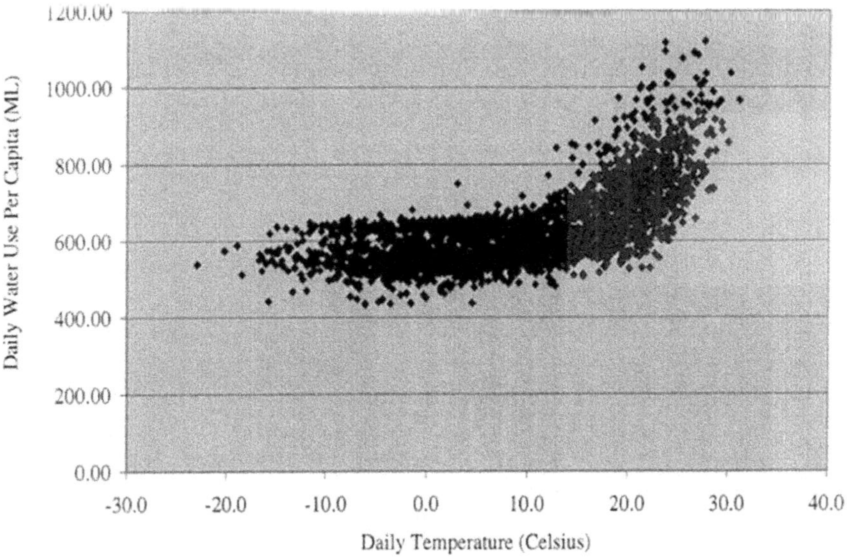

Fig. 4. Daily Per Capita water use and daily mean temperature for the city of Toronto, January–December, 1988–1995.

4.3. *Impact of temperature on water use*

A scattergram is constructed for the temperature and water use, showing that the daily water use (January–December) tends to vary directly with temperature (Fig. 4). Three distinct patterns emerge from the scattergram:

- Daily water use per capita is fairly low and steady when the temperature is below 15°C;
- Daily water use per capita increases when the temperature exceeds 15°C; and
- The water use peaks when the temperature is between 20°C and 30°C.

The impact of temperature on daily water use during the summer months (May–August) is further confirmed by the regression analysis (Table 2). The R square is 0.32, the regression coefficient is 13.73, the F statistic is 475.51, and the t-statistic is 21.8. The regression coefficient of temperature is statistically significant and has the expected sign.

To further understand the relationship between temperature, rainfall and water demand, a multiple linear regression model of the relationship between rainfall and temperature with water use was fitted to the data of

Table 2. Regression analysis between water use and temperature in the city of Toronto, May–August, 1988–1995.

	Coefficients	Standard Error	t-Stat	P-value
Intercept	445.79	12.61	35.34	3E-177
Temp (C)	13.73	0.62	21.8	3E-86

Table 3. Multiple regression analysis of relationship between water use, temperature and rainfall in the city of Toronto, May – August, 1988–1995.

	Coefficients	Standard Error	t-Statistic	P-value
Intercept	453.18	12.33	36.73	1.6E–186
Temperature	13.79	0.61	22.46	1.57E–90
Rainfall	−3.74	0.51	−7.23	9.2E–13

the growing season from 1988–1995 (Table 3).

$$Ci = 453.18 + 13.80Ti - 3.75Ri \tag{2}$$

where,

Ci = Daily water use per capita in Mega Litre (ML) (May–August, 1988–1995)

Ti = Daily average temperature in °C; and

Ri = Daily total rainfall in mm.

Both the regression coefficients of temperature and rainfall are highly significant, as is the overall regression. The R square is 0.36. The two variables explain 36% of the variation. The overall F statistic is 276.4 which shows that the relationship between temperature, rainfall and water use is significant at the 99.9% level of confidence. For every additional degree in temperature, on average, an extra 13.79 ML will be used.

Table 3 shows that the t-statistic for temperature is 22.46, which is statistically significant at the 99.9% confidence level. This indicated that the daily average temperature has a significant influence on the water use for the summer months in the City of Toronto. The regression coefficient for rainfall which is −3.74, this indicated that the daily total rainfall has a negative impact on water use. For every additional mm of rain, on average, 3.7 ML less of water will be used. The water use decreased with increasing rainfall, as one would expect because of significantly less lawn watering.

5. Conclusions

The study examines the relationship between weather variables and daily water use. The impact of temperature and rainfall on the daily water use for the growing season in 1988–1995, the city of Toronto, shows the following patterns:

- Higher temperature increases the water use significantly;
- Water use peaks when the temperature is between 20°C to 30°C;
- Water use decreases with increasing rainfall. Irrespective of rainfall, water use in the city of Toronto never goes below the intercept of 500 ML; and
- The peak days of water use start by middle of June and begin to decline by the end of August.
- For every additional mm of rain, on average, 3.7 ML less of water will be used.
- For every additional degree in temperature, on average, an extra 13.79 ML of water will be used.

Therefore, this study is important for water resource planners and administrators because, as identified by the studies mentioned above, water use is highly correlated with daily mean temperature and daily rainfall. It enables the estimation of seasonal changes in water use in the City of Toronto, providing the foundation to devise planning strategies to reduce water use and/or to increase supplies.

References

1. A. A. Asibey, L. C. Nkemdirim and D. L. Draper, The impacts of climatic variables on seasonal water consumption in calgary, Alberta, *Canadian Water Resources Journal.*
2. H. Auld, M. Loielle, B. Smith, and T. Alsopp, The climate of Metropolitan Toronto, Climatological Studies No. 41, Environment Canada.
3. E. E. Crease, and J. E. Robinson, Urban water systems for sustainable development, *Canadian Water resources Journal* **21**(3) (1996) 210.
4. N. R. Draper, and H. Smith, *Applied Regression Analysis* (John Wiley & Sons, Inc., 1996).
5. Environment Canada, The State of Canada's Environment, Government of Canada, Ottawa, 1991.
6. A. P. Grima, *Residential Water Demand: Alternative Choices for Management* (University of Toronto Press, 1972).
7. J. E. Lewis, Climatic Change and its Effects on Water Resources for Canada: A Review, *Canadian Water Resources Journal* **14**(1) (1989).

8. S.-P. Miaou, A class of time series urban water demand models with nonlinear climatic effects, *Water Resources Research* **26**(2) (1990) 169–178.
9. G. Smith, Ontario's Water Policy: From policy vision to plan implementation, *Canadian Water Resources Journal* **15**(2) (1990).
10. D. H. Waller, R. S. Scott, C. Gates, and D. B. Moore, Canadian Municipal Water Conservation Initiatives, Intergovernmental Committee on Urban and Regional Research (ICURR), Toronto, 1997.

Advances in Geosciences
Vol. 17: Hydrological Science (2008)
Eds. Namsik Park et al.

AN APPLICATION OF CONDITIONAL DIFFERENTIAL EVOLUTION ALGORITHM FOR SEARCHING RESERVOIR RULE CURVES

ANONGRIT KANGRANG* and RATTANA HORMWICHIAN†
Faculty of Engineering, Mahasarakham University, Kantarawichai, Maha Sarakham, 44150, Thailand
**anongrit@hotmail.com*
†rat_hom@yahoo.com

Rule curves of a reservoir system are necessary monthly guides to those responsible for reservoir operation to achieve the minimum of water shortage in the long run. The curves indicate either the interval (i.e. upper and lower bounds) of required water levels or desired storage volumes of each reservoir at any particular month. This paper proposed a conditional differential evolution (CDE) algorithm to connect with simulation model for searching the optimal reservoir rule curves. A minimum average water shortage was used as the objective function for searching procedure. The proposed model was applied to determine the optimal rule curves of the Lampao Reservoir (in the Northeast Region of Thailand). The results showed that the pattern of the obtained rule curves similar to the existing rule curve. Then all obtained rule curves were used to simulate the Lampao reservoir system with the 500 samples of synthetic inflow. Furthermore, these results of the obtained rule curves were compared to the obtained rule curves of simulation model. The results indicated that the situations of water shortage and excess release water of using the rule curves of CDE are smaller than the situations of using the existing rule curves.

1. Introduction

Nowadays, water deficit and flood plan are still serious problems in Thailand especially in the Northeast region. Integrated water resources management including demand and supply management is accepted method to solve these problems. These managements are addressed in the possible practice and high efficiency.[1] In the supply management, a reservoir simulation model is widely used to analyze the behavior of reservoir system on the computer.

Rule curves of a reservoir system are necessary monthly guides to those responsible for reservoir operation to achieve the minimum of water shortage in the long run. The curves indicate either the interval (i.e. upper and lower bounds) of required water levels or desired storage volumes of each reservoir at any particular month. The searching of the optimal rule curves is a non-linear optimization problem. The optimization techniques applying to search the optimal rule curves included simulation model, dynamic programming (DP), and genetic algorithm (GA).

In the past, the rule curves are obtained from reservoir simulation model by trial error process.[2] This method is straightforward and applicable for both single multiple reservoirs. However, the reservoir simulation method does not guarantee to yield the optimal rule curves because of the experienced person who adjusted the rule curves.

Dynamic programming (DP) is another optimization technique that applied to search the non-linear problems of water resource as well as to search the optimal rule curves.[3] However, there are limited conditions on these search techniques such as the dimension problem of DP.

Genetic algorithms (GAs) is another search technique that applied to search optimal rule curves of the reservoir system.[4,5] The best part of GAs model is that it can handle any type of objective function of the search. In addition, the applied GAs can handle any condition of reservoir simulation such as initial reservoir capacity and the period of inflow record. The appropriate objective function for searching the curves is average water shortage. Also, a smoothing function constraint is required to include into the proposed GAs for fitting the rule curves.[6] However, there are limited conditions on these search techniques such as the complicate process of GAs.

A differential evolution (DE) is search optimal technique that used widely in artificial intelligence fields, for reaching a goal state from a starting node. Choice of next node and starting node can be varied to give a list of related algorithms.[7–9] An alternative technique to reduce the fluctuation of rule curves in the output process is to limit the boundary of searching. To reduce searching boundary can activate the process to reach the optimal solution fast.

This paper proposes a conditional differential evolution (CDE) algorithm to connect with simulation model for searching the optimal reservoir rule curves. A conditional constraint was applied to the search process for reducing the fluctuation of the obtained rule curves. A minimum average water shortage was adopted be the objective function of the search process. Comparison results of the conditional differential evolution (CDE) and the simulation model were presented to demonstrate the effectiveness of the proposed model at the end of the paper. The proposed model was applied to determine the optimal rule curves of the Lampao Reservoir (in the Northeast region of Thailand).

2. Model Formulation

2.1. *Reservoir simulation model*

This study developed reservoir simulation model to describe the behavior of the reservoir system. This reservoir simulation model was constructed on the concept[10] of HEC-3 and it can be used to simulate the reservoir operation effectively. The reservoir operating policies are based on the reservoir rule curves and the principles of water balance concept. The reservoir system operated along the standard operating policy as expressed in Eq. (1):

$$R_{\upsilon,\tau} = \begin{cases} D_\tau + W_{\upsilon,\tau} - y_\tau, & \text{for } W_{\upsilon,\tau} \geq y_\tau + D_\tau \\ D_\tau, \quad \text{for } x_\tau \leq \ W_{\upsilon,\tau} < y_\tau + D_\tau \\ D_\tau + W_{\upsilon,\tau} - x_\tau, & \text{for } x_\tau - D_\tau \leq W_{\upsilon,\tau} < x_\tau \\ 0, \quad \text{otherwise,} \end{cases} \tag{1}$$

which $R_{\nu,\tau}$ is the release discharges form the reservoir during year ν and period τ ($\tau = 1$ to 12, representing January to December); D_τ is the water requirement of month τ, $x_\tau$is lower rule curve of month τ; y_τ is upper rule curve of month τ; and $W_{\nu,\tau}$ is the available water calculated by simple water balance as described in Eq. (2):

$$W_{\upsilon,\tau+1} = S_{\upsilon,\tau} + Q_{\upsilon,\tau} - R_{\upsilon,\tau} - E_\tau - DS \tag{2}$$

where $S_{\upsilon,\tau}$ is the stored water at the end of month τ; $Q_{\upsilon,\tau}$ is monthly reservoir inflow; E_τ is average value of evaporation loss; and DS is the minimum reservoir storage capacity (the capacity of dead storage). In the Eq. (1), if available water is in a range of the upper and lower rule level, then demands are satisfied in full. If available water over the top of the upper rules level, then the water is spilled from the reservoir to downstream river in order to maintain water level at upper rule level. If available water is below the lower rule level, release water is reduced. The policy usually reserves the available water ($W_{\nu,\tau}$) for reducing the risk of water shortage in the future, when $0 \leq W_{\upsilon,\tau} < x_\tau - D_\tau$.

The release water of reservoir were used to calculate the situations of water shortage and excess release water such as the number of failure year, the number of excess release water, as well as the average annual shortage. These results will be then recorded for using in developed CDE model.

2.2. *Development of conditional differential evolution model*

The developed CDE to connect simulation model are described as follows. The CDE requires encoding schemes that transforms the decision variables into chromosome. Then, the genetic operations (reproduction, mutation and crossover) are performed. These CDE operations will generate new sets of chromosomes. The most common encoding schemes use binary strings. In this study, each decision variable represents a monthly level of the rule curves of reservoirs.

After the chromosomes (rule curves) of the initial population have been determined, the release water is calculated by the simulation model using these rule curves. Then, the release water is used to calculate the objective function for evaluating CDE fitness. Next, the reproduction including selection, mutation and crossover is performed for creating a new rule curve parameters in next generation. This procedure is repeated until the criterion is satisfied as described in Fig. 1. There are 24 parameters (monthly level of the upper and lower rule curves of reservoirs) for one reservoir which are represented by the chromosomes. The criterion in this model is consideration of the variation in system performance with population size. The system performance is improved when the population size is increased. Then, the optimal value is met when the performance is constant.

The objective function of searching the optimal rule curve is the minimum of average water shortage (Aver–MCM/year)[6] obtaining from the simulation model which described as follows:

$$Aver = \frac{1}{n}\sum_{\upsilon=1}^{n} Sh_\upsilon \tag{3}$$

where n is the total number of considered year Sh_υ is water deficit during year υ.

To reduce the fluctuation of the obtained rule curves, the boundary of the search for each generation is limited. The range of searching for lower and upper rule curves is fixed base on the previous rule curves. These ranges cover the existing rule curves for the old reservoir that similar to pattern of

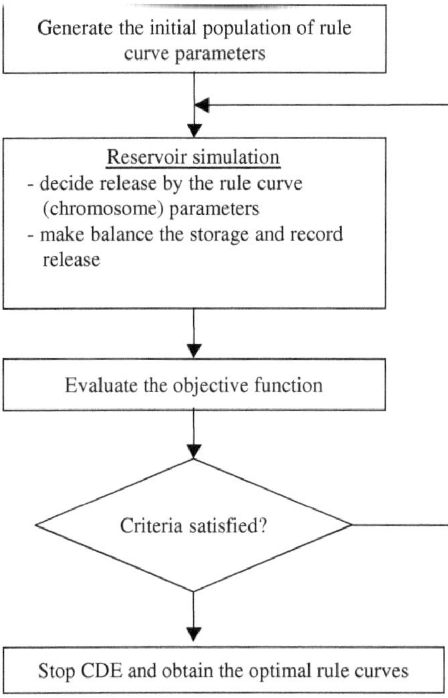

Fig. 1. Integration of CDE and simulation model.

existing rule curves. For the new reservoir, these ranges are inner the active storage for the new reservoir that dead storage for lower bound and normal high water level for upper bound.

3. Illustrative Application

The proposed CDE model was applied to search the optimal rule curve of the Lampao Reservoir that located in the Chi River Basin (in the Northeast Region of Thailand).

Figure 2 shows the locations of the Lampao Reservoir. In the following, the obtained assessment results of the considered water-deficit and excess-release properties for existing (HEC-3) and CDE cases were presented. As shown in Fig. 3, the schematic diagram of flows within the total drainage basin of the Lampao reservoir system. The Lampao Reservoir has the capacity of 1,400 MCM (million cubic meters or 10^6 m^3). This reservoir is located on the Pao River. The verification needs monthly flow records and the other related data such as monthly water requirements supplied by

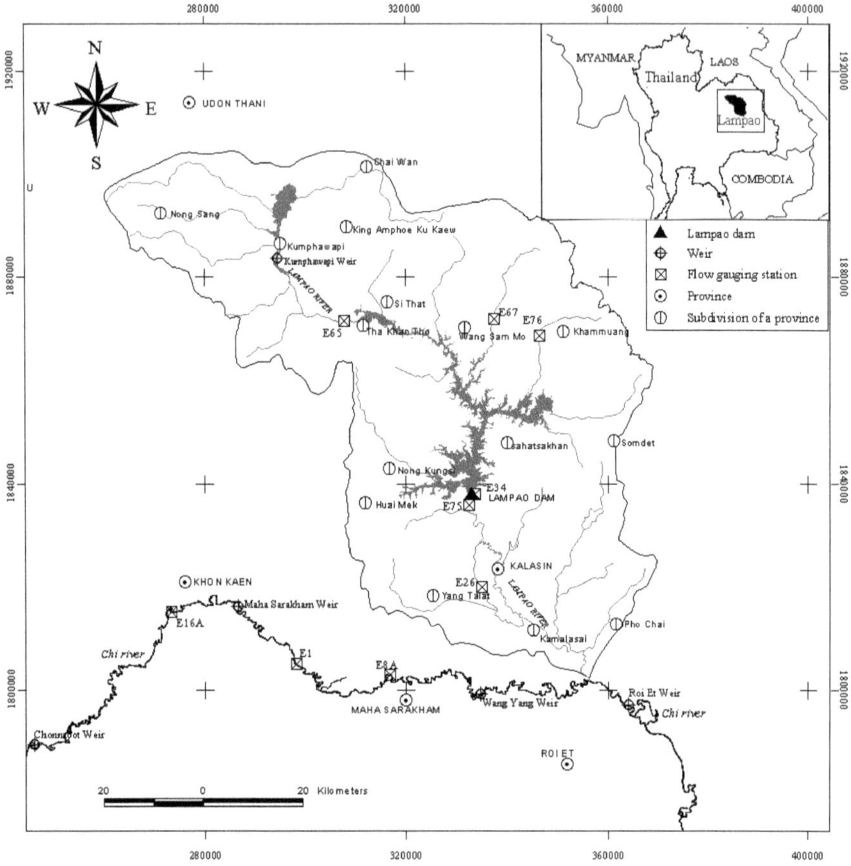

Fig. 2. Location map of the lampao reservoir.

the reservoirs, their characteristic curves, and monthly evaporation rate. For the inflow record data, sequences of 23-year (1986–2008) monthly-flow records of reservoir were used. The locations of the flow gauging stations are presented in Fig. 2. The 500 samples of generated inflow were used to evaluate the proposed model. The stochastic inflows were generated by the MAR (1).[11] The other average hydrological data for each month included series of evaporation losses and precipitation of the reservoirs were used for the simulation model. The considered water-requirement information of the studied basin was collected from the report of the Royal Irrigation Department of Thailand (RID).

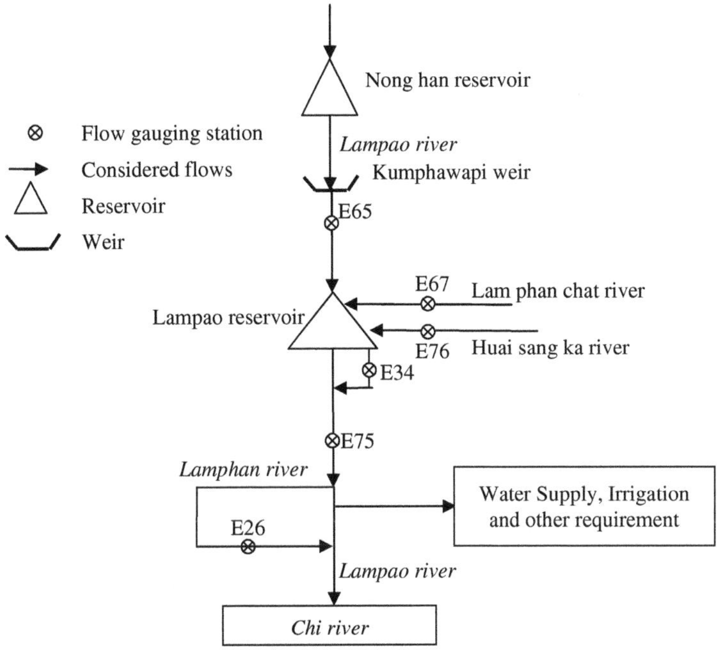

Fig. 3. Schematic diagram of flows in the Lampao River Basin.

4. Results and Discussion

Figure 4 shows the optimal rule curves of CDE compared with the rule curves of simulation model (HEC-3). The pattern of the new rule curves is similar to the existing curves of the simulation. However, the lower rule curves of both reservoirs during dry season (Jan–May) are the lower than theirs existing curves. Beside, the upper rule curves of using CDE are higher than the existing curves. Then obtained rule curves were used to simulate the Lampao River Basin system. The monthly inflow were generated by SVD (MAR 1)[11] for evaluating water shortage and flood frequency. The results are shown in Table 1. The results show the circumstances of water shortage and flood frequency (frequency of water shortage, average water shortage, the frequency of excess water and the average water release). The frequency of water shortage, the average water shortage and the maximum water shortage of rule curve from CDE model are 0.411 ± 0.062 time/years, 102 ± 18 MCM/year and 361 ± 104 MCM/year respectively. The flood frequency of excess water release, the average excess water release and the

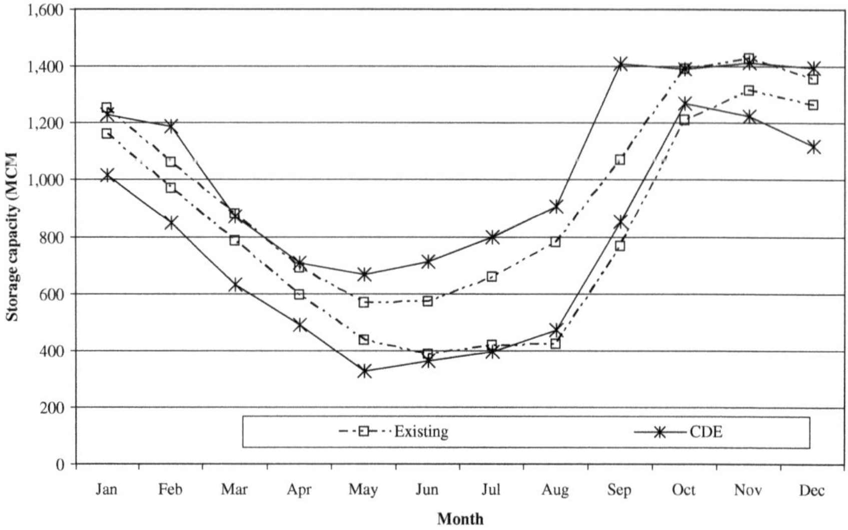

Fig. 4. Optimal rule curves of the Lampao reservoir.

Table 1. Frequency, magnitude and duration of water shortage and excess release of the systems.

Situations	Rule Curves		Frequency (times/year)	Magnitude (MCM/year) Average	Magnitude (MCM/year) Maximum	Duration (year) Average	Duration (year) Maximum
Water shortage	Existing	μ	0.926	327.2	669.9	13.0	16.4
		σ	0.049	20.6	85.6	6.3	4.8
	CDE	μ	0.411	102.4	361.2	2.6	4.1
		σ	0.062	18.0	103.5	0.9	1.8
Excess release water	Existing	μ	0.979	922.6	2738.1	18.2	20.1
		σ	0.029	25.2	565.7	6.3	4.3
	CDE	μ	0.764	571.4	2343.0	4.1	7.6
		σ	0.052	24.4	542.7	1.9	2.8

Note: μ = mean, σ = standard deviation

maximum excess release of rule curve's CDE are 0.764 ± 0.052 time/years, 571 ± 24 MCM/year and $2{,}343 \pm 543$ MCM/year respectively.

The results also indicated that most situations of shortage and excess release for CDE model are smaller than the situations of HEC-3 such as the maximum water shortage, the frequency of water shortage, flood frequency of excess water release, and the maximum excess release.

5. Conclusion

Rule curves are necessary guides for long term reservoir operation. The optimization techniques applied to search the optimal rule curves include simulation model, dynamic programming and genetic algorithm. This paper proposed a conditional differential evolution (CDE) connected simulation model to search the optimal rule curve. A minimum average water shortage was applied as the objective function of the search process. The limited bound of searching is used as the conditional constraint to reduce the fluctuation of the obtained rule curve. The developed CDE model was applied to determine the optimal rule curves of the Lampao Reservoir (in the Northeast Region of Thailand). The results showed that the pattern of the obtained rule curves similar to the existing rule curve. Then these obtained rule curves were used to simulate the reservoirs system by the synthetic inflow. The results indicated that water shortage and average duration of water shortage of CDE model are smaller than the situations of HEC-3. The excess release's situations of the CDE model are less than situations of HEC-3 for all properties.

Acknowledgments

The authors would like to express their appreciation to the Faculty of Engineering, Mahasarakham University and the Mahasarakham University for financial support, the Royal Irrigation Department of Thailand for supporting data.

References

1. J. C. I. Dooge, Future management of water resources, in *Proceedings of the Institution of Civil Engineers, Water and Maritime Engineering* **154**(3) (2002) 229–238.
2. S. K. Jain, M. K. Goel, and P. K. Agarwal, Reservoir operation study of sabamati system, India, *Journal of Water Resources Planning and Management* **124**(1) (1998) 31–38.
3. A. O. Esogbue, *Dynamic Programming for Optimal Water Resources Systems Analysis* (Prentice Hall, Englewood Cliffs, New Jersey, 1989), p. 435.
4. L. Chen, Real coded genetic algorithm optimization of long term reservoir operation, *Journal of the American Water Resources Association* **39**(5) (2003) 1157–1165.
5. J. F. Chang, L. Chen, and C. L. Chang, Optimizing reservoir operating rule curves by genetic algorithms, *Hydrological Processes* **19** (2005) 2277–2289.

6. A. Kangrang, and C. Chaleeraktrakoon, Genetic algorithms connected simulation with smoothing function for searching rule curves, *American Journal of Applied Sciences* **4**(2) (2007) 73–79.
7. R. Storn, and K. Price, Differential evolution—a simple and efficient heuristic for global optimization over continuous spaces, *Journal Global Optimization* **11** (1997) 341–354.
8. P. Kaelo, and M. M. Ali, A numerical study of some modified differential evolution algorithms, *European Journal of Operational Research* **171** (2006) 674–692.
9. D. Zaharie, Critical values for the control parameters of differential evolution algorithms, eds. M. Radek and O. Pavel, *Proceedings of MENDEL 2002, Eighth International Mendel Conference on Soft Computing*, Brno, Czech Republic, 2002, pp. 62–67.
10. US Army Corps of Engineers., HEC-3 reservoir system analysis for conservation, user's manual. Hydrologic Engineering Center, Davis, California, USA, 1974.
11. C. Chaleeraktrakoon, Stochastic procedure for generating seasonal flows, *Journal of Hydrologic Engineering, ASCE* **4**(4) (1999) 337–343.

Advances in Geosciences
Vol. 17: Hydrological Science (2008)
Eds. Namsik Park et al.

WATER RESOURCES MANAGEMENT IN TURKEY AS A CASE STUDY SOUTHEASTERN ANATOLIA PROJECT (GAP)

BÜLENT AÇMA
Anadolu University, Department of Economics,
Unit of Southeastern Anatolia Project (GAP),
Eskisehir, 26470 Turkey
bacma@anadolu.edu.tr

The Southeastern Anatolia Project (GAP), one of the most important projects for develop remarkable natural resources of the world, is accepted as a change for getting benefit from rich water and agricultural resources of the Southeastern Anatolia Region.

The GAP Project has been considered as a regional development projects through years, but the dimensions of sustainability, protection of environment and participatory have been attached to the master of the project in recent years.

When the GAP Project is completed, the Upper Mesopotomia, the centers of many civilisation, will re-again its importance as it had in the ancient times, and will be alive a center of civilisation. Moreover, when the problem of water shortage and water supplies in the world for the future is kept in mind, the importance of Southeastern Anatolia's water supplies will be doubled. For this reason, the GAP Project, developed by depending on water and natural resources of the region, will have an important place in the world.

The aim of this study is to introduce the region with rich natural resources and the GAP Project. For this reason, firstly, the natural potential of the region will be introduced. Second, The GAP Project will be presented in detailes. In the third stage, the projects being processed for protecting the natural sources and environment will be analyzed. In the last stage, strategies and policies to develop and to protect the natural resources of the region in short, mid, and long terms will be proposed.

1. Introduction

The Southeastern Anatolia Project (GAP), which is one of the most important development projects in the world, is a popular subject in both Turkey and especially in Southeastern Anatolia Region. Because of being underdeveloped region and of having unvalued rich sources, (see,

Appendices 1–3) the importance of the GAP Project is receiving more and more importance.

The GAP had originally started as an energy production and irrigation project seeking to utilize the rich land and water resources of the region. It was later converted into an integrated regional development project upon the completion of the GAP Master Plan in 1989. The Master Plan is an overall guide for the course that regional development will follow and for plans, programs and projects to be developed on more specific terms. The basic development scenario adopted by the Master Plan is to transform the region as a agriculture based industrial center (GAP Administration, 1991).

At present, the GAP is a human centered and integrated regional development project carried out along with the principles of sustainable human development. The development envisaged under the GAP has the goal of creating opportunities for the people of the region fully materialize their preferences and economic potentials. Other than dams, hydroelectric plants and irrigation schemes over the rivers of Euphrates and Tigris, the concept of "Southeastern Anatolia Project" is conceived as a regional development drive aiming the multi-faceted and sustainable socio-economic development of the region on the basis of a multi-sectoral and integrated approach which covers such diverse areas as urban, rural and agricultural infrastructure, transportation, industry, education, health, housing, tourism and investments in many other fields.[4]

2. The Natural Potential of the Region

The State Hydraulic Works (DSI) is engaged in the program for developing land and water resources in the region. The program consists of two parts each of them covering projects related to the basins of either the Euphrates or the Tigris (see Appendices 4 and 5). The program envisages the construction of 22 dams, 19 hydraulic power plants and an irrigation system that will bring 1.82 million hectares of land under irrigation. Upon the completion of the project, 29% of the total water potential of Turkey will be managed through the facilities on the Euphrates and the Tigris, which together flow more than 52.9 billion cubic meters of water a year. The planned irrigation area corresponds to 20% of total irrigable land in Turkey and annual energy production to 22% of total electric energy potential in Turkey.[7]

The GAP region extends over an area of 75,000 km^2 and a wide range of crops each requiring different climatic conditions are raised in this area including olive, pistachio, hazelnut and persimmon. The region has 3.2 million hectares of land fit for crop culture. Forested areas make up 1.3 million hectares while 2.3 million hectares of land consists of pastures and ranges.[12]

The GAP focuses on efficient utilization of these natural resources. For the first time in Turkey the management, operation and maintenance of new irrigation systems have been directly transferred to Irrigation Districts, which are organizations formed by local farmers. In 1998, the region accounted for 41.6% of the total cotton output of Turkey. Favorable climatic conditions in the region make it possible to reap two crops a year. The region is also quite fit for animal husbandry. In this context, research projects led by the GAP Administration focus on genetic improvement and development of advanced breeding techniques.

According to studies made, upon the completion of irrigation projects in GAP, the area brought under irrigation will be equal in size to the total area so far brought under irrigation by the State. This will naturally bring along significant changes in agricultural output and crop design. Such irrigation-led crops like soybean, groundnut, corn, sunflower and fodder crops will be the basis of flourishing agro-industries.[2]

Gradual implementation of the GAP Project, which envisages irrigation on 1,82 million hectares of land and creation of new water reservoirs will significantly alter the land use and water regime in the region. Also in this process, population movements, rapid urbanization and industrialization will bring along new transformations in both rural and urban areas. Other than advantages to be reaped after irrigation, there are some other points and problems to be considered: Problems emerging as a result of excessive and uninformed practices of irrigation; effects of climate change on crop farming and plant cover in the region; corresponding changes in the flora and fauna; erosion and adverse effects of uncontrolled growth on natural, historical and cultural properties. All these make it necessary to reconsider the project with a view to the dimensions of culture and the environment.[1]

In terms of natural resources, Southeastern Anatolia is one of the most unique areas in Turkey. It is the border gate through which species peculiar to steppe and semi-desert areas enter Turkey and the region rooms in two different living environments, which is not found in other regions of the country. These are:

i) Banks of the Euphrates and the Tigris, their flood plains and major tributaries of these two rivers,
ii) Steppe and semi-desert areas especially in the southern parts of the region.

For this dual (wetland and desert) nature it is possible to observe many endemic species of plants, animals, birds and fish in the region. The project that targets agricultural and economic change and thus social transformation through water resources management is transforming steppes into irrigated farmlands and river beds into dam lakes. While ensuring increases in agricultural output, this transformation, however, may lead to the disappearance of natural living environments of many waterborne and steppe species. Considering this risk, the GAP Administration has since 2001 been conducting the "Project for Exploring Biological Diversity in the Region". The "Wild Life Project" was launched in 2002 in the context of efforts for rehabilitation.[8]

3. History of Gap Project

The Southeastern Anatolia Project (GAP) is a multi-sector and integrated regional development effort approached in the context of sustainable development. Its basic objectives include the improvement of living standards and income levels of people so as to eliminate regional development disparities and contributing to such national goals as social stability and economic growth by enhancing productivity and employment opportunities in the rural sector. The project area covers in the basins of the Euphrates and Tigris and in Upper Mesopotamia.

The GAP had originally been planned in the 70s consisting of projects for irrigation and hydraulic energy production on the Euphrates and Tigris, but transformed into a multi-sector social and economic development program for the region in the 80s. The development program encompasses such sectors as irrigation, hydraulic energy, agriculture, rural and urban infrastructure, forestry, education and health. The water resources development component of the program envisages the construction of 22 dams and 19 hydraulic power plants and irrigation of 1.82 million hectares of land. The total cost of the project is estimated as 32 billion US $. The total installed capacity of power plants is 7,476 MW and projected annual energy production reaches 27 billion kWh.

The project rests upon the philosophy sustainable human development, which aims to create an environment in which future generations can benefit and develop. The basic strategies of the project include fairness in development, participation, environmental protection, employment generation, spatial planning and infrastructure development.

The agricultural development objectives of the GAP include the following: Raising levels of income in the rural sector; providing inputs for industrial enterprises in the region; creating employment opportunities so as to minimize out-migration and encouraging export oriented production in the region.[3]

4. Background of Southeastern Anatolia Project

In the region of Southeastern Anatolia, the area encompassing the provinces of Adiyaman, Batman, Diyarbakir, Gaziantep, Kilis, Mardin, Siirt, Sanliurfa and Sirnak is referred to as the "GAP Region." This region is surrounded by Syria to the south and Iraq to the southeast has a surface area of 75,358, which corresponds to 9.7% of the total surface area of the country. 20% of a total of 8.5 million hectares of irrigable land in Turkey is in the GAP region made up of wide plains in the basins of Lower Euphrates and the Tigris.

This region, also called as "Fertile Crescent" or "Upper Mesopotamia", is known as the cradle of antic civilizations. For long in history, the region served as a bridge from Anatolia and Mesopotamia.

The Euphrates and the Tigris, the two major rivers in Turkey flow in this region. Both of these rivers originate from Eastern Anatolia and flow to the Persian Gulf. Southeastern Anatolia receives less precipitation than the other regions of the country. It was thus considered to develop water resources in the basins of these two rich rivers for irrigation and energy generation purposes and to bring these two rivers with irregular flow regimes under control.

The first idea and decision to utilize the waters of these two rivers rationally belong to Atatürk, the founder of the Republic. In those years when the country was making great efforts for change and modernizations in all fields including material and spiritual, the need for electrical energy came to the fore as urgent and priority issue. Then, upon the order of Atatürk, the Electricity Studies Administration was founded in 1936 to investigate issues on how rivers in the country could be utilized for energy production. The Administration began its detailed studies with the "Keban Project" and established observation stations to assess the flow and other characteristics of the Euphrates.

Geological and topographical surveys in the Keban Pass started in 1938. The Administration gave weight to drillings along the Euphrates and the Tigris in the period 1950–60. Observing that some new needs and requirements emerged, the General Directorate of State Hydraulic Works (DSİ) was established in 1954. This gave rise to the idea of conducting basin scale studies in Turkey and consequently the DS started its work in 26 different water basins of the country.

As a result of work carried out by Diyarbakir based "Euphrates Planning Authority" established in 1961, the "Euphrates Basin Development Report" came out as the main document assessing the irrigation and energy potential of the Euphrates. In 1966, the "Lower Euphrates Development Report" followed. The Diyarbakir Regional Directorate of DS conducted a similar work for the Tigris basin.[6]

All these made clear how the basins of the Euphrates and Tigris could be developed and utilized and, in 1977, these two basin projects were given the single title of "Southeastern Anatolia Project." Then, in 1986, the State Planning Organization was given the mandate of addressing development activities in Southeastern Anatolia within the framework of integrated regional planning and ensuring the coordination of these activities.

The Southeastern Anatolia Project Regional Development Administration was instituted upon the Government Decree no. 388 in Force of Law, published in the Official Journal no. 20344 dated 6 November 1989. This law assigns the following mandate to the new organization: To ensure the rapid development of and investments in areas covered by the Southeastern Anatolia Project; to deliver or cause to be delivered services for this purpose including those in the fields of planning, infrastructure, grant of licenses; housing, industry, mining, agriculture, energy and transportation; to take or cause to be taken relevant measures to raise the educational level of people living in the region and to ensure coordination among different organizations and agencies involved in these activities.

The GAP Higher Council is the highest decision making body of the organization that examines and finalizes all plans, projects and programs. This council presided over by the Prime Minister or a State Minister appointed by him or her and composed of the State Minister in Charge of GAP, State Minister in Charge of SPO (State Planning Organization) and the Minister of Public Works and Settlement.

The GAP Administration has its Head Office in Ankara and a Regional Directorate in Sanliurfa. At present, the concept "Southeastern Anatolia Project" is conceived, beyond dams, hydraulic power plants and irrigation facilities over the rivers Euphrates and Tigris, as a multi-sector, integrated and sustainable development effort also comprising investments in such diverse areas as urban and rural infrastructure, agricultural facilities, transportation, industry, education, housing, health, tourism, etc.

Upon the completion of the project, through facilities over the rivers Euphrates and Tigris, which together flow more than 50 billion cubic meters of water a year, 28% of the total water potential of the country will have been taken under control, 1.8 million hectares of land will have been brought under irrigation and annually 27 billion kWh of electric energy will have been produced on an installed capacity of 7,485 MW.

Expected high potential in both industry and agriculture will increase the income level of the region fivefold and create employment for 3.8 million people in the region whose population is projected to reach more than 10 million in 2010.

The core duty of the GAP Administration is to plan for and realize all efforts and activities geared to the development of the region in the context of a "comprehensive regional planning approach" that covers all economic and social sectors in consistency with the objectives, targets and strategies of regional development. Such a comprehensive approach to planning is expected to guide decision makers in relation to development directions and magnitudes, enable them to link each project component to others and to draw concrete frames by assessing economic and social investments in time and space. The GAP Master Plan is the main guide used for these purposes. Project implementation will improve urban infrastructure and enhance the population absorbing capacity of urban centers. Furthermore exports from the region will be promoted by mobilizing the resources of the region and ensuring a sustained economic growth (GAP Master Plan, 2002).

This is the largest investment initiative even launched in Turkey with a total value of 32 billion US$ at 1997 prices. As of the end of 2008, 19 billion US$ has been spend for the project.

The Republic of Turkey assigns great importance to the elimination of inter-regional disparities in the process of social and economic development. This emphasis derives not only from aspirations for a fair and equitable process of development but also from the sound assumption that inducement of the potential of relatively underdeveloped regions will contribute significantly to such targets as economic growth export promotion and social stability. In sum, the GAP is to reinstate civilization to the Upper Mesopotamia.

13 projects were prepared to develop the rich water resources (the rivers of Euphrates and Tigris) of the region for irrigation and power generation purposes. 7 of these projects are in the basin of the Euphrates and 6 in that of the Tigris. These projects envisage the construction of 22 dams and 19 power plants, generation of 27 billion kWh of energy a year over an installed capacity of 7,485 MW. Parallel to or as a result of this development in water and land resources, there will be improvements and corresponding developments in the sectors of agriculture, industry, rural-urban infrastructure and education and health services.

The GAP Master Plan which draws the general frame for regional development paid particular attention to linking investments for the development of water and land resources to a time schedule by considering both financial and technical capacities, projected for changes in economic and social sectors, employment, population and its spatial distribution induced by this development; assessed at macro level the needs for education, health, housing and urban infrastructure and expressed annual need for funds up to year 2010.

The Master Plan formulates the basic objective as "raising the development indicators of the region up to country standards in shortest possible time", defines economic and social targets serving to this objective and adopts the basic strategy of "transforming the region into an export center for agriculture based goods".

During the time elapsing since the preparation of the GAP Master Plan in 1989, rather rapid changes have taken place in the region, country and in Middle East and these led to delay in reaching targets. The UN embargo on Iraq and terrorist activities in the region had their negative effects on development efforts and region's trade with Middle Eastern countries. Furthermore, imbalances in public financing in Turkey caused problems in meeting the financing needs of the project sufficiently and in time. Meanwhile, the international community has added new dimensions and concepts to the definition of "development." Coming to the fore during this period were such concerns and concepts as the environment, sustainability and participation, which were either overlooked or totally absent in the original Master Plan. All these developments and urgent needs of the region for development made it necessary to revise the GAP Regional Development Plan with a new understanding.

The most salient principle in the new plan is human development. This approach is expected to instigate changes in three important areas and make them effective:

- Rational utilization of public resources and potential,
- Enhancing people's participation, and
- Catching up with targets in the field of human development.

The macro frame of the GAP Regional Development Plan (GAP-RDP) is drawn by the Long Term Strategy (Decree no. 697 dated 27.06.2000 and the 8th Five-Year Development Plan.

This frame has recently been further enriched by efforts waged under the "Program for Transition to a Strengthened Economy" prepared as a part of the process for Turkey's accession to the European Union.

Irrigation in Harran Plain started in 1995. As of the end of 2007, 272,072 hectares of land has been brought under irrigation by the DSİ. In terms of physical realization, 12% of irrigation projects are already in operation, 8% under construction, 25% at the stage of contracting out and 55% are being planned. There have already been significant increases in agricultural output. The value to be created by irrigation is estimated as 3 billion $.

5. Effects of the Gap Project Upon the Socio-Economic Structures of the Region

The Region is also named as the "Fertile Crescent" or "Upper Mesopotamia", and known to be the cradle of civilization in human history. Throughout history, the Region has served as a bridge ensuring passage from Anatolia to Mezopotamia. The Tigris and the Euphrates, two important rivers of Turkey flow through the Region. Both originating from the Eastern Anatolia, these two rivers reach sea in the Persian Gulf. Southeastern Anatolia receives less precipitation compared to the other regions of the country. Hence the idea was to utilize the rich water potential of these two rivers for irrigation and energy production purposes, and to regulate the otherwise irregular flow of both rivers.[15]

The integrated project covers not only multi-purpose dam and irrigation schemes but also investments in such development related areas as agriculture, energy, transportation, telecommunication, health, education and urban and rural infrastructure building.

The basic development scenario of the GAP Master Plan[13] is to transform the region into a "base" for agro-industrial products. In more concrete terms, the GAP envisages the following:

1) irrigation of 1.82 million hectares of land,
2) production of 27 billions kWh energy,
3) 106 percent increase in per capita income, and
4) generation of employment for 3.8 million people

When the project is completed, facilities on the Euphrates and Tigris, which together flow more than 52.94 billion cubic meters of water a year, will bring under control 28.5% of the total water potential of Turkey. 1.82 million hectares of land that is an area four times as large as Cukurova will be brought under irrigation and 27 billion kWh of energy will be produced with an installed capacity of 7,500 MW. The area to be brought under irrigation corresponds to 20% of total land economically irrigable in Turkey and energy production to 22% of the total hydraulic energy potential of Turkey.

The GAP is the largest investment ever launched in Turkey for regional development. Upon the completion of irrigation projects, 1.82 million hectares of land will have been brought under irrigation. As of the end of 2007, 272,972 of land has been brought under irrigation by the DSI. In terms of physical realization, 12% of irrigation projects is already in operation, 8% is under construction, 25% is at the stage of contracting out and finally 55% is at planning stage. When all these projects are completed, the new area brought under irrigation will be equal to the total size of land so far opened to irrigation by the State. Thus, this high agricultural and industrial potential created by the GAP will significantly contribute to the economic output of the region and generate employment for about 3.8 million people. Irrigation will naturally increase agricultural output as well.

As of the year 1985, which constitutes the base year of the GAP Master Plan, the per capita regional product was 47% of the same figure for the country as a whole. In 2000, this rose to 52%. In the period 1987–2000, the rate of annual increase in GNP was 3.63% compared to 3.45% for the region.

6. Water Management and Civilization in the Region

The Project aims at removing a socioeconomic "gap" between the project area and more developed regions in Turkey. The Region's economy is dominated by the agricultural sector and agriculture is largely practiced under rain-fed conditions. The Region's contribution to Turkey's gross domestic product (GDP) is a now 4%, and to the national value-added in the industrial sector is an-even lower 2%.

The Region, on the other hand, is rich in soil and water resources. The Tigris and Euphrates Rivers together provide about 28% of all national water supply by rivers; the economically irrigable land area in the region amounts to 20% of that for the whole country.

In order to improve the economy of the Region, mobilization of the water and soil resources on a regional scale has been planned, within the framework of integrated, multi-sectoral regional development. As such the 32 billion-dollar project comprises not only of water resources projects, but also" of investments in all development-related sectors such as agriculture, energy, transportation, telecommunications, healthcare, education, urban and rural infrastructure, in an integrated manner. The project is one of the largest of its kind in the world. The water resources development program aims at producing 27 billion kilowatt-hours of hydroelectric energy and irrigating 1.82 million hectares of land. A total of 22 dams, including the sixth largest-volume dam in the world (Atatürk Dam), 19 hydropower plants, the anlıurfa Irrigation Tunnel System the Iargest of its kind, numerous irrigation networks, canal systems constitute the physical groundwork in water resources.

Water resources development of this size and scale is bound to have effects and implications that go far beyond irrigation-related activities, touching every facet of life and involving all social and economic sectors. On-farm development, farmer training and extension programs, agricultural input provision, credit and marketing arrangements, agro-processing, related rural infrastructure, operation and maintenance of the extensive irrigation system, environmental protection, preservation of historical and cultural heritage, social attitudes and expectations are some of the issues that need to be addressed in water sector.

Project planning and implementation are done based on a Master Plan and on an Action Plan. Macro-level planning and management, coordination, monitoring-evaluation and implementation in selected areas are carried out by the Regional Development Administration.

A comprehensive investment program is under implementation: Out of the total public sector finance requirement of 32 billion US dollars, 19 billion US dollars has already been invested, one billion dollar is being spent this year .

The project will double the nation's hydroelectric production, increase the irrigated areas by 50%, more than quadruple the gross regional product, and more than double the per-capita income in the region.

7. Strategies and Resolution for Sustainable Development

As it is well known fact that all development projects should be carried out under the principles of sustainability. The GAP Administration and UNDP accepted this fact and conducted a seminar on in March, 1997. Since then, the studies done by the GAP Administration on sustainable development can be summarized as (GAP Administration, September, 1999):

Sustainability is the basic development approach of the GAP. Its basic components are:

1) Sustainability of irrigation,
2) Environmental sustainability,
3) Sustainability of agricultural production,
4) Social sustainability, and
5) Spatial sustainability.

Sustainability and human devleopment in GAP include;

1) Sustainability of natural sources and the environment,
2) Sustainability in agriculture (irrigation, soil and land uses),
3) Sustainability of economic dimension (develop, growth, employment and marketing),
4) Spatial sustainability (carrying capacity), and
5) Social sustainability (development of human resources, participating, equity) (GAP Administration, September, 1999).

The GAP Sustainable Development Programme supports the implementation of 29 sub-projects presented below under five headings:

- Encouragement of social sustainability and improvement of social services,
- Encouragement of agricultural sustainability and improvement of rural productivity,
- Encouragement of local entrepreneurship and industrial development for economic viability.

Encouragement of sustainable human settlements:

- Sustainable utilisation of natural resources at an optimal level.[10]

The projects listed below are actually interrelated within the framework of sustainable human development. The list includes a grouping for analytical purposes.

1. Projects for Environmental Protection: The outputs of these projects can be taken under two main headings.
   (i) to create the conditions which allow future generations utilise natural resources;
   (ii) to create healthy and safe living conditions for people.
   - Management-operation-maintenance of GAP irrigation systems (GAP-MOM),
   - Regulation of irrigation water and introduction of methods and technologies which save water,
   - Pilot implementation for a modern irrigation technology,
   - Urban waste water treatment,
   - Re-use of water after irrigation,
   - Eastern Anatolia erosion control and watershed rehabilitation project,
   - Protection of bio-diversity,

- Environmental work in the GAP region — The Euphrates Basin,
- Environmental work in the GAP region — The Tigris Basin,
- Assessment of climatic conditions in the GAP region for the present and near future.

2. Urban Planning and Infrastructure Projects: These projects are developed and materialised with the participation of people. Participation assures the social feasibility of projects and helps the promotion of a democratic culture in urban management and life.
   - Participation to urban zoning and planning,
   - Resettlement and employment of people affected by dam lakes,
   - Ataturk Dam Lake sub-region development plan,
   - Eco-city and eco-village planning and development works in pilot areas.
3. Projects for the Elimination of Disadvantaged Positions and Poverty: These projects aim at gender balanced development; closing income gaps; better access to health, education and social services; and awareness building.
   - Establishment of multi-purpose community centres for women (CATOM),
   - Enhancement of the income level of people living in areas out of the coverage of irrigation,
   - Distribution of treasury land to landless peasants,
   - Establishment of centres for supporting and guiding entrepreneurs (GAP-GIDEM).
4. Human Resources Development and Institutional Arrangements: These projects enhance efficiency and income and also set the conditions of internal and external participation.
   - Transfer of irrigation systems to farmers,
   - Institutional capacity building,
   - Farmers' training and extension,
   - Agricultural research and development projects,
   - Consolidation of farm lands.[9]

8. Conclusion

The GAP Project, which is one of the most important development projects in the world, is based on a sustaniable development and is human oriented. Actually, in recent years the concepts of sustainable development and regional development have become a complementary concept to each other.

In order to develop the Southeastern Anatolia, which has underdeveloped socio-economic structure and unvalued natural resources up to now, only physical development is not satisfactory. The most important thing to do is to realize the regional development process. But this process must be accorded with both socio-economic and cultural structure.

As for the level it reached today, the GAP has already brought along significant changes in the life of people living there.

The GAP is important for not only development of natural resources and environment, but also a unique model of international significance in terms of water management for human and social-economic development.

The GAP is a technological and sociological monument of civilization waiting future generations.

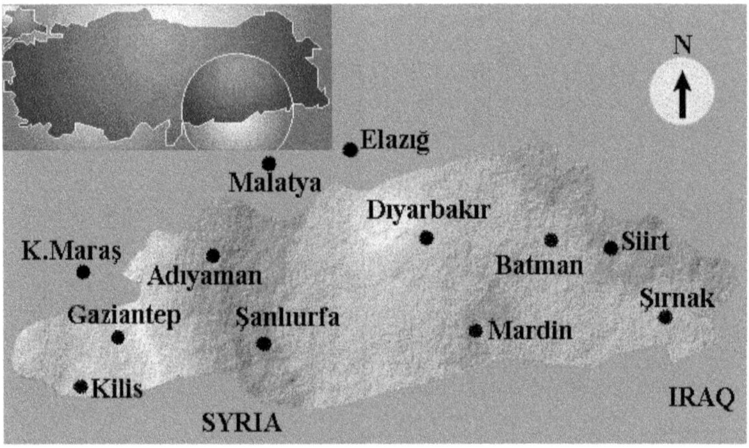

Appendix 1. Map of Southeastern Anatolia Region.

Appendix 2. Map of General Turkey.

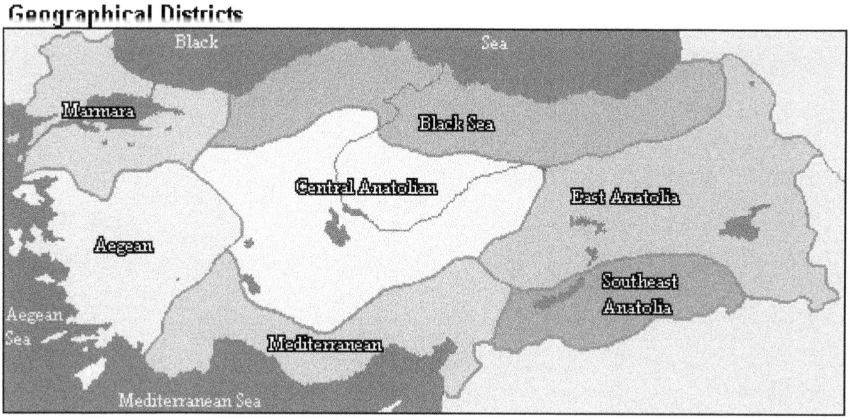

Appendix 3. Appendix 3. Regions of Turkey's.

Appendix 4. Appendix 4. Ataturk Dam.

Appendix 5. Appendix 5. Karakaya Dam.

References

1. B. Açma, New Approaches to regional development: Southeastern Anatolia Project (GAP) in Turkey, Conference Young Scientists-OPEN MINDS, in Lodz, Poland, 2003a, pp. 13–14.
2. B. Açma, A new approach for planning and environmental policies in regional development: A case study Southeastern Anatolia Project (GAP) in Turkey, *Sixth International Symposium & Exhibition on Environmental Contamination in Central & Eastern Europe and the Commonwealth of Independent States*, Prague, Czech Republic, 2003b, pp. 1–4.
3. B. Açma, New horizons for natural resource economics in Asia: The Southeastern Anatolia Region and Southeastern Anatolia Project (GAP) in Turkey, The International Convention of Asia Scholars (ICAS), Singapore, 2003c, pp. 19–22.
4. B. Açma, Sustainable regional development: The GAP Project in Turkey, Invited paper in Internatonal Atlantic Economic Conferences, Athens, Greece, 2001, pp. 13–20.
5. B. Acma, Industrialization strategy of Southeastern Anatolia Project in Turkey, Invited paper in APDR Conferences, Ponta Delgade University of Azores, Portugal **29**(2) (2000).
6. DSI Technicial Bulletin, Ankara, Turkey, 2006.
7. GAP Administration, Latest State in GAP, Republic of Turkey Prime Ministry, Ankara, September 2008.

8. GAP Administration, GAP Regional Development Plan, Republic of Turkey Prime Ministry, in Turkish, Ankara, 2002.
9. GAP Administration and UNDP, Sustainable Development Programme in GAP, Republic of Turkey Prime Ministry, in Turkish, Ankara, September 1997.
10. GAP Administration, GAP Project an Innovative Approach to Integrated Sustainable Regional Development, Republic of Turkey Prime Ministry, Ankara, April 1996.
11. GAP Administration, GAP Status Report, Republic of Turkey Prime Ministry, in Turkish, Ankara, 1993a.
12. GAP Administration, GAP Action Plan, Republic of Turkey Prime Ministry, in Turkish, Ankara, 1993b.
13. GAP Administration, Social Trends ant Attitudes Towards Transformation in GAP Region, Republic of Turkey Prime Ministry, in Turkish, Ankara, 1993c.
14. Unver, Olcay and K. Ravij, Grupta, Water Resources Management, METU Press, Ankara, 2002.
15. Unver, I. Olcay, GAP: A Pioneering model for Water-Based Regional Development, Republic of Turkey Prime Ministry, in Turkish, Ankara, 1999.

Advances in Geosciences
Vol. 17: Hydrological Science (2008)
Eds. Namsik Park *et al.*

GROUND WATER LEVEL CRITERIA FOR SUPPORTING LINED CONCRETE CANAL DESIGN

SURASAK KLUNGSUPAVIPAT* and SUWATANA CHITTALADAKORN†
Department of Water Resources Engineering, Faculty of Engineering, Kasetsart University, 50, Phaholyothin Rd. Lardyao, Chatuchak, Bangkok 10903, Thailand
**surasak2545@yahoo.com*
†fengswc@ku.ac.th

Irrigation canal system usually benefits from the gravity open channel flow. The main purpose of the canal system is to serve irrigation area which is usually in a plain area. The canal shall provide enough level or water head, while the lined concrete canal is normally used for conveying water to protect water loss of the system. However, in many cases, those irrigation canal systems are disturbed by the subsurface water which is unavoidably resulted in many unpredictable problems, especially the damages of lined concrete. Normally, the lined concrete has no reinforcement therefore it is unable to resist the tension stress. When the subsurface water pressure is higher than that of water in the canal system, it generates over tension stress on the concrete, resulting in lined concrete cracking. The seriousness of those cracks mainly depends on the force of subsurface water pressure and the sizes of concrete lined.

This study presents "Ground Water Level Criteria for Supporting Lined Concrete Canal Design" that can classify the areas of high subsurface water pressure which need to be careful of the further concrete crack. A case study at Lampao irrigation canal system, Kalasin, Thailand is presented of the seriousness of concrete lining damages. The analysis results of the subsurface water pressure in the area have showed accurately the position of damages as compared to the investigated results from the site. The design criteria of lined canal was therefore developed to deal with those problems from the ground water, including the criteria for designing the appropriate types, sizes, and spaces of those subsurface drainages needed for the safety of the lined canal. The contribution of this study is to help the canal system designer to provide enough lined concrete strength by which it could save further maintenance costs.

*Ph.D. student.
†Associate Professor. Ph.D.

1. Introduction

An irrigation canal transfers the water in to agriculture areas. The concrete pipeline is paved to protect the water loss through to the end of canal. The main line canal functions to carry the water longitudinally as designed. If damaged, the water leakage will cause the loss of energy, and subsequently the water transfer will not be carried out as designed.

The main lined concrete canal is designed to keep on the contour line. It is to be dug and filled through the topography to keep the canal its constant level. The position of the excavation from the existing the ground may obstruct the flow and be damaged by the ground water.

The past design did not analyze the ground water flow for reducing its pressure because they did not know the position and magnitude of the pressure. If these factors had been known they can seek for the better prevention measure.

The problem for irrigation systems throughout the world is the constructing canals in cut areas, where a hillside has been excavated as shown in Fig. l(a), and the cut slope may fail because the phreatic line is very near the soil surface. The first problem is stabilized the cut slope, while the second problem is avoided canal bank sloughing that would result in failure of the canal lining. Figure 1(b), the groundwater is higher than the canal water surface. This is generally the result of not having design criteria for subsurface drainage of the lined canals.

The location of the research at the Lam Pao irrigation project is located in Kalasin province in the northeast Thailand, about 500 km from of Bangkok (see Fig. 2). The operation of the Lam Pao dam and reservoir began in 1968. The distributing system by two main conveyance canals was designed for supplying the water. The total length of the Right Main Canal (RMC) and Left Main Canal (LMC) are 90 and 32 km, respectively.

In the Fig. 3 shown the failure problem in RMC lined canal occurring between km 13+000 and km 40+000 every year. In this case, the problem has currently been caused by subsurface drainage.

Skogerboe and Merkley (1996) in the book of Irrigation maintenance and operations learning process presented the failure concrete lining from subsurface water pressure behind the line can cause cracks and bulging with eventual eruption of the line. Rahimi and Barootkoob (2002) showed the failure of concrete cracking in low flow to medium plastic from swelling soils in the Khoozestan province of Iran. Hartani (2003) studied the characteristics of drainage of sloping lands with Variable Recharge Model

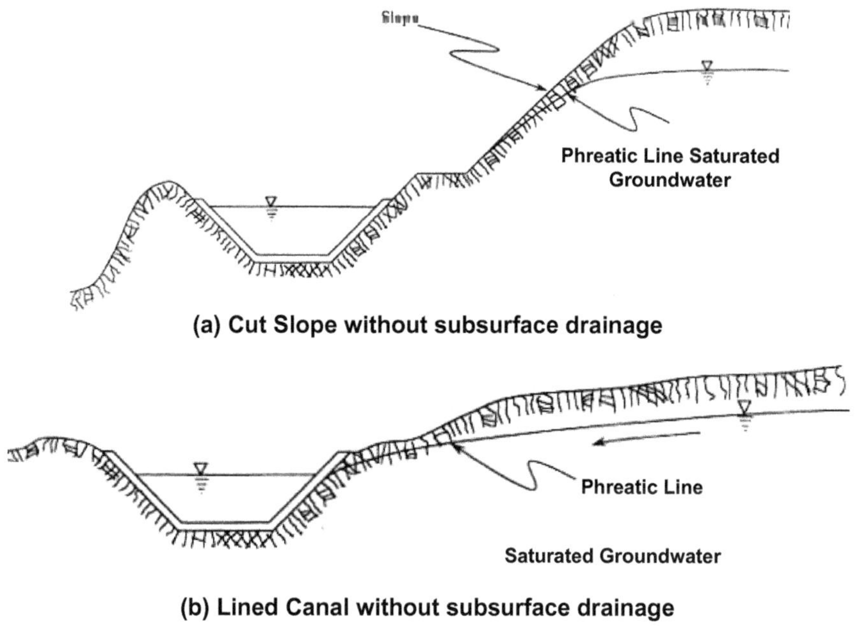

Fig. 1. Typical Schematics of Cut slope and Line canal.

Validation and Applications. Semi-analytical transient equations describing the behavior of water tables in the subsurface drained soils when drains rest on a sloping impervious barrier have been derived previously and represent the bases of the SIDRA model. Moustafa (1988) developed the spatial modeling of soil properties for subsurface drainage projects. Subsurface drainage system and the proper design of such a system require sufficient and reliable soil data, which are inherently variable in both space and time. Helmi (1998) revealed the problem of draining heavy soil of the low hydraulic conductivity overlying an impermeable layer by means of the combined system of the tile and mole drains. Katibeh (2004) developed the simulation mathematical model of Seepage from Lined Canal using Finite-Element Method. Imam (1991) developed methodology to design of the irrigation canals by integrating approach. Maskey (2002) developed the model groundwater remediation strategy using the global optimization algorithms. Maged Hussein (2001) expressed the development and application of geographic information system based utilities for a fully coupled surface-subsurface distributed parameters hydrologic model. Naveh (2004) studied MODFLOW to managing groundwater levels in

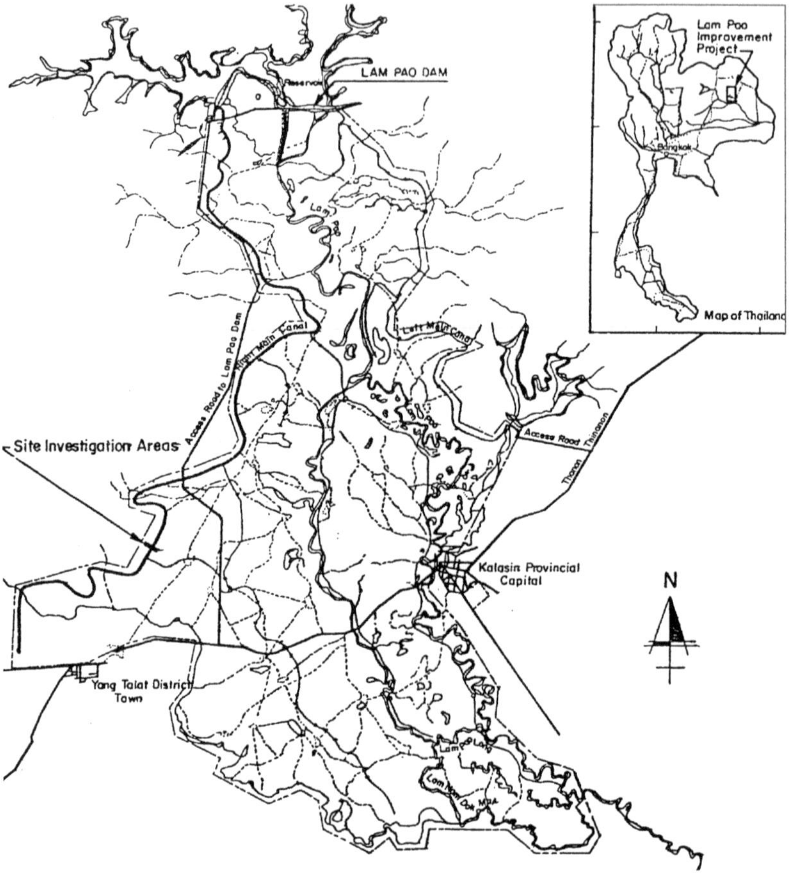

Fig. 2. Site investigation area.

an agricultural area with peat soils. The Hula Decision Support System (HDSS) is designed to aid the Hula site operators in managing groundwater levels in the Hula Lake region of Israel. Stormont (2005) studied the impact of unsaturated flow on pavement edgedrain performance. Reddi (2000) studied the permeability reduction of soil filters due to physical clogging. Wray *et al.*, (2005) studied three-dimensional model for moisture and volume changes prediction in expansive soils. The design criteria from ASCE has been developed into Standard Guidelines for the Design of Urban Subsurface Drainage, Standard Guidelines for Installation of Urban Subsurface Drainage, and Standard Guidelines for Operation and Maintenance of Urban Subsurface Drainage.

Fig. 3. Failure investigation.

2. Objective

The main objective of this research is to develop the software to be used by the designer in designing the protection measure where possibility of damage from ground water pressure including 3 components as shown below:

2.1. *Analyze failure index*

2.2. *Criteria for designing*

2.3. *Developed software for designing*

3. Methodology

The framework of this research, as can be seen in Fig. 4, contains many parts which are the important elements such as the investigation of failure problems, soil data, construction drawing, and the ground water level. Study parameters are effective from causal factor for decision of the

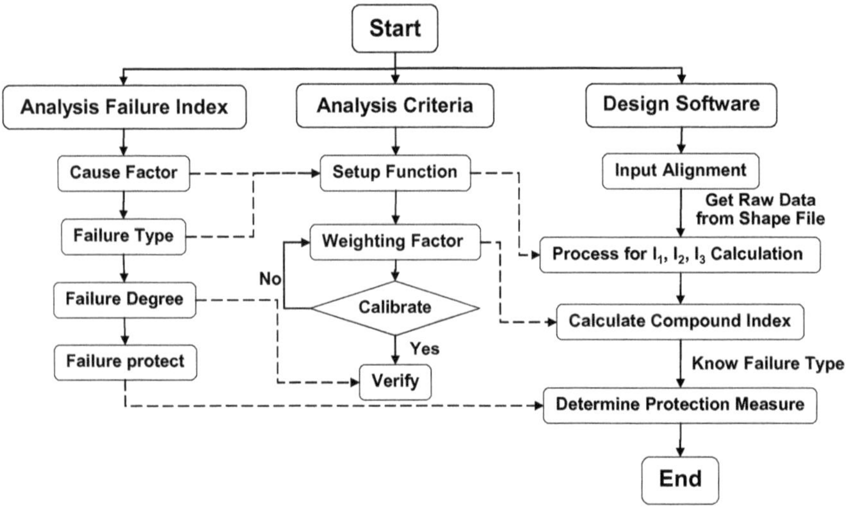

Fig. 4. Framework of research.

index identify problem. There are the uses of weighting factors to adjust corresponding investigated failure and study the protection measured by reducing the pressure from the ground water level. The software has been developed to be helpful for the designer to design another project.

The methodology of research for developing the software for decision making has been designed for 3 steps as follows;

3.1. *Failure index analysis*

The failure index analysis is to classify the damage along the way of canal. The study area is Lampao irrigation project in Thailand. The component of failure index composes of cause factor, failure type, failure degree and failure protect.

3.1.1. *Cause factor*

Various factors are the causes of the damage of canal; this shall include ground water level, soil type, excavated depth, side slope, settlement, construction, concrete strength, velocity erosion. In this research the 3 major parameters of concerning from ground water pressure are ground water level, soil type, and excavated depth see Fig. 5.

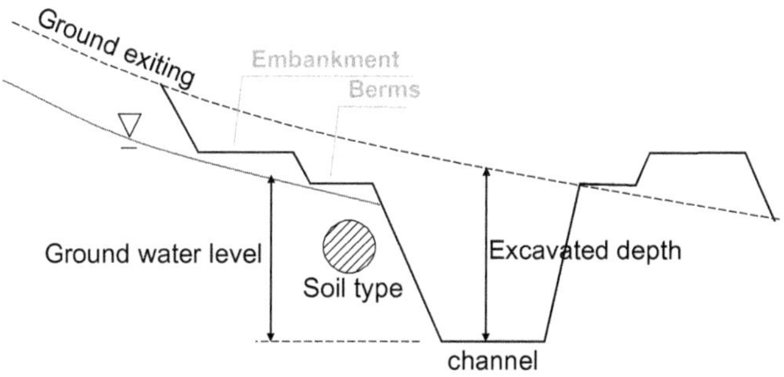

Fig. 5. Schematic of cause factor.

Fig. 6. Failure type.

3.1.2. *Failure type*

The failure type see in Fig. 6 is mainly caused by 3 factors as follow;

1. Overland flow past on the top bank canal.
2. Pressure from ground water level beside the wall of canal.
3. Concrete is expiry.

This research focuses on the failure from ground water pressure which firstly is behavior of cracking then soil particle move out from hole until the lined canal broken will be studied; see Fig. 7.

3.1.3. *Failure degree*

The failure degree is defined using the quantity of damage by consideration for every 100 meters from the penetration of damage along the way which is shown as an example in Fig. 8.

Fig. 7. The failure from ground water pressure.

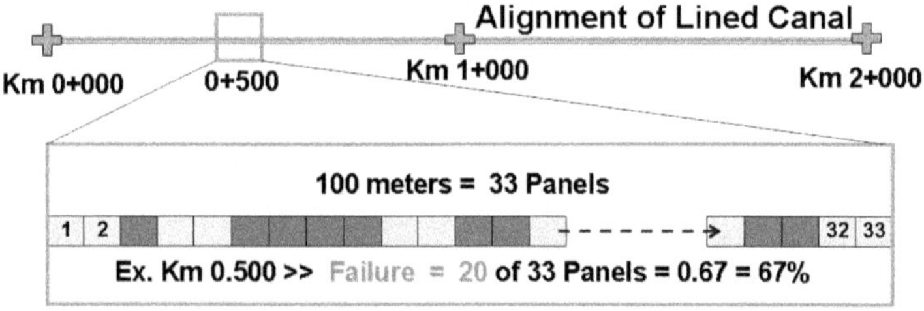

Fig. 8. Consider failure every 100 meters.

3.1.4. *Failure protection*

The failure protection can be selected as a type of subsurface drainage appropriate with the failure index. There are 5 types of failure protection which is shown as an example in Fig. 9.

3.2. *Analysis criteria*

In the study area, the concrete lined canal was damaged from the ground water pressure the causes come from different physical parameters including, the soil types, the ground water levels, and the excavation deeply.

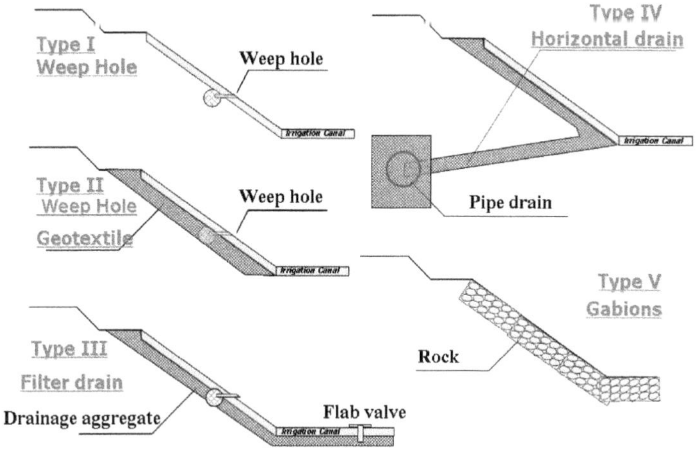

Fig. 9. Type of failure protection.

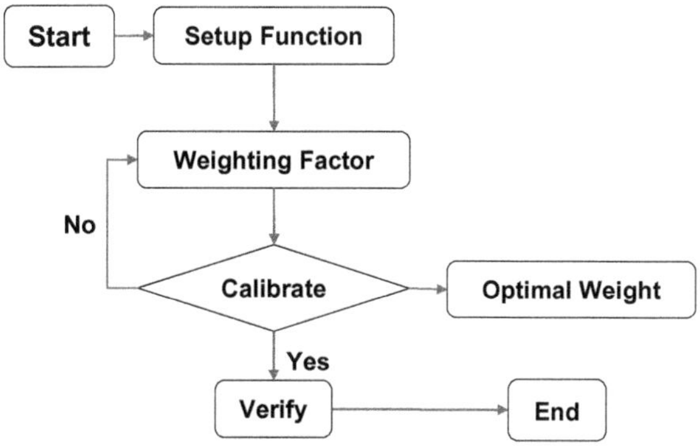

Fig. 10. Steps of analysis criteria.

The parameters must distribute in the same scale, by using the integer number between 0–100 to represent the minimum to the maximum values. The analysis criterion is the best solution from the comparison of the compound index with failure index for every considered position in any calibration zone. The analysis criteria are used for calculated solution at every position from indexes see Fig. 10.

3.2.1. *Compound index*

$$\text{Index} = \mathrm{I}_1(\mathrm{w}_1) + \mathrm{I}_2(\mathrm{w}_2) + \mathrm{I}_3(\mathrm{w}_3) \tag{1}$$

$$\text{Weighting} = \mathrm{w}_1 + \mathrm{w}_2 + \mathrm{w}_3 = 1 \tag{2}$$

Where I_1 = Soil type index (0–100: Clay to Sand), I_2 = G.W. Level index (0–100: Below to above bed canal), I_3 = Excavated depth index (0–100: canal depth below ground level).

3.2.2. *Weighting factor*

The weighting factors used to adjust the failure index come from the calculation fitting on the failure index from investigation.

Decision Variables are setting parameter of w_1, w_2, and w_3 as a group of decision variables.

Objective function

$$\text{Max.Z} = \text{R}^2 \tag{3}$$

Constraints

$$\begin{aligned} w_1, w_2, w_3 &>= 0 \\ w_1, w_2, w_3 &<= 1 \\ w_1 + w_2 + w_3 &= 1 \\ I_1, I_2, I_3 &>= 0 \\ I_1, I_2, I_3 &<= 100 \end{aligned}$$

The determination of coefficient as.

$$R^2 = \left\{ \frac{\sum(XY) - \frac{\left(\sum X\right)\left(\sum Y\right)}{n}}{\sqrt{\left[\sum X^2 - \frac{\left(\sum X^2\right)}{n}\right]\left[\sum Y^2 - \frac{\left(\sum Y^2\right)}{n}\right]}} \right\}^2 \tag{4}$$

Where X = Model Failure Index, Y = Observe Failure Index.

3.2.3. *Failure index*

For the failure index is the analysis from damage evaluation every in 100 meters along the canal, and changed scale to percentage for comparing with the compound index in each location.

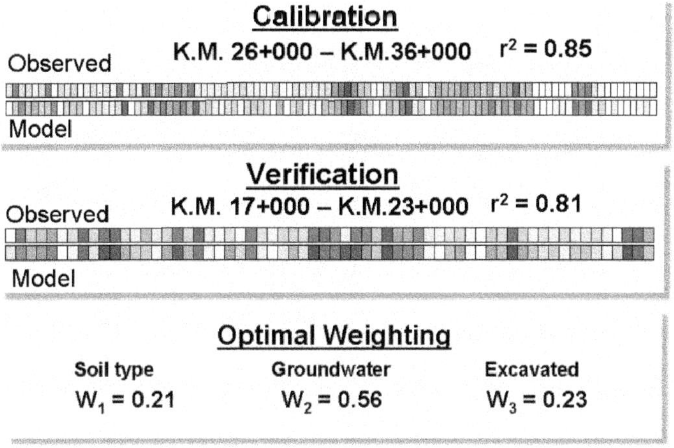

Fig. 11. Calibration and verification model.

3.2.4. *Calibration and verification*

The result of failure to compare observed and model conform the investigation by determination coefficient as 0.85 from calibration and 0.81 from verification see Fig. 11.

3.3. *Design software*

The design of decision support software for the designer to explore where on the alignment of canal which the damage is from ground water pressure and the type of drainage can be applied. The software designed is based on the Geographic Information System (GIS). By using the alignment of canal buffer at any parameter layers and recomputed for the index before analysis to failure index it would happen anywhere along the canal system. The flowchart show Fig. 12.

The software is named concrete lined canal which is designed based on the GIS tool. After the opening of the program, the user must open shape files form soil type, ground water level and contour around the boundary considering. This model has 5 steps as follow.

1. Input alignment, user tacking the Irrigation alignment from start point to end point, together with Kilometer see Fig. 13 after tacking finished, user must input bed level in the table; see Fig. 14.

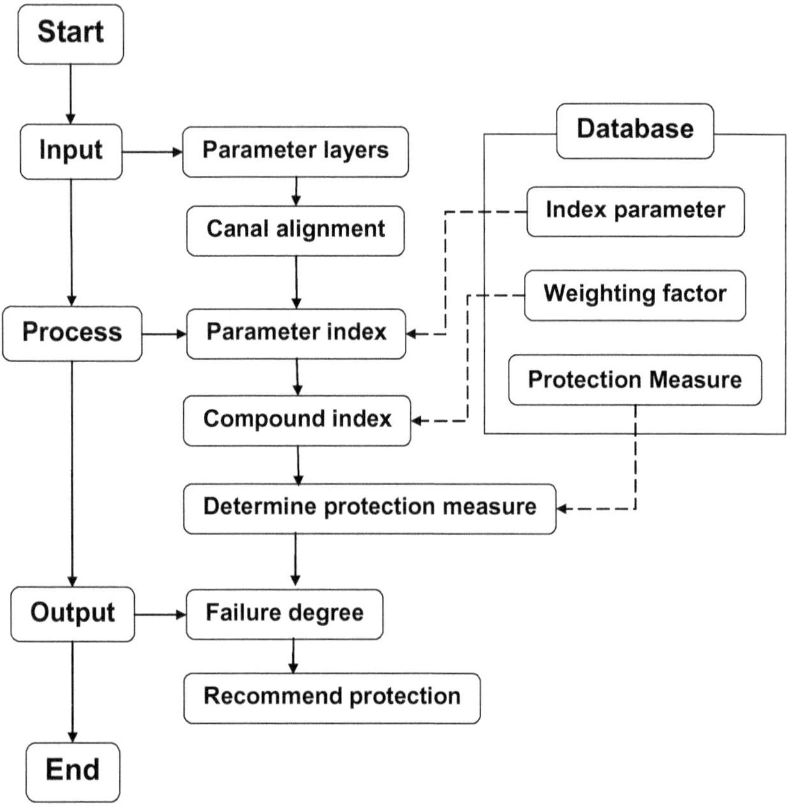

Fig. 12. Flowchart to develop software.

2. Get raw data from shape file, Fig. 15 shows all layer parameters and operated by buffering module to get the raw data along the alignment see Fig. 16.
3. Process for i_1, i_2, i_3 calculation, see Fig. 17, shows the database from analysis criteria for adjust raw data to index parameter and weighting factor. Figure 18, the result shows processing parameter indexes from raw data.
4. Calculate compound indexes and failure indexes, the software will automatically calculate for the compound index form function. Then the result shows the model failure indexes on the last column of the table; see Fig. 19.
5. Determining protection measure, the stored data from the database is keeping for classifying the protection measure. The color scale ranks

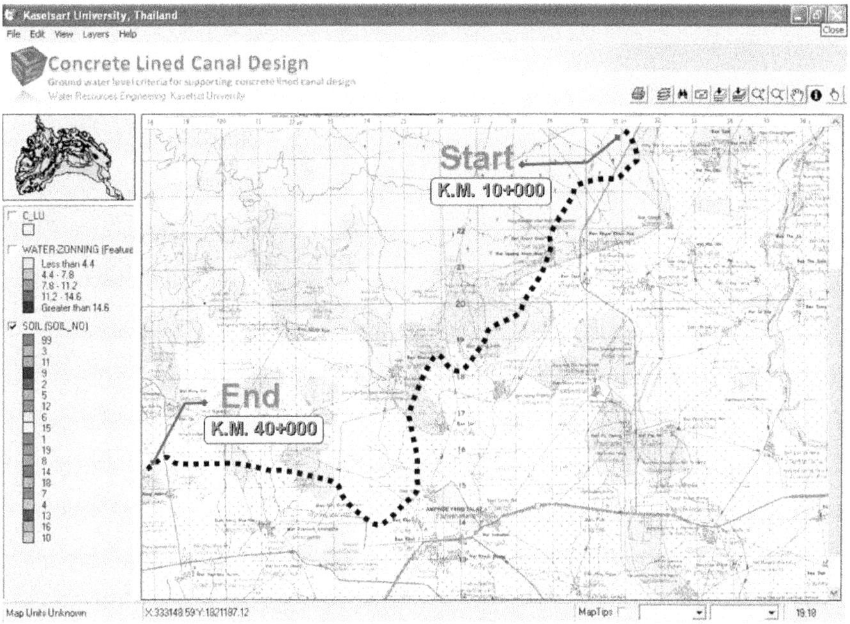

Fig. 13. Input alignment.

Km	(User Input) Bed Level (msl)	(User Input) Observe Failure Index
10.000	160.0	.20
10.100	159.9	.20
10.200	159.7	.40
10.300	159.6	.35
10.400	159.3	.40
10.500	159.0	.65
↓	↓	↓
39.900	145.0	.80
40.000	145.9	.90

Fig. 14. Get raw data from shape file.

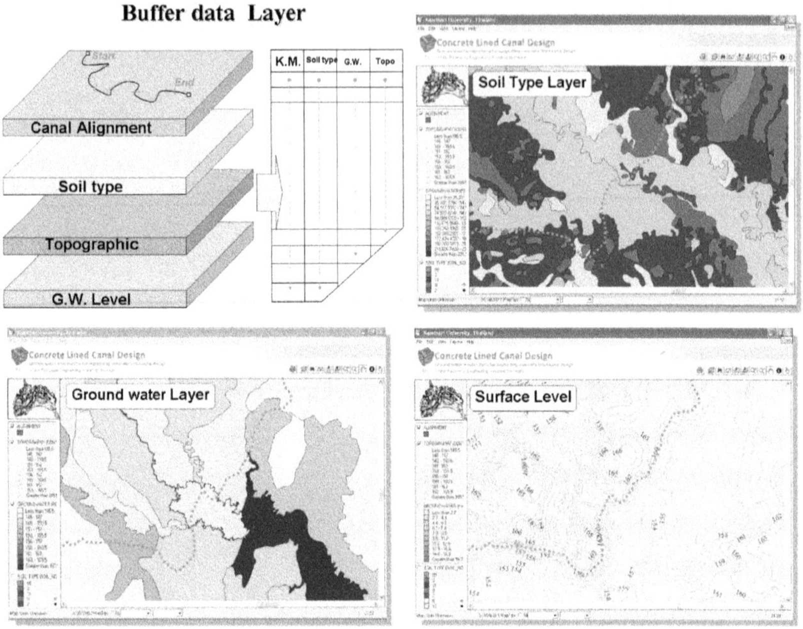

Fig. 15. Buffering data layer concept.

Km	Soil type	Ground water Level (msl)	Surface level (msl)
10.000	SC	160.0	163
10.100	CL	160.9	164
10.200	CL	160.7	163.5
10.300	CL	163.6	166.6
10.400	SM	166.3	169.3
10.500	SM	167.0	170.0
↓	↓	↓	↓
39.900	SC	146.0	145.0
40.000	SC	144.9	145.9

Fig. 16. Table buffering data.

Parameter index

Soil type	I_1
CL	0
SC	50
SM	100

Ground water	I_2
< 0	0
0 - 3	50
> 3	100

Excavated	I_3
< 0	0
0 - 5	50
> 5	100

Weighting factor

Soil	Ground water	Excavated
w1	w2	w3
0.21	0.56	0.23

Fig. 17. Parameter indexes and weighting factor.

Km	Soil type	Ground water Level	Excavated
10.000	50	0	50
10.100	0	50	50
10.200	0	50	50
10.300	0	100	100
10.400	100	100	100
10.500	100	100	100
↓	↓	↓	↓
39.900	50	50	0
40.000	50	0	0

Fig. 18. Process for i1, i2, i3 calculation.

Km	I_1	w_1	I_2	w2	I_3	w_3	Model Failure Index %
10.000	50	0.21	0	0.56	50	0.23	22
10.100	0	0.21	50	0.56	50	0.23	39.5
10.200	0	0.21	50	0.56	50	0.23	39.5
10.300	0	0.21	100	0.56	100	0.23	79
10.400	100	0.21	100	0.56	100	0.23	100
10.500	100	0.21	100	0.56	100	0.23	100
↓	↓	↓	↓	↓	↓	↓	↓
39.900	50	0.21	50	0.56	0	0.23	38.5
40.000	50	0.21	0	0.56	0	0.23	10.5

Fig. 19. Compound index and failure index.

Color Scale	Percent	Failure Level	Protection measure
	0%	No Failure	-
	1% - 20%	Little	Weep Holes (W.H.)
	20% - 40%	Moderate	(W.H.) + Geotextile
	40% - 60%	High	Filter drain
	60% - 80%	Very High	Horizontal drain
	80% - 100%	Severe	Gabions, flume

Fig. 20. Scale rank order from database.

white to red represented the percent of failure, Failure level and protection measure; see Fig. 20. Model did automatically determine protection measure. The table shows color scale bar on the row; for clearer view see Fig. 21.

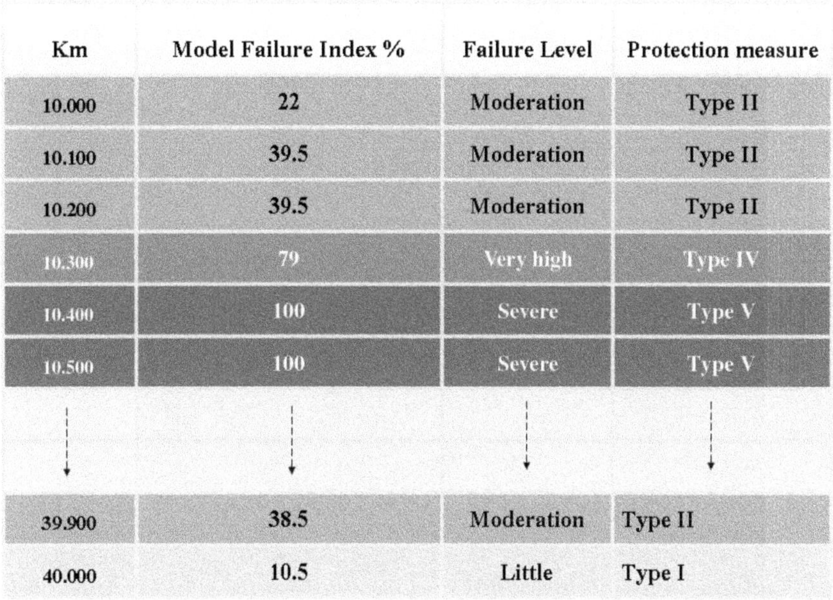

Km	Model Failure Index %	Failure Level	Protection measure
10.000	22	Moderation	Type II
10.100	39.5	Moderation	Type II
10.200	39.5	Moderation	Type II
10.300	79	Very high	Type IV
10.400	100	Severe	Type V
10.500	100	Severe	Type V
↓	↓	↓	↓
39.900	38.5	Moderation	Type II
40.000	10.5	Little	Type I

Fig. 21. Table result recommends the protection measure.

4. Study Case

The study case for comparing the software model result and the investigation result: The right main canal (RMC) of Lampao irrigation project in Kalasin province Thailand, The canal is about 90 kilometers long with 3.50 meters deep and 1.5 side slope could transfer 38 cubic meters of water per second. The canal is often damaged from ground water pressure then the investigation failure lined canal to the case study see Fig. 22 show the observation well and ground water contour map.

The results from software model are shown in the following Fig. 23. After finishing data input, the model will calculate and show the result of protection measures needed.

5. Results

The result shows decision support system for analysis failure index and selected subsurface drainage which were appropriate to the failure position by the software completely.

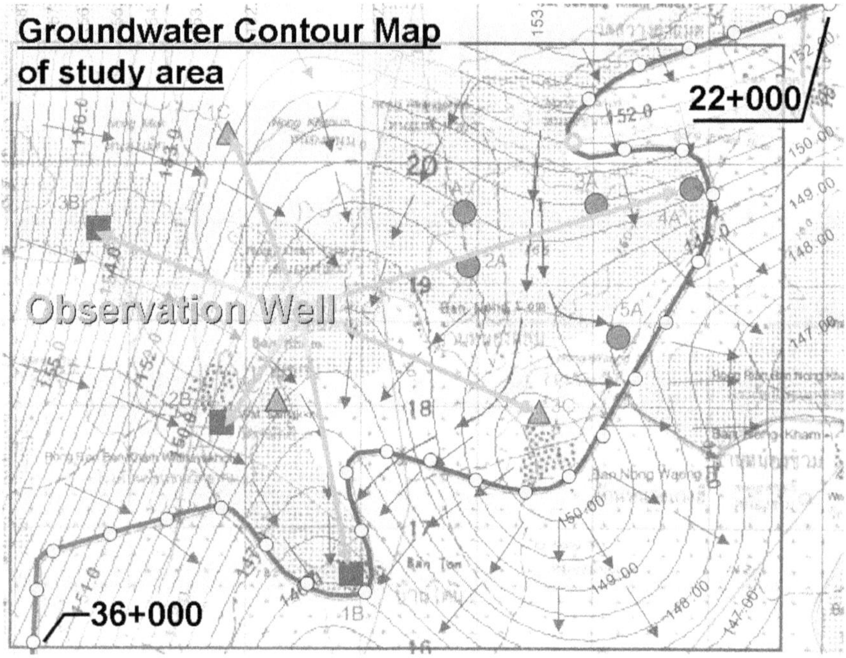

Fig. 22. Ground water map in study area.

K.M.	26.1	26.2	26.3	26.4	26.5	26.6	26.7	26.8	26.9	27.0	27.1	27.2	27.3	27.4	27.5	27.6	27.7	27.8	27.9	28.0
Type	II	IV	II	III	II	II	IV	-	II	II	II	-	II	II	I	I	II	V	-	II

K.M.	28.1	28.2	28.3	28.4	28.5	28.6	28.7	28.8	28.9	29.0	29.1	29.2	29.3	29.4	29.5	29.6	29.7	29.8	29.9	30.0
Type	II	V	III	IV	III	III	IV	III	V	II	II	I	I	II	II	I	II	II	II	II

K.M.	30.1	30.2	30.3	30.4	30.5	30.6	30.7	30.8	30.9	31.0	31.1	31.2	31.3	31.4	31.5	31.6	31.7	31.8	31.9	32.0
Type	-	III	-	II	II	I	I	-	II	II	III	IV	V	IV	III	II	I	-	III	II

K.M.	32.1	32.2	32.3	32.4	32.5	32.6	32.7	32.8	32.9	33.0	33.1	33.2	33.3	33.4	33.5	[illegible]	[illegible]	[illegible]	[illegible]	34.0
Type	IV	V	I	II	-	III	III	IV	III	II	III	II	III	III	[illegible]	[illegible]	[illegible]	[illegible]	[illegible]	[illegible]

K.M.	34.1	34.2	34.3	34.4	34.5	34.6	34.7	34.8	34.9	35.0	35.1	35.2	35.3	35.4	[illegible]	[illegible]	[illegible]	[illegible]	[illegible]	[illegible]
Type	III	-	-	-	I	I	I	IV	III	V	I	II	-	I	-	[illegible]	[illegible]	[illegible]	[illegible]	-

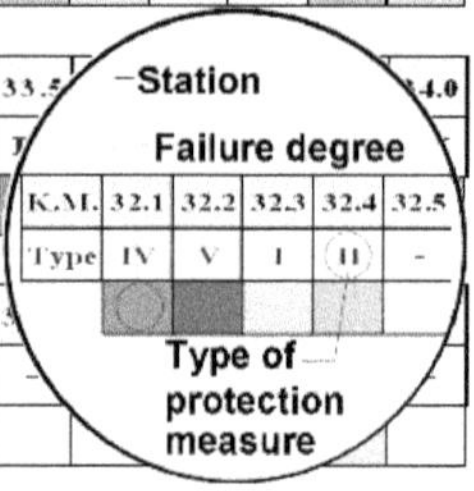

Fig. 23. The results analyze by software.

6. Conclusions

By using failure index the decision support model was appropriately developed for canal design and alignment in cut slope condition.

The next version which will be further developed is to include the slope stability index and seepage index.

References

1. H. R. A. S. Barootkoob, Concrete canal lining cracking in low to medium plastic soil, *Journal of Irrigation and Drainage Engineering* **51** (2002) 141–153.
2. K. Forrester, Subsurface drainage for slope stabilization, 2001, 208 pages.
3. V. Gaylord and G. P. M. Skogerboe, Irrigation maintenance and operations learning process, 1996, 358 pages.
4. T. Hartani, Drainage of sloping lands with variable recharge model validation and applications, *Journal of Irrigation and Drainage Engineering* **129**(4) (2003) 284–290.
5. H. M. Hathoot, Theory of pipe drainage assisted by mole drainage. *Journal of Irrigation and Drainage Engineering* **124**(2) (1998) 102–107.
6. M. Hussein, Development and application of geographic information system based utilities for a fully coupled surface-subsurface distributed parameters hydrologic model, *Conference Proceeding Part of: Bridging the Gap: Meeting the World's Water and Environmental Resources Challenges*, 2001.
7. E. H. Imam, Design of irrigation canals: integrated approach. *Journal of Irrigation and Drainage Engineering* **117** (1991) 852–869.
8. H. Katibeh, From lined canal using finite-element method. *Journal of Irrigation and Drainage Engineering* **130**(5) (2004) 441–444.
9. S. Maskey, Groundwater remediation strategy using global optimization algorithms, *Journal of Water Resources Planning and Management* **130**(3) (2002) 243–254.
10. M. M. Moustafa, Modeling of soil properties for subsurface drainage projects, *Journal of Irrigation and Drainage Engineering* **124**(4) (1998) 218–228.
11. N. Naveh, Managing groundwater levels in an agricultural area whit peat soils, *Journal of Water Resources Planning and Management* **130**(3) (2004) 243–254.
12. L. N. Reddi, Permeability reduction of soil filters due to physical clogging, *Journal of Geotechnical and Geoenvironmental Engineering* **126**(3) (2000) 236–246.
13. J. C. Stormont, Impact of unsaturated flow on pavement edgedrain performance, *Journal of Transportation Engineering* **131**(1) (2005) 46–53.
14. W. K. Wray, Three-dimensional model for moisture and volume changes prediction in expansive soils, *Journal of Geotechnical and Geoenvironmental Engineering* **131**(3) (2005) 311–324.

Advances in Geosciences
Vol. 17: Hydrological Science (2008)
Eds. Namsik Park *et al.*

INTELLIGENT MODEL FOR PIPE NETWORK MANAGEMENT

SUWATANA CHITTALADAKORN
Associate Professor, Ph.D., Department of Water Resources Engineering, Faculty of Engineering, Kasetsart University, 50, Phaholyothin Rd., Lardyao, Chatuchak, Bangkok 10903, Thailand.
fengswc@ku.ac.th

PAKORN DITTHAKIT
Lecturer, D.Eng., School of Engineering and Resources Management, Walailak University, 222 Thaiburi, Thasala District, Nakhon Si Thammarat 80160, Thailand.
dpakorn@hotmail.com

The stochastic optimization method, namely Differential Evolution (DE), was applied to develop a computer program for pipe network management comprising three options: new pipe system design, pipe replacement plan, and optimal pump scheduling. The developed program was linked to a hydraulic simulation model for a pressurized pipe network, namely EPANET. To demonstrate model's applicability, it was tested by study cases of a water distribution network of Nakhon Sawan province and the transmission conduit and tunnel system of Metropolitan Waterworks Authority (MWA), Thailand. It was firstly found that in case of the new pipe system design the developed program could optimally select the pipe and pump sizes under several hydraulic constraints. Secondly, in case of the pipe replacement plan the program showed satisfying performance to determine the most appropriate pipelines to be replaced under specified constraints with the replacement cost close to the limited budget. Finally, the optimal pump scheduling module was capable to minimize the operational cost of electricity under the constraints of hydraulic conditions and Time of Use (TOU) electricity tariff rate.

1. Introduction

With the rapid growing of urban areas at present, water demands become more and more increasing. Several government offices such as Metropolitan Waterworks Authority (MWA), Provincial Waterworks Authority (PWA), and local administration offices, etc are responsible to efficiently administer and operate pipe networks for public services under limited budget. The pipe network system is a fundamental infrastructure for community. This system is composed of pipe, pump, storage tank, reservoir, etc. The efficient pipe network system must be able to sufficiently supply an amount of water for both household and industrial consumers under limited investment and desired pressure constraints. Pipe network management, i.e. new pipe system design, pipe replacement plan, and optimal pump scheduling become much more complex problems. For this reason, it is necessary to develop intelligent model for pipe network management to improve performance of pipe network management. In this study, the intelligent model for pipe network management was developed. It includes three modules as follows: (1) the pipe network design module was developed for new pipe network design for fixed and operation cost minimization under hydraulic constraints; (2) pipe replacement plan module was developed to replace the old pipes with the new ones by considering pipe age, pressure, flow velocity, and its priority for pipe replacement under limited budget; and (3) pump scheduling module was developed to plan pump operation scheduling of transmission plant to minimize the operational cost of electricity corresponding to water demand and hydraulic constraints. Such model was developed based on the linked simulation-optimization approach. The public domain of closed conduit hydraulic model namely EPANET and stochastic optimization method namely Differential Evolution (DE) were utilized for model development.

Many researches have been conducted for pipe network management. Duan *et al.* (1990) studied reliability-based of pumping system design and pump operations. Pezeshk and Helweg (1996) applied Adaptive Search Algorithm (ASA) and KYPIPE model for On-Line optimal pump operations due to its faster computing method compared with genetic algorithm and simulated annealing. Nitivattananon *et al.* (1996) used dynamic programming for determining optimal pump operation scheduling for short and long time periods. Frank and Steve (1999) developed economic model for pipe replacement planning. The model could predict leakage of pipe network system from existing pipe leakage database for evaluation of improvement budget. Milan (2000) developed design and improvement method and reliability audit of pipe network using genetic algorithm. Sakarya and Mays (2000) applied Augmented Lagrangian Penalty method for determining optimal pump operations of transmission system with water quality considerations. Ditthakit (2001) applied simulated annealing approach for pie network design. Kleiner *et al.* (2001) applied dynamic programming for pipe network improvement planning. The model was able to select optimal method and duration for pipe improvement. Ariamalar *et al.* (2002) studied pipe network improvement by considering both fixing and replacing pipe measures. Moradi-Jalal *et al.* (2004) developed WAPIRRA model based on genetic algorithm to select pump size, number of pump, pump scheduling in irrigation system.

This paper presents the theories and features used for developing the intelligent model for pipe network management. It was then demonstrated model's capabilities for various study cases from existing system of a water distribution network.

2. Different Evolution Algorithm

The Differential Evolution Algorithm (DE) is selected in this study for seeking the optimal solution responding to the desired objective function. This algorithm is capable for searching nearest solution rapidly and directly. It is classified to be the stochastic search space algorithm like genetic algorithms (GAs), but it is less complex than genetic algorithms. DE uses real number instead of binary number transformation, i.e. needs no bit — String encoding; hence it is more faster speed for finding solution, especially considering the discrete decision variable like pipe and pump sizes in the optimization problems. DE also fits to non-linear problems and non-differentiate function. DE is successfully applied in various optimization problems and water resource engineering problems, for example the pipe design and the replacement of water distribution system (Charocnongart, 2005; Prempree, 2005), reservoir operation (Thasaduak and Chittaladakorn, 2005), and water pumping system (Babu and Angira, 2005). The principle procedure for approaching the solution by Differential Evolution proposed by Price and Storn (1998) can be generally described as follow.

2.1. *Initial population*

In the first step, a set of numbers of initial population is randomly generated from specified boundary constraints using the uniformly distribution random numbers to cover the entire solution space. Each of those has its cost value or fitness value.

$$P_{G=0} = x_j^{(L)} + rand_j\,[0,1] \cdot \left(x_j^{(U)} - x_j^{(L)}\right), \quad j \in [1, D] \tag{1}$$

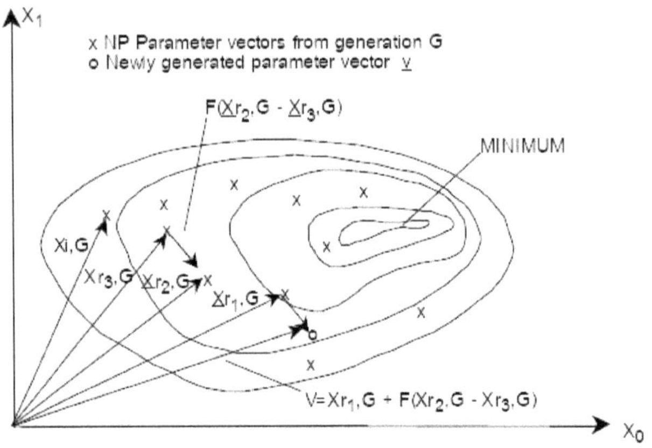

Source: Price and Storn (1998).

Fig. 1. Procedure to search the mutant vector from 2 variables population.

where $rand_j$ $[0,1]$ = a uniformly distributed random value in range $[0,1]$; $x_j^{(U)}$, $x_j^{(L)}$ = upper and lower boundary of each D real parameters.

2.2. *Mutation*

The next step is to mate couple of population, calculate the different value between them, and multiply the different value by weighting factor or mutation factor (F). Then a newly difference population vector is obtained. Figure 1 shows the example of procedure to search the Mutant Vector from 2 variables population. The sub procedures are as follows.

1) Defining the Target Vector ($X_{i,G}$).
2) Randomly select three Vectors ($X_{r1,G}, X_{r2,G}, X_{r3,G}$) from the initial population which are different from Target Vector.
3) Calculate the Mutant Vector ($V_{i,G+1}$) from the relation:

$$V_{i,G+1} = X_{r1,G} + F(X_{r2,G} - X_{r3,G}) \tag{2}$$

where $i = 1, 2, 3, \ldots, NP$; $X_{i,G}$ = target vector; $V_{i,G+1}$ = mutant vector; $X_{r1,G}, X_{r2,G}, X_{r3,G}$ = random vector; F = weighting factor.

2.3. *Crossover or recombination*

This process is performed to increase the diversity of the perturbed parameter vectors. As shown in Fig. 2, the trial vector, $U_{i,G+1}$ is produced

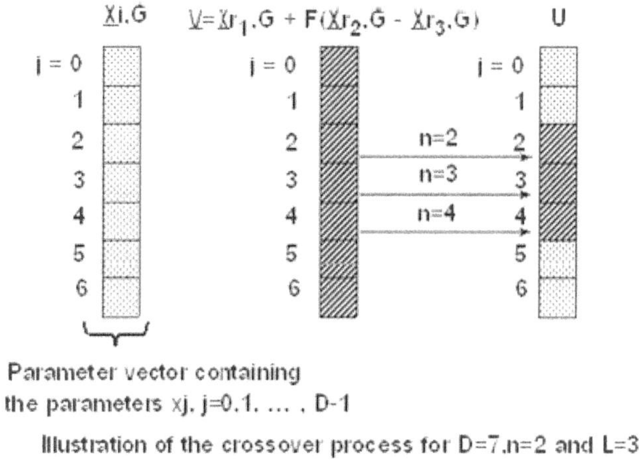

Source: Price and Storn (1998).

Fig. 2. Procedure to crossover step of the target vector and mutant vector which its value D = 7.

by duplicating some elements of the mutant vector, $V_{i,G+1}$ or some elements of the target vector, $X_{i,,G}$ depending on criteria presented in mathematical form as:

$$\begin{aligned} U_{i,G+1} &= V_{ji,G+1} \quad \text{if } (\text{randb}(j) \leq CR) \quad \textit{or} \quad j = \text{rnbr}(i) \\ U_{i,G+1} &= X_{ji,G} \quad \text{if } (\text{randb}(j) > CR) \quad \textit{or} \quad j \neq \text{rnbr}(i) \end{aligned} \tag{3}$$

where $U_{i,G+1}$ = trial vector; $V_{i,G+1}$ = mutant vector; $X_{i,G}$ = target vector; $\text{randb}(j)$ = the jth evaluation of a uniform random number generator with outcome $\in [0,1]$; $\text{rnbr}(i)$ = a randomly chosen index $\in 1, 2, 3, \ldots, D$; CR = the crossover constant $\in [0,1]$.

2.4. *Selection*

The next population generation $(G + 1)$ is created from the newly better population by comparing the function value of trial vector with target vector. If the objective function value of trial vector is better than that of target vector, the next population generation $(G+1)$ will be replaced by the trial vector, and vice versa.

The process of mutation, crossover, and selection is routinely repeated until a termination criterion, for example the maximum number of generation is satisfied. The algorithm then terminates proving the best point

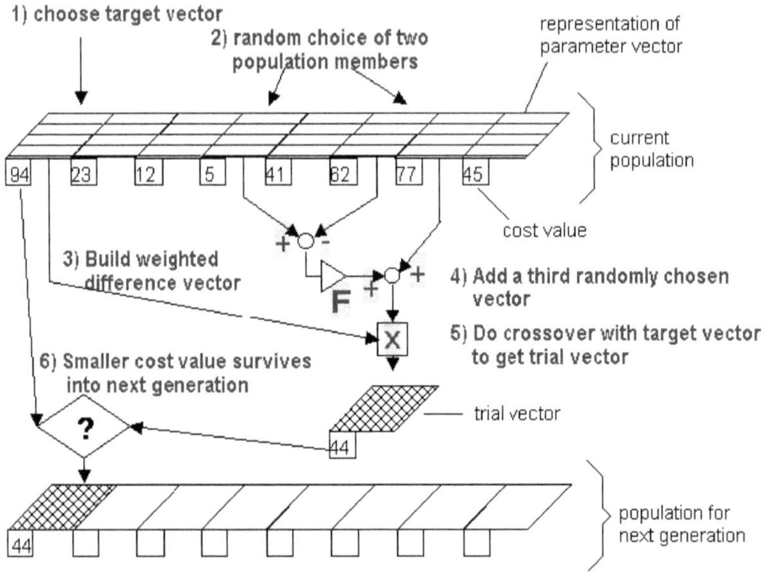

Source: Price and Storn (1998).

Fig. 3. Procedure of solution seeking by Differential Evolution algorithm.

that has been explored over all the generations. Figure 3 shows differential evolution procedure.

3. EPANET Model

EPANET is public domain developed by United States Environmental Protection Agency (US EPA). It can perform extended period simulation of hydraulic and water-quality behavior within pressurized pipe networks. A network can consist of pipes, nodes (pipe junctions), pumps, valves and storage tanks or reservoirs. EPANET tracks the flow of water in each pipe, the pressure at each node, the height of water in each tank, and the concentration of a chemical species throughout the network during a simulation period comprised of multiple time steps. EPANET's Windows user interface provides a visual network editor that simplifies the process of building piping network models and editing their properties. In addition, the EPANET Programmer's Toolkit is a dynamic link library (DLL) of functions that allow developers to customize EPANET's computational engine for their own specific needs.

4. Model Development and Features

4.1. *Model framework*

The intelligent model for pipe network management is the linked simulation-optimization model as shown in Fig. 4. It was developed using Delphi programming language. The Different Evolution and EPANET model serve as optimization and hydraulic simulation models, respectively. The EPANET model is routinely invoked during running process to check hydraulic constraints. The model consists of three sub-modules: pipe network design module, pipe replacement planning module, and pump scheduling module. To run the model, it requires pipe network and pump information, hydraulic constraints and economic data as well as optimization's parameters.

4.2. *Objective functions and constraints*

The objective function and constraints were designed for three sub-modules as follows.

4.2.1. *Pipe network design module*

The objective of this module is to select the optimal pipe and pump sizes to minimize the fixed and operational cost under hydraulic constraints. The considered decision variables are, therefore, pipe and pump sizes.

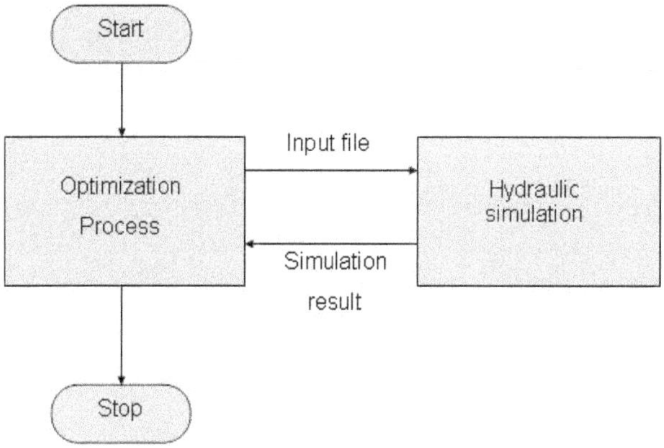

Fig. 4. Linked simulation-optimization model.

The penalty method was developed to prevent the violation of hydraulic constraints of the final obtained solution. The objective function and constraints can be mathematically expressed as:

$$\begin{aligned}\text{Minimize cost} = {} & \text{Network cost} + \text{Operation cost} \\ & + \text{Penalty cost}\end{aligned} \tag{4}$$

Subject to:

$$\sum Q_{\text{in}} - \sum Q_{\text{out}} = Q_e \tag{5}$$

$$\sum_{K=1}^{N} \Delta H_K = 0 \tag{6}$$

$$d_{\min} \leq d_i \leq d_{\max}; i = 1, 2, \ldots, \text{no. of pipes} \tag{7}$$

$$P_{\min} \leq P_i \leq P_{\max}; i = 1, 2, \ldots, \text{no. of nodes} \tag{8}$$

$$V_{\min} \leq V_i \leq V_{\max}; i = 1, 2, \ldots, \text{no. of pipes} \tag{9}$$

$$HL_i \leq HL_{\max}; i = 1, 2, \ldots, \text{no. of pipes} \tag{10}$$

The Eq. (5) shows the conservation mass constraint, where Q_{in}, Q_{out}, and Q_e are the quantities of node inflow, node outflow, and node water demand, respectively. The Eq. (6) is energy conservation constraint, that is, the sum of head losses around a pipe loop is equal to zero. The Eqs. (5)–(7) are the constraints of pipe size, required node pressure, limited pipe flow velocity, and limited head loss, respectively. The network cost is the sum of pipe and pump cost and the operation cost is electricity cost of pumping system. The penalty cost is the sum of the product of penalty coefficient (usually between 0.5 to 4 times of network cost) and the number of pipes and nodes which violate the hydraulic constraints. It can be mathematically presented as:

$$\text{Penalty cost} = \sum_{i=1}^{R} C_i r_i \tag{11}$$

where R = the number of considered constraints; C_i = penalty coefficient; r_i = the number of pipes and nodes which violate the hydraulic constraints.

4.2.2. *Pipe replacement planning module*

This module was developed to effectively budget for pipe replacement planning. The old pipes, which are replaced with the new ones, are the pipe age over 25 years, small in sizes that cause the pipe flow velocity over 3 m/s. In addition, they depend on their urgent priority, for example traffic improvement, increase in water consumer, and pipe leakage. Again, the penalty cost is used to prevent the violation of hydraulic constraints of the final obtained solution. The objective function is to minimize the replacement cost, and the decision variable considered herein is the type of pipes ID to be replaced:- no replacement, replacement with the existing size, and replacement with the larger size. The objective function and constraints can be mathematically expressed as:

$$\text{Minimize cost} = \sum_{i=1}^{n} \text{replacement cost}_i + \sum_{i=1}^{n} \text{penalty cost}_i \tag{12}$$

Subject to:

$$AP_i > 25 \tag{13}$$

$$VP_i > 3 \tag{14}$$

$$SP_i = UP_i \tag{15}$$

$$\text{Replacement cost} \leq \text{Budget} \tag{16}$$

where AP_i = age of pipe (years); VP_i = flow velocity in pipe (m/s); SP_i = pipe selection results:- 0 = no replacement, 1 = replacement with the existing size, 2 = replacement with the larger size; UP_i = urgent priority:- 0 = no rapidity, 1 = rapidity: the existing size replaced, 2 = rapidity: the larger size replaced.

The Eqs. (13) and (14) represent the constraint for selecting to replace the pipes ID having the pipe age over 25 years and pipe flow velocity over 3 m/s, respectively. The function of Eq. (15) is to verify agreement between the assigned urgent priority and the replaced pipe selection results. In addition, the replacement cost obtained from model's selection must not exceed the available budget as presented in Eq. (16). If the a certain pipe ID to be replaced satisfy the constraints, Eq. (13)–(15), and the replacement cost of the whole pipe system does not exceed the available budget, the penalty cost is equal to zero. Conversely, if any pipes ID to be replaced do not satisfy only one constraint or more constraints, Eqs. (13)–(15), or the replacement cost of the whole pipe system is more than the available budget, the penalty cost become a huge number of cost. In general, the more violation situation of the constraints happen, the more penalty cost is obtained. With this process, the final obtained solution escapes from the violation of determined constraints.

4.2.3. *Pump scheduling module*

The purpose of pump scheduling module is to use for planning the pump system operations with the electricity cost minimization under hydraulic constraints. The decision variable herein is pump speed with respect to control horizon. It can present the objection function and constraints as follows.

$$\text{Minimize cost} = \text{Eletricity cost} + \text{Penalty cost} \tag{17}$$

Subject to:

$$\sum Q_{\text{in}} - \sum Q_{\text{out}} = Q_e \tag{18}$$

$$\sum_{K=1}^{N} \Delta H_K = 0 \tag{19}$$

$$T_{\min} \leq T_i \leq T_{\max};\, i = 1, 2, \ldots, \textit{control horizon} \tag{20}$$

$$T_{i,\text{initial}} = T_{i,\text{last}};\, i = 1, 2, \ldots, \textit{control horizon} \tag{21}$$

Eqs. (18) and (19) are the same equations as discussed above in Eqs. (5) and (6), respectively. Equation (20) shows that the water levels in the clear water tanks are between the minimum and maximum of storage level. Equation (21) shows that the water levels in clear water tanks at the initial simulation time are equal to those at the last simulation time. For the penalty cost, it is considered as the sum of product of penalty coefficient and the number of clear water tanks which violate the hydraulic constraints. It can be mathematically presented as:

$$\text{Penalty cost} = \sum_{i=1}^{W} C_i K_i \tag{22}$$

where W = the number of considered constraints; C_i = penalty coefficient; K_i = the number of clear water tanks which violate the hydraulic constraints.

5. Model Application and Results

To demonstrate the model's applicability, it was investigated by two study cases of water distribution network of Nakhon Sawan province and the transmission conduit and tunnel system of Metropolitan Waterworks Authority (MWA), Thailand. The water distribution network of Nakhon

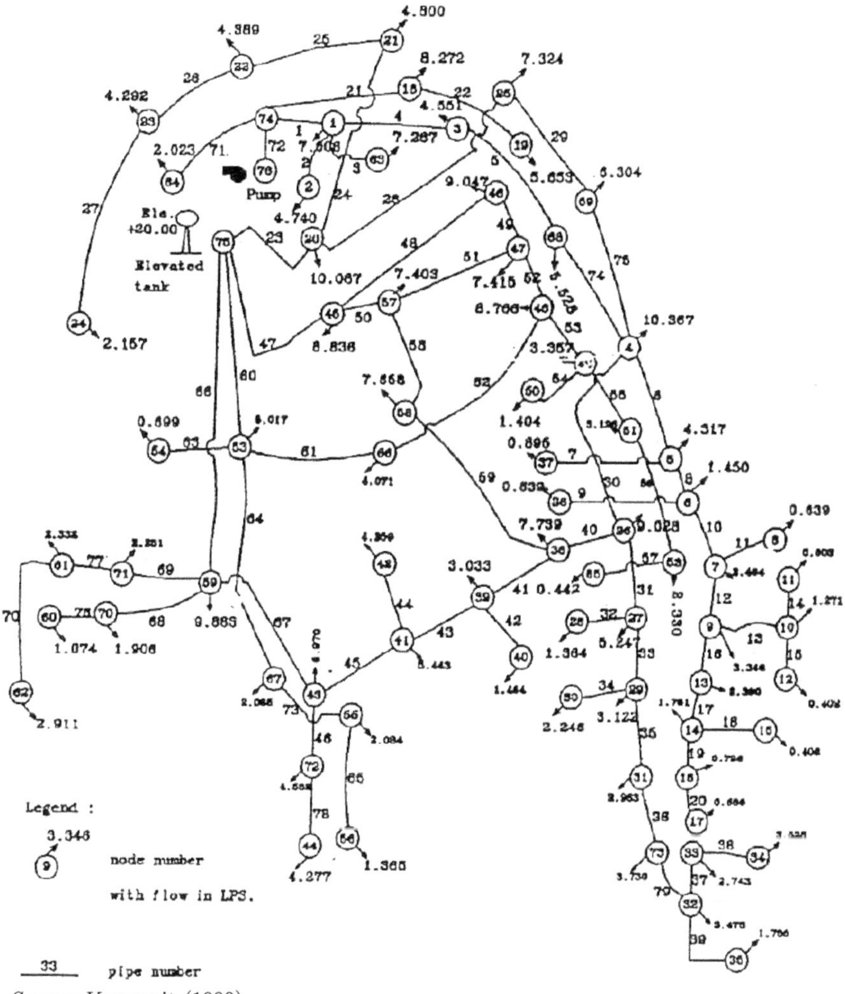

Source: Kosunwit (1990).

Fig. 5. Schematic diagram of water distribution network of Nakhon Sawan province.

Sawan province as presented in Fig. 5 consists of 79 pipes, 76 nodes, 2 sources: a storage tank and a pumping station. The transmission conduit and tunnel system of Metropolitan Waterworks Authority (MWA) as shown in Fig. 6 have large capacity of approximately 3,000,000 m^3/day. This system is supported by 13 pumps in the 3 transmission pumping stations. The study results can be discussed for each module as follows.

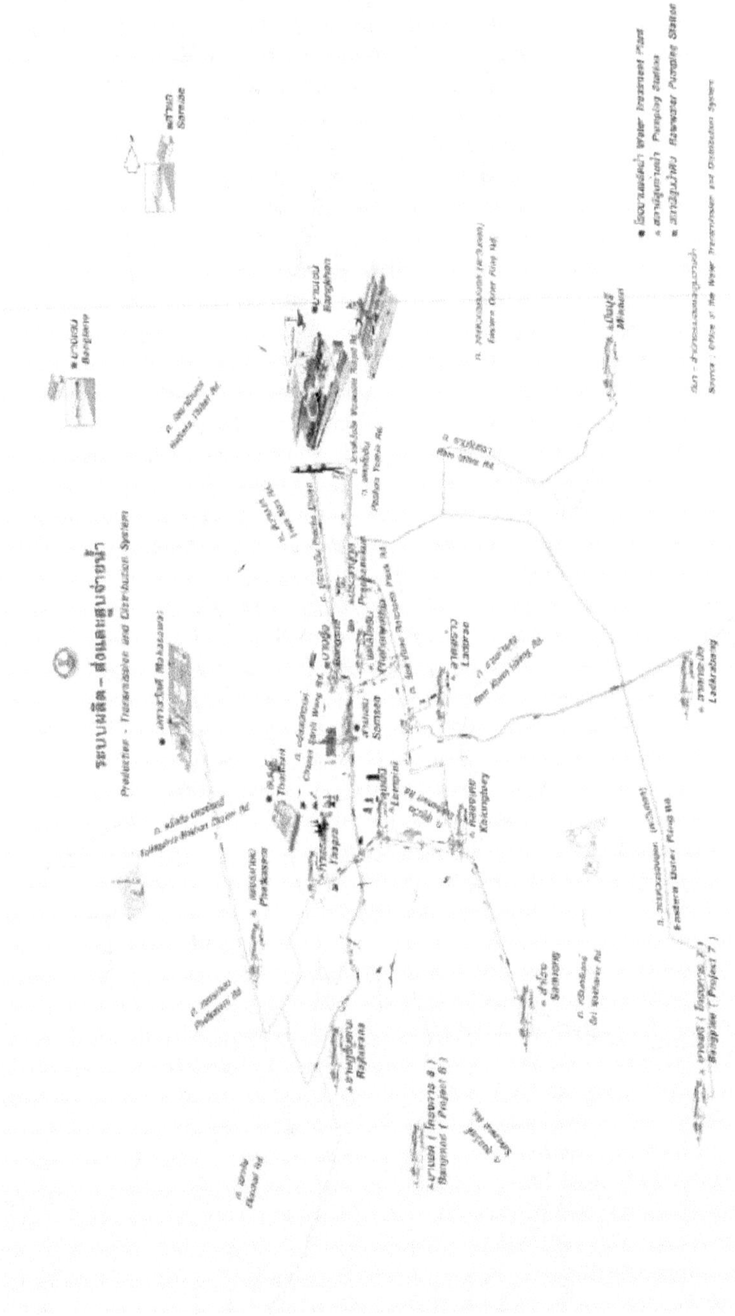

Source: Wiriyagulopas (2006).

Fig. 6. Schematic diagram of MWA.

5.1. *Pipe network design*

For model testing purpose, the used Hazen-William roughness coefficient was 140, peaking factor was 1.50, and the hydraulic constraints comprise:- (1) the allowable maximum and minimum pressure were 10 and 60 m, respectively; (2) the allowable flow velocity was in the range of 0.3 to 1.50 m/s; and (3) head loss was less than 10 m/km. Three study cases were experimented:- (1) to select optimal pipe size only; (2) to select optimal pipe size with operation cost consideration; (3) to select optimal pipe and pump sizes with operation cost consideration. The study results can be concluded as follows. In the first case, the performance comparison of three optimization approaches:- traditional, Simulated Annealing (SA), and DE approaches for selecting pipe size were undertaken. The results are presented in Table 1. It was found that the pipe cost obtained by using DE could be more saved 10.63, 10.17 percent than those by using traditional and simulated annealing methods, respectively. Furthermore, it was found that the obtained results correspond to all determined hydraulic constraints.

The performance comparison among three study cases as described above are presents in Table 2. It was found that in the first case the lowest pipe cost of 34,246,080 Thai Baht was found, but the highest operational cost and annual total costs of 380,817 and 3,914,688 Thai Baht were obtained, respectively. In the second case, the pipe cost of 34,353,230 Thai Baht was higher than that of the first case of 0.33%, but it gave lower operational cost (350,871 Thai Baht/year) than the first case of 7.86% and the annual total cost of 3,89,160 Thai Baht. This shows that the developed

Table 1. Performance comparison of selecting pipe size among three optimization approaches.

	Length (m)		
Pipe size (mm)	Traditional	SA	DE
40	680	680	680
55	1,970	1,970	1,670
65	4,020	4,020	3,120
80	4,100	4,100	5,000
100	4,980	4,980	4,330
150	11,070	11,070	10,770
200	4,600	5,000	11,250
250	7,200	6,800	5,600
300	6,940	6,940	3,140
Total pipe cost (Thai Baht)	38,321,305	38,125,305	34,246,080

Table 2. The performance comparison among three study cases.

List	Case 1	Case 2	Case 3
1. Pipe cost [1]	34,246,080	34,353,230	34,433,165
2. Pump cost [1]	120,000	120,000	85,000
3. Operational cost [2]	380,817	350,871	285,823
4. Annual total cost [1]	3,914,688	3,896,160	3,824,914

[1] unit = Thai Baht; [2] unit = Thai Baht/year.

model could appropriately select the pipe size leading to a decrease in operational cost under determined hydraulic constraints. For the third case, the pipe cost was higher than that of the first and second cases of 0.55% and 0.22%, respectively, but it could select smaller pump size and could save pump cost more than 29.17% compared to the first and second cases. In addition, it could save more annual operational cost of 24.94% and 18.54% than those of the first and second cases, respectively. When compared the results of the pipe network design among thee study cases, it was found that of the third case was the most appropriate way for detailed design and construction. From the results above, it shows that the developed model can select the optimal pipe and pump sizes leading to more save operational cost under determined hydraulic constraints.

5.2. *Pipe replacement planning*

The water distribution network of Nakhon Sawan province was also used as a case study. The limited budget of 12 million Thai Baht was assumed for the testing purpose. It was found that some replaced pipes did not correspond with the assigned urgent priority and some pipes having the age over 25 years did not be replaced. This may because the determined penalty cost of both constraints might be not appropriate and it is rather difficult to determine the urgent priority for the complicated pipe network. For future research, the model should be developed to fill up this research gap. However, when investigating pipe flow velocity constraint, there is no pipe having pipe flow velocity more than the criteria of 3 m/s. And the replacement cost of 11,999,702 Thai Baht was less than the available budget of 12,000,000 Thai Baht. The study results showed that the model could select the most appropriate pipelines to be replaced with the reasonable replacement cost closed to the given budgets. Consequently, those selected pipelines can be used as a guideline for planner to efficiently plan for pipe replacement schedule with the most beneficial use of available budget.

5.3. *Optimal pump scheduling*

The transmission conduit and tunnel system of Metropolitan Waterworks Authority (MWA) was used as a study case in this section. The model was investigated in four study cases as follows:- (1) constant flow rate to distribution pumping station; (2) varied flow rate to distribution pumping station depending on TOU (3); varied flow rate to distribution pumping station depending on TOU with increasing the storage capacity of 50%; and (4) all 3 transmissions pumps in the west bank, Mahasawat Transmission Pumping Station, shutting down. The results showed that the proposed model could appropriately set pumping schedule for first three cases. The electricity power consumption during high tariff rate at on peak duration could be decreased 17.84, 23.56, and 32.36% and the total electricity cost could be decreased 15.95, 18.00, and 20.22%, respectively. In addition, it showed that the third case gave the most saved cost and its total cost was 106,226.67 Thai Baht per day or approximately 39 million Thai Baht per year. For the forth case, the simulation results showed that the 10 remained pumps in the east bank transmission pumping station could not enough capacity to transmit water to the west bank, Phetkasem and Ratburana Distribution Pumping Stations. As the results of this study, it could be concluded that the most appropriate scheme was the second case. This is because the investment cost would be much lower due to no need increasing the capacity of the reservoirs, although it could save less electricity cost than that of the third case.

6. Conclusion

This paper presents the application of differential evolution, which is the stochastic optimization technique, for developing the intelligent model for pipe network management. The model include three sub-modules:- pipe network design module, pipe replacement planning module, and pump scheduling module. This developed model was linked between EPANET toolkit, which is pressurized pipe network hydraulic simulation model, and Differential Evolution method. The model was verified with water distribution network of Nakhon Sawan province and the transmission conduit and tunnel system of Metropolitan Waterworks Authority (MWA), Thailand. The results can be concluded as follows:- (1) Differential Evolution can be efficiently applied for water distribution problem; (2) EPANET toolkit is efficient and can be applied for pressurized pipe network; (3) the selection of the optimal DE's parameters and penalty coefficient is important to obtain the nearest optimal solution; (4) the developed model is random based search, so the obtained results from each run may be not the same, but it close to the optimal solution; and 5) this model can appropriately select pipe size, plan pipe replacement, and schedule pump operations under the determined hydraulic constraints.

Acknowledgments

The writers wish to express sincere gratitude to Kasetsart University Research and Development Institute (KURDI) for financial support for this research project. The second writer is thankful to Institute of Research and Development Walailak University for providing financial help for attending oral presentation at 5th Asia Oceania Geosciences Society Conference, AOGS 2008, Busan, Korea. Finally, the writers are grateful to anonymous reviewers for their great recommendations to improve the quality of this paper.

References

1. S. Ariamalar, M. C. Robert, and S. Mano, Costs for water supply distribution system rehabilitation, *Environmental protection agency*, available at: *http://www.epa.gov/ORD/NRMRL/Pubs/600ja02406/600ja02406.pdf*, October 18, 2004 (2002).
2. B. V. Babu, and R. Angira, Optimization of water pumping system using differential evolution strategies, *Proceedings of the Second International Conference on Computational Intelligence, Robotics, and Autonomous Systems (CIRAS-2003)* Singapore, 2003.
3. P. Charoenongart, Optimal planning for pipe replacement of distribution network using differential evoltuion algorithm, MS Thesis, Kasetsart University, 2005.
4. P. Ditthakit, Optimization of pipe sizes for loop water distribution by simulated annealing method, MS Thesis, Kasetsart University, 2001.
5. N. Duan, L. W. Mays and K. E. Lansey, Optimal reliability-based design of pumping and distribution systems, *J. Hydraul. Eng.* **116** (1990) 249–268.
6. G. Frank, and H. Steve, Economic modeling for prioritizing pipe replacement programs, *Weston Solutions, Inc.*, available at: *http://www.westonsolutions.com/Publications/Grablutz1.PDF*, October 15, 2004–1999.
7. Y. Kleiner, B. J. Adams, and J. S. Roger, Water distribution network renewal planning. *Journal of Computing in Civil Engineering* **15**, 1 (2001) 15–26.
8. J. Kosunwit, The study of water distribution system design criteria, MS Thesis, Kasetsart University, 1990.

9. C. Milan, Advantages of using genetic algorithms over deterministic methods in optimal design of the water networks rehabilitation, *Algoritmy* (2000) 293–300.
10. M. Moradi-Jalal, S. I. Rodin and M. A. Mario, Use of genetic algorithm in optimization of irrigation pumping station, *Journal of Irrigation and Drainage Engineering* **130** (2004) 357–365.
11. V. Nitivattananon, E. C. Sadowski and R. G. Quimpo, Optimization of water supply system operation, *J. Water Resour. Plan. Manage* **122** (1996) 374–384.
12. A. Prempree, Optimal design of water distribution piping system using differential evolution algorithm, MS Thesis, Kasetsart University, 2005.
13. K. Price, and R. Storn, Differential Evolution a simple and efficient adaptive scheme for global optimization over continuous spaces, Technical Report TR-95-012, ICSI, 1998.
14. A. B. A. Sakarya, and L. W. Mays, Optimal operation of water distribution pumpsconsidering water quality, *J. Water Resour. Plan. Manage.* **126** (2000) 210–220.
15. S. Thasaduak and S. Chittaladakorn, *Efficiency Evaluation of Differential Evolution Algorithms for Resesrvoir Operation, The 10th Nat. Conf. on Civil Eng*, Thammasart University, Chon Buri, 2005 pp. WRE108-WRE113.
16. P. Wiriyagulopas, The study of optimal pump scheduling for water distribution system using stochastic optimization method, MS Thesis, Kasetsart University, 2006.

Advances in Geosciences
Vol. 17: Hydrological Science (2008)
Eds. Namsik Park *et al.*

DECISION SUPPORT MODEL FOR OPTIMAL MANAGEMENT OF COASTAL GATE

PAKORN DITTHAKIT

Lecturer. D.Eng., School of Engineering and Resources Management,
Walailak University, 222 Thaiburi,
Thasala District, Nakhon Si Thammarat 80160, Thailand.
dpakorn@hotmail.com

SUWATANA CHITTALADAKORN

Associate Professor. Ph. D., Department of Water Resources Engineering,
Faculty of Engineering, Kasetsart University,
50, Phaholyothin Rd. Lardyao, Chatuchak, Bangkok 10903, Thailand.
fengswc@ku.ac.th

The coastal areas are intensely settled by human beings owing to their fertility of natural resources. However, at present those areas are facing with water scarcity problems: inadequate water and poor water quality as a result of saltwater intrusion and inappropriate land-use management. To solve these problems, several measures have been exploited. The coastal gate construction is a structural measure widely performed in several countries. This manner requires the plan for suitably operating coastal gates. Coastal gate operation is a complicated task and usually concerns with the management of multiple purposes, which are generally conflicted one another. This paper delineates the methodology and used theories for developing decision support modeling for coastal gate operation scheduling. The developed model was based on coupling simulation and optimization model. The weighting optimization technique based on Differential Evolution (DE) was selected herein for solving multiple objective problems. The hydrodynamic and water quality models were repeatedly invoked during searching the optimal gate operations. In addition, two forecasting models:- Auto Regressive model (AR model) and Harmonic Analysis model (HA model) were applied for forecasting water levels and tide levels, respectively. To demonstrate the applicability of the developed model, it was applied to plan the operations for hypothetical system of Pak Phanang coastal gate system, located in Nakhon Si Thammarat province, southern part of Thailand. It was found that the proposed model could satisfyingly assist decision-makers for operating coastal gates under various environmental, ecological and hydraulic conditions.

1. Introduction

The coastal areas are the world's most important and intensely used of all areas settled by humans. Coastal areas attract the human population because they are focal point for economic growth.[4] They are fertile areas with natural resources and also support several agricultural activities. However, with more population increase at present, most natural resources, especially forests and water sources, have been considerably deteriorated leading to serious adverse effects on environment. For example, during dry season there is inadequate freshwater for agricultural and domestic consumptions due to insufficient storages and salt water intrusion. In addition, the water quality problem often occurs in the portion of river flowing through community. On the contrary, during wet season the occurrence of flooding leads to loss of human lives and property.

To control the salinity level at specified locations, several researches proposed the development of coupling simulation-optimization model to determine the optimal upstream freshwater inflows or upstream reservoir releases into bays and estuaries.[2,8,10,16,17] Prabnarong and Kaewrat[11] reviewed the environmental changes since the operation of the Uthokawiphatprasit watergate over the Pak Phanang River in the southern east coast of Thailand. As recent years, the coastal gates, the most widespread structural measures to alleviate such problems in coastal areas, have been constructed in several countries.[6,7,14] However, the operations of coastal gate to simultaneously control several water quality and quantity parameters become a complicated task. It hence requires decision support tool for assist the decision makers.

A number of mathematical models have been developed and used extensively in the recent years as decision support tools for coastal gate operations. Hoanh *et al.*[7] used GIS and hydraulic model to manage conflicting demands on water quality for shrimp and rice production in the Mekong River Delta, Vietnam. Dufour *et al.*[6] presented integrated water management system for operation of the Bang Pakong diversion dam, Thailand. RID[14] developed the River Operation Model (ROM), which is the main forecasting and mathematical simulation models applied in this work, for integrated water management in the Pak Phanang River Basin Development Project, Thailand. To determine the coastal gate operation scheduling, all models as reviewed above used trial & error approach. This method is extensive time-consuming process, especially for the complex systems. It also requires the considerable gate regulator's experience, and is not guaranteed to obtain the optimal solution. In addition most of the existing mathematical simulation model cannot determine the optimal multi-purpose gate operation like coastal gate, which require control both water quality and water quantity criteria simultaneously. It therefore could develop the decision support model which can help decision-makers for optimally determining coastal gate operations.

In present article, the novel method for management of coastal gate operations through simulation-optimization model is proposed. The main features and model development of the proposed model are also described. Finally, the model's applicability is demonstrated using a hypothetical system of coastal gate system of Pak Phanang river basin, located in Nakhon Si Thammarat province, southern part of Thailand.

2. Overview of Model Features

The model features consist of three main modules: forecasting model, mathematical simulation model, and optimization model. The more detailed information for each feature can be described as follows.

2.1. *Forecasting model*

In the planning and scheduling of coastal gate operations, it requires the efficient time series forecasting models. Hence, Auto Regressive & Updating Procedure model (AR model) and Harmonic Analysis model (HA model) were selected herein for forecasting the time series data of hourly upstream flows and tide levels, respectively. These forecasted data will be subsequently used as boundary input data for hydrodynamic model.

2.1.1. *Auto regressive model (AR model)*

The forecasting procedure of water levels and rainfall time series data by AR model comprise of the four main steps: (1) measuring and sending real-time data, (2) the hydrological and hydraulic model, (3) rainfall prediction, and (4) updating procedure or data assimilation procedure. By this method, the predicting efficiency depends upon the selected model quality, the rainfall forecasting accuracy, and the chosen updating procedure efficiency. The updating procedure is feedback procedure, which uses the measured real-time data for adjusting the initial state of system and the forecasting results. In general, the updating procedure has several types: input variables, model states, model parameter, and output variables.[14] In this study, the updating output variables procedure is selected since it is traditional method and appropriate for conceptual rainfall-runoff model like Tank Model, a selected rainfall-runoff relationship model in this study.

2.1.2. *Harmonic analysis model (HA model)*

The tide phenomena results from the relative motion influence of the earth, moon and sun as well as the oceanography effect. Generally, the tide characteristic is a certain periodic wave. Thus it can be analyzed using harmonic analysis. The harmonic analysis, applied for forecasting tide about 1 to 2 days ahead, consists of two stages. Firstly, the measured tide at previous day $(t-1)$, obtained from telemetering station, is used to determine the values of mean amplitude and phase of specified constituent. Secondly, these two parameters and the time values (t) are substituted in harmonic equation, and then the forecasted tide is obtained for subsequent time period.

2.2. *Mathematical simulation model*

2.2.1. *Hydrodynamic model*

The hydrodynamic model is used to model the flow behavior of water in the river network and flood plain. The model uses implicit finite difference scheme for solving the One-Dimensional Saint-Venant's Equation. The hydrodynamic model comprises of five sub-modules, namely:- Node-Branch (NB), Flood Plain (FP), Structure (STR), Gate Operation (GOP), and Rating Curve (RC). It can thus analyze flow over a variety of hydraulic structures such as gates, weirs, bridges, and culverts. The GOP model offers the possibility to control hydraulic structures (i.e. pumps, gates, etc.) by choosing different controlled strategies. Any number of conditions can be specified for each strategy. For a given control strategy all the required conditions must satisfy the strategy to be executed. These conditions can be set on hydraulic variables, i.e. water levels, discharges, etc. Often, the conditions of a control strategy describe hydraulic conditions at specified locations inside the model area as well as time-dependent conditions, e.g. the seasonally variation of tide.

2.2.2. *Water quality model*

The water quality model is used to model the mass transportation in watercourse by solving an advection-dispersion equation. This model can be used to study mass transportation for several water quality parameters, e.g.

salinity, dissolved oxygen, biological oxygen demand, and pH. Typically, the dissolved oxygen and biological oxygen demand are analyzed simultaneously because the organic decomposition process requires oxygen from reaeration process. The water quality model used in this study comprises of three sub-models, namely:- BOD/DO Model, pH Model, and Salinity Model. To run this model, it requires the water level and discharge data computed from hydrodynamic model.

2.2.3. *Rainfall-runoff model*

A rainfall-runoff model will be needed to develop hyetographs and simulate runoff for historical, hypothetical, forecast, and real-time storm events. This range of storm event includes all the hydrologic input data needed to conduct operational planning and basin planning. The hydrology model will allow selection of several methods to compute runoff volumes and model direct runoff. It will also be suitable for simulation in urban, rural and mixed land use catchments. To allow the use of the model in basin planning, the simulation engine will need to have the capability of simulation runoff for future land use conditions. The tank model, a lumped conceptual model, is adopted herein as rainfall–runoff relationship model. These data are subsequently used as upstream boundary data for hydrodynamic model.

2.2.4. *Water demand model*

A dynamic water demand model will be required for proper implementation of coastal gate management. The model should be prepared and updated as needed depending on information available. Required frequency of updates would be based on the latest National Statistics Office population, industrial. The model will give estimates describing water demand variation, by zone and by season. The water demand model is used to calculate the weekly and monthly irrigation water demand and return flow. The irrigation areas are considered as irrigation blocks. The required input information is irrigation area, cropping activity, effective rainfall, potential evapotranspiration, crop coefficient, other demands, and rainfall. These data are subsequently used as input data for hydrodynamic model.

2.3. *Optimization model*

The Differential Evolution algorithm (DE) is selected in this study for seeking the best solution responding to the desired objective function. This algorithm is capable of searching the nearest solution rapidly and directly. It is classified to be the stochastic search space algorithm like genetic algorithms (GAs), but it is less complex than genetic algorithms. Furthermore, it uses real number instead of binary number transformation, needs no bit — String encoding, and fits to non-linear problems and non-differentiate function. DE is successfully applied in various optimization problems and water resource engineering problems, for example the pipe design and the replacement of water distribution system,[3,12] reservoir operation,[15] and water pumping system,[1] and irrigation canal design.[5] The principle procedure for approaching the solution by Differential Evolution created by Price and Storn[13] can be generally described as follow.

2.3.1. *Initial population*

In the first step, a set of numbers of initial population is randomly generated from specified boundary constraints using the uniformly distribution random numbers to cover the entire solution space. Each of those has its cost value or fitness value.

$$P_{G=0} = x_j^{(L)} + rand_j\,[0,1] \cdot \left(x_j^{(U)} - x_j^{(L)}\right),\ \mathrm{j} \in [1, D] \tag{1}$$

where $rand_j[0,1]$ = a uniformly distributed random value in range $[0,1]; x_j^{(U)}, x_j^{(L)}$ = upper and lower boundary of each D real parameters.

2.3.2. *Mutation*

The next step is to mate couple of population, calculate the different value between them, and multiply the different value by weighting factor or mutation factor (F). Then a newly difference population vector is obtained. Figure 1 shows the example of procedure to search the Mutant Vector from 2 variables population. The sub procedures are as follows.

1) Defining the Target Vector $(X_{i,G})$
2) Randomly select three Vectors $(X_{r1,G}, X_{r2,G}, X_{r3,G})$ from the initial population which are different from Target Vector
3) Calculate the Mutant Vector $(V_{i,G+1})$ from the relation:

$$V_{i,G+1} = X_{r1,G} + F(X_{r2,G} - X_{r3,G}) \tag{2}$$

where i = 1,2,3,...,NP; $X_{i,G}$= target vector; $V_{i,G+1}$= mutant vector; $X_{r1,G}, X_{r2,G}$, $X_{r3,G}$ = random vector; F = weighting factor.

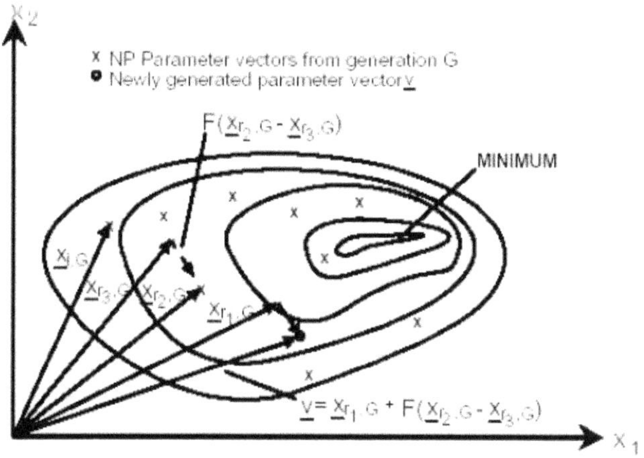

Fig. 1. Procedure to search the Mutant Vector from 2 variables population.

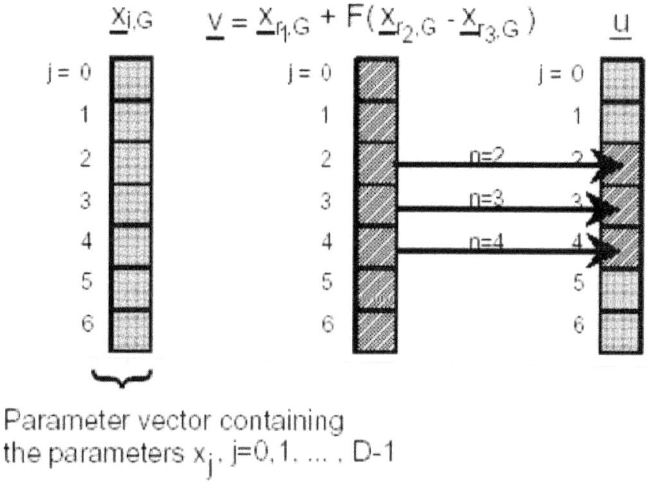

Source: Price and Storn (1998)

Fig. 2. Procedure in Crossover step of the Target Vector and Mutant Vector which its value D = 7

2.3.3. *Crossover or recombination*

This process is performed to increase the diversity of the perturbed parameter vectors. As shown in Fig. 2, the trial vector, $U_{i,G+1}$ is produced by duplicating some elements of the mutant vector, $V_{i,G+1}$ or some elements of the target vector, $X_{i,,G}$depending on the criteria presented in mathematical form as:

$$\begin{aligned} U_{i,G+1} &= V_{i,G+1}\text{if (randb(j)} \leq \text{CR) or j} = \text{rnbr(i)} \\ U_{i,G+1} &= X_{i,G}\text{if (randb(j)} > \text{CR) or j} \neq \text{rnbr(i)} \end{aligned} \tag{3}$$

where $U_{,G+1}$ = trial vector; $V_{i,G+1}$ = mutant vector; $X_{i,G}$ = target vector; randb(j) = the jth evaluation of a uniform random number generator with outcome $\in [0, 1]$; rnbr(i) = a randomly chosen index $\in$ 1,2,3,...,D; CR = the crossover constant $\in [0, 1]$.

2.3.4. *Selection*

The next population generation (G + 1) is created from the newly better population by comparing the objective function value of trial vector with target vector. If the objective function value of trial vector is better than that of target vector, the next population generation (G+1) will be replaced by the trial vector, and vice versa.

The process of mutation, crossover, and selection is routinely repeated until a termination criterion, for example the maximum number of generation is satisfied. The algorithm then terminates proving the best point that has been explored over all the generations.

3. Model Development

3.1. *Objective function and constraints*

The multi-objective optimization based on weighting method was designed. This mechanism was used to ensure that if failure cannot be prevented, the least important interests fail first and the most important ones fail last (Lobbrecht *et al.*, 2005). The desired conditions are to prevent poor water quality and water scarcity. In this study, the control variables or decision variables are gate opening. And the state variables are water levels, dissolved oxygen concentrations, and salinity concentrations at specified locations. The objective function is to maximize satisfaction function. The mathematical formulation for one time step Δt of the control horizon to simultaneously satisfy all requirements can be expressed as follows

$$\text{Maximize Z } = \sum_{i=1}^{m} W_{ai} \sum_{j=1}^{n} R_{di,j} S_{i,j} \left(\bar{x}, \bar{u}\right) \tag{4}$$

whoro $\bar{x}$ — timo dopondont stato variablos; $\bar{u}$ — time-dependent control variables; $S_{i,j}(\bar{x}, \bar{u})$ = satisfaction function j for water subsystem i; W_{ai} = weighting factor for an area; and $R_{di,j}$ = relative importance of interesting parameter j in water subsystem i within a particular area.

Since there were three state variables:- water levels, salinity concentrations, and dissolved oxygen concentrations considered in this study and each parameter has different rang value, all these three parameters were normalized to be the same range value using the simple linear transformation function as mathematically expressed in equation 5. Considering the equation 5, it shows that the maximum satisfaction function value is 10, which is obtained when the value of $x_{calculated}$ and $x_{desired}$ are equal. The satisfaction function value will decreases where the different value between $x_{calculated}$ and $x_{desired}$ is more than a half of the different value between x_u and x_l.

$$S_{i,j}(\bar{x}, \bar{u}) = 10 - \frac{|x_{calculated} - x_{desired}|}{0.5(x_u - x_l)} \tag{5}$$

where $x_{calculated}$ = the calculated state variables; $x_{desired}$ = the desired state variables; and x_u, x_l = upper and lower limit on the state variables.

The problem is subject to three constraints as follows.

(1) Hydraulic constraint is defined by the Saint-Venant equations for one-dimensional gradually varied unsteady flow that describe the flow in the different components of the river system as expressed.

$$H(h, Q, g) \tag{6}$$

where h = the vector of water surface elevations; Q = the vector of discharge; and g = the matrix of gate setting. All given matrix forms are considered in the time and space dimensions.

(2) Water quality constraint is defined by the Advection-Diffusion equation. To run water quality model, it requires the output data (flows and water levels) of hydrodynamic model. The results of water quality model are the concentrations of water quality parameters at each space and time.

$$W(wq, g) \tag{7}$$

where wq = the water quality parameters; and G = the matrix of gate setting.

(3) Physical and operational bounds on gate operations

$$\underline{g} \leq g \leq \bar{g} \tag{8}$$

where $\bar{g}$,$\underline{g}$ = the upper and lower bounds for gate setting. The bounds on gate setting are intended primarily to reflect the physical limitations on the gate operations. This may be very useful, especially for the cases where sharp changes in the gate operations (i.e. sudden opening and closures), are not desirable or physically impossible.

3.2. *Coupling simulation and optimization model*

The overall framework of coupling simulation and optimization model, namely CoastalGate model is presented in Fig. 3. The procedure begins with obtaining input data for optimization, hydrodynamic, and water quality models. Through this simulator-optimizer formulation, the problem is solved efficiently by incorporating the mathematical simulation model into a procedure when a set of gate settings (g) or control vector is chosen. The simulation model is run, which is subject to the selected control vector, to solve the hydraulic constraint set (H) for the water levels, and flows, and to solve water quality constraint set (W) for the salinity concentrations and dissolved oxygen concentrations. Then the multiple-objective function is evaluated. Furthermore, the violation of physical and operational bounds on gate operations is checked and the alternate optimal solutions, which often occur during running this model in the case of non-orifice flow, are considered. These procedures are repeated with a new data set of gate settings until a convergence criterion is satisfied and no bound constraints are violated.

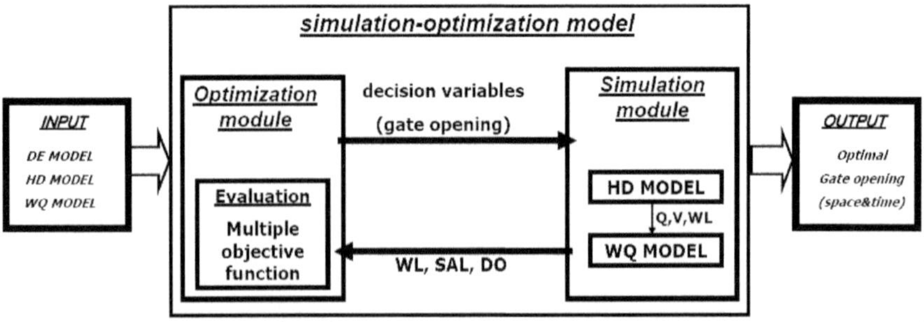

Fig. 3. The overall framework of coupling simulation and optimization model.

4. Model Demonstration

4.1. *Description of study area and used data*

To demonstrate the applicability of the developed model, it was applied for planning hourly gate operations in Pak Phanang River Basin (PPRB) Development Project, Nakorn Si Thammarat province, Thailand. Figure 4 shows schematic diagram of coastal gate system in PPRB. The system consists of thirteen gates, namely Uthokawiphatprasit (G1), Sua Hueng (G2), Pakrawa (G3), emergency gate (G4), Thapraya (G5), Klong Khong (G6), Chian Yai (G7), Bang Sai (G8), Sukoom (G9), Praekmueng (G10), navigation gate (G11), Klong Lad (G12), and Nha Goat (G13). The first ten

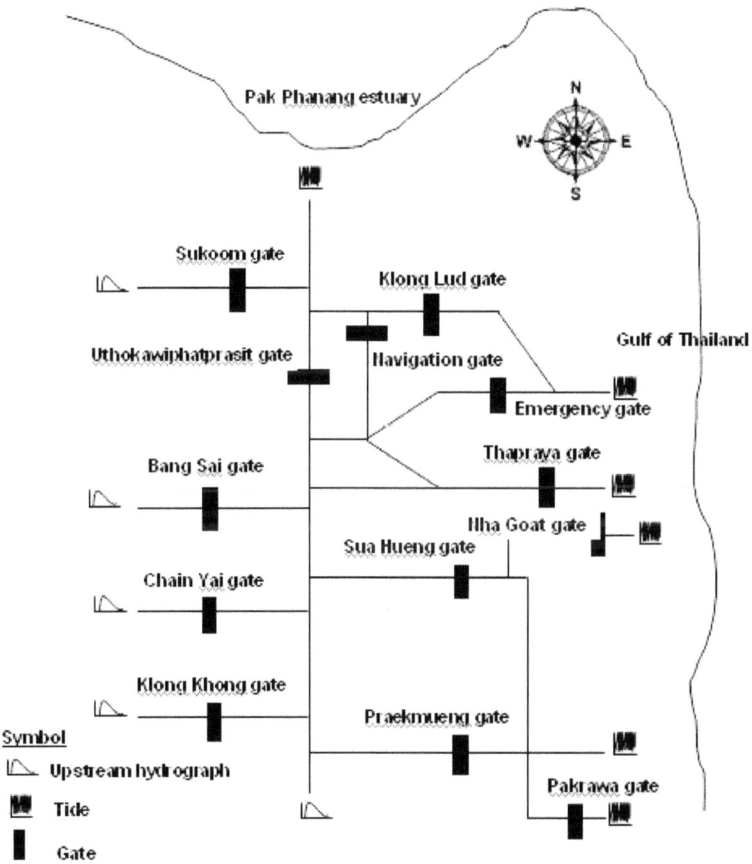

Fig. 4. Schematic diagram showing coastal gate system of PPRB.

gates were considered herein as control variables. The navigation gate was closed throughout control horizon. The other two gates were controlled by comparing the water level between upstream and downstream gate. Except Uthokawiphatprasit gate, which is the adjustable-crest weir, the other nine gates are sluice gate.

To run model, the data for all models were prepared. The complete list of all data in meteorology, hydraulics, structure, and water quality used in hydrodynamic and water quality model are available from RID (2004). The upstream and downstream boundary data for hydrodynamic model were generated for the operation period of 24 hours in advance on 24/03/2005 during dry season using Forecasting model and hydrological database obtained from the telemetering system.

The current policy for gate system operation during dry season is follows: (1) for freshwater storage purpose, G1 and G10 are closed when water level is below 0.3 m MSL. G2, G3, G4, and G5 are closed all time. G6, G7, G8 and G9 are opened when water levels in Pak Phanang River are higher than those in each canal; (2) for salinity control purpose, G1, G10 and G13 are operated depending on water levels at upstream and downstream gate. In general, the gates are closed when downstream water levels are higher than upstream water levels and they is opened when there is sufficient freshwater to prevent saltwater intrusion. The salinity concentration for upstream and downstream gate should be $\leq$ 2 ppt, and 20–26 ppt, respectively; (3) for dissolved oxygen control purpose, G1, G10 and G13 are operated depending on water levels at upstream and downstream gate the same as salinity control purpose. In addition, G12 is controlled corresponding to the difference of tide levels between Pak Phanang estuary and Gulf of Thailand to alleviate water quality problem of Pak Phanang community located downstream Uthokawiphatprasit gate.

In this study, the current policy for gate system operation was considered as a baseline scenario (BAS). Four study cases were analyzed the effect of decision-maker defined parameters on coastal gate operations. The first case is BAS and the other three cases are full optimization scenarios (FOP#1, FOP#2, and FOP#3, respectively), which the different weighing factor sets are given. The gate opening and crest level of adjustable-crest weir are integer and adjusted not more than 2 meters for each run time due to gate's speed limit. Hence, the total search space is 10^5 combinations:- 10 gates and each gate takes 5 possible values. Since hourly time step of control horizon was considered, the selected control points is in vicinity of control structures to assure that gate operations has an effect on the control points.

4.2. *Results and discussion*

Before applying the Different Evolution method for finding the optimal solution, it requires the determination of the optimal four parameters: maximum number of generations, population size (NP), weighting factor (F), and crossover constant (CR) for the concerning problem. These parameters differ from problem to problem depending on the characteristics of each problem. In general, the sensitivity analysis by varying all these parameters is used for determining the suitable parameters. In this study, it was found that the maximum number of generations, population size, weighting factor, and crossover constant are 3, 50, 0.8, and 0.8 respectively The reason why select the maximum number of generations of 3, and population size of 50 because it gave the maximum satisfaction function value and save more computational time compared to other cases, which use larger value. Likewise, the selected values of weighting factor (F) of 0.8, and crossover constant (CR) of 0.8 gave the maximum satisfaction function value.

The example of the optimal coastal gate operation scheduling for 24 hours in advance of Uthokawiphatprasit gate for three full optimization scenarios is presented in Fig. 5. It was found that no a sharp change occur, i.e. gate changes did not exceed the given allowable maximum change. Therefore, the preventing method in violation of physical bounds on gate operations is acceptable.

The values of satisfaction function for BAS, FOP#1, FOP#2, and FOP#3 were 1176, 1184, 1192, and 1206, respectively. This showed that all

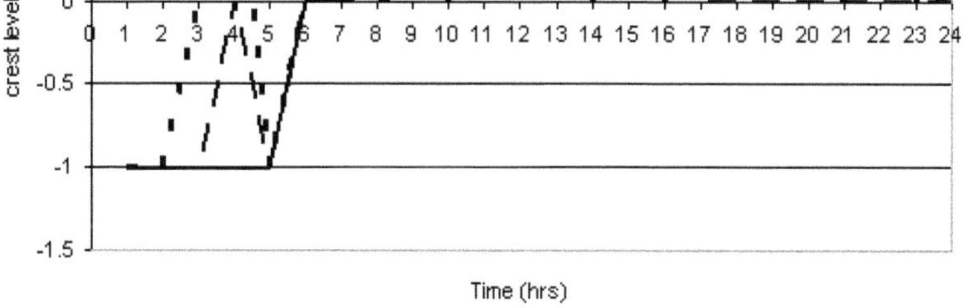

Fig. 5. Schematic diagram showing coastal gate system of PPRB.

three full optimization scenarios gave better results than baseline scenario, and FOP#3 gave maximum value of satisfaction function. The example of state variables at control points of Uthokawiphatprasit gate are presented in Fig. 6. The average upstream water level for BAS, FOP#1, FOP#2,

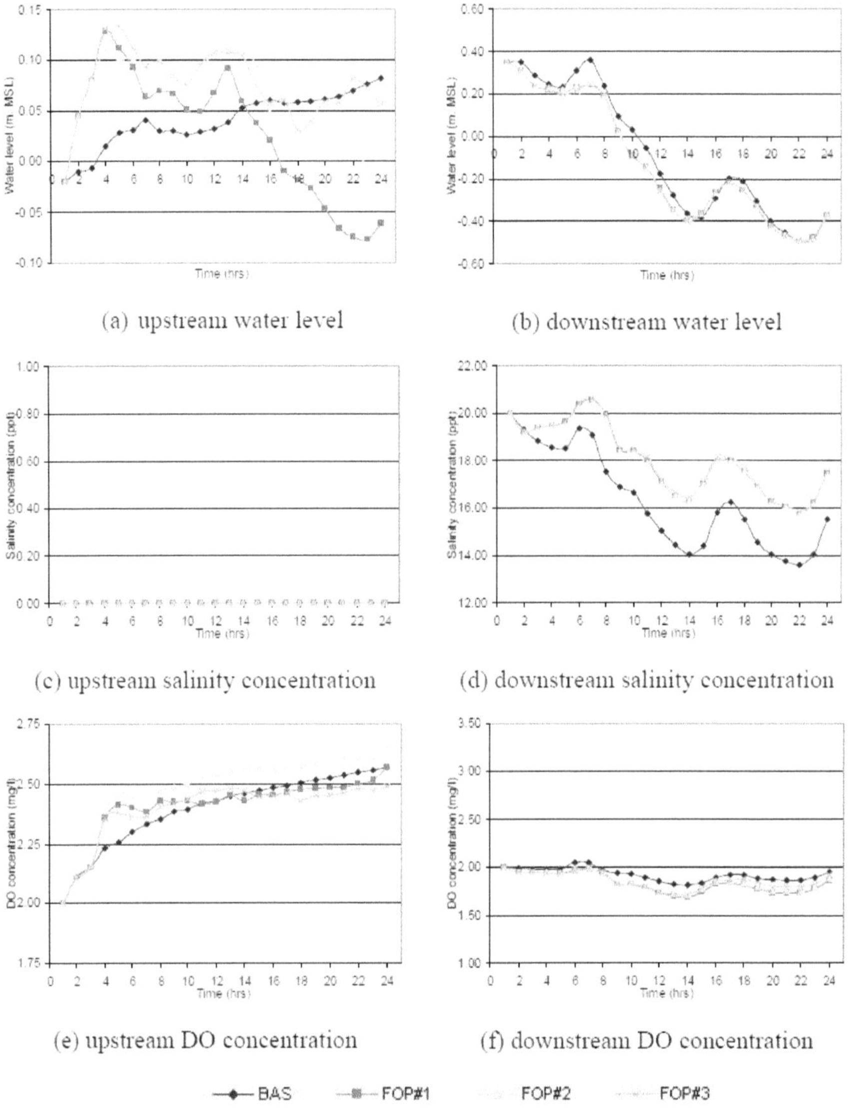

Fig. 6. State variables versus control horizon at checkpoints of Uthokawiphatpraist gate.

and FOP#3 were 0.040, 0.026, 0.060, and 0.076 m, respectively. This showed that the average upstream water level for FOP#3 was the closest to its desired target of 0.3 m. The downstream water level for FOP#1, FOP#2, and FOP#3 scenarios were similar, but those for BAS were a little higher. Although gate operation scheduling guided by the model allowed seawater through the upstream river to raise upstream water levels, the salinity concentrations at chosen upstream control point for BAS, FOP#1, FOP#2, and FOP#3 scenarios were acceptable with salinity concentrations of 0.0 ppt. The salinity concentrations at downstream control point for three full optimization scenarios were similar, i.e. the average salinity concentrations were 18, 18, and 18.1 ppt, respectively. Those values were closer to its desired value of 23 ppt compared to the baseline scenario, which gave the average salinity concentration of 16.3 ppt. The trend of dissolved oxygen concentrations at upstream control point for all scenarios is gradually increasing and FOP#2 gave the maximum average dissolved oxygen concentrations. The trend of the change of dissolved oxygen concentrations at downstream control point for all scenarios was the same. However, the gate operated by BAS gave the maximum of the average dissolved oxygen concentrations.

5. Conclusion and Recommendation for Future Work

This paper presents the development of decision support model for optimal management of coastal, namely CoastalGate Model. The features of model consist of three main parts:- Forecasting Model, Mathematical Simulation Model, and Optimization Model. The weighting optimization technique based on Differential Evolution (DE) was selected herein for solving multiple objective problems. The proposed model is illustrated by using Pak Phanang coastal gate system as a case study.

It was found that the Differential Evolution (DE) method as a global optimizer in surface water system is useful for the use of obtaining optimum solution of coastal gate operation. The developed simulation-optimization model could help decision makers for planning coastal gate system operation under various environmental, ecological and hydraulic conditions. Due to multiple optimization problems, the results depend on the determination of preferred weighting factor and the relative importance of the interesting parameter. The execution time was approximately 50 minutes for one run time step on a 160 MHz Pentium PC. This is because in solving the problem it is necessary to routinely invoke mathematical simulation model several times, especially calling water quality model, which requires rather much computation time. Hence, this developed model is quite appropriate for planning coastal gate operation in advance than using for real-time control tasks, which need making a decision in very short time interval. However, the further development of faster computing equipment and more efficient numerical methods for HQ and WQ models will extend the capability of this proposed methodology.

This proposed model can be implemented for automatic gate system. For semi-automatic gate and manual gate systems, the longer time control, such as daily gate operation may be required. In addition, the determination of suitable weighting factor, which is the characteristics of particular area, could be studied for future work. This may use multi-criteria decision making process, i.e. Analytic Hierachy Process (AHP).

Acknowledgments

The writers wish to express sincere gratitude to the Graduate School, Kasetsart University (KU) for financial support for this research project. The second writer is thankful to Institute of Research and Development Walailak University for providing financial help for attending oral presentation at 5th Asia Oceania Geosciences Society Conference, AOGS 2008, Busan, Korea. Finally, the writers are grateful to anonymous reviewers for their great recommendations to improve the quality of this paper.

References

1. B. V. Babu and R. Angira, Optimization of water pumping system using differential evolution strategies, *Proceedings of the Second International Conference on Computational Intelligence, Robotics, and Autonomous Systems (CIRAS-2003).* Singapore, 2003.
2. Y. Bao and L. W. Mays, New methodology for optimization of freshwater inflows to estuaries, *Journal of Water Resources Planning and Management* **120**(2) (1994) 199–217.
3. P. Charoenongart, Optimal planning for pipe replacement of distribution network using differential evoltuion algorithm, MS Thesis, Kasetsart University, 2005.
4. J. R. Clark, *Coastal Seas: The Conservation Challenge* (Blackwell Science Ltd, Oxford, 1998), p. 134.
5. P. Ditthakit and S. Chittaladakorn, Irrigation canal design by differential evolution approach, *The 12th Nat. Conf. on Civil Eng.* Naresuan University, Pitsanulok, Thailand, (2007) pp. WRE112-WRE117.

6. A. Dufour, S. Prajamwong, B. Camire and S. Asselin, Integrated water management system feasibility study for the bang pakong river basin, *Proceedings of the World Water and Environmental Resources Congress held in Philadelphia*, Pennsylvania, June 24–26, 2003.
7. C. T. Hoanh, T. P. Tuong, S. P. Kam, N. D. Phong, N. V. Ngoc and E. Lehmann, Using GIS-linked hydraulic model for managing water quality conflict for shrimp and rice production in the Mekong River Delta, *Vietnam Proceeding of MODSIM 2001, International Congress on Modeling and Simulation*, Canberra, Australia, 10–13 December, 2001. Volume 1: Natural Systems (Part one). The Modeling and Simulation Society of Australia and New Zealand Inc., 2001, pp. 221–226.
8. G. Li and L. W. Mays, Differential dynamic programming for estuarine management, *Journal of Water Resources Planning and Management* **121**, 6 (1995) 455–462.
9. A. H. Lobbrecht, Y. B. Dibike and D. P. Solomatine, Neural networks and fuzzy systems in model based control of the overwaard polder, *Journal of Water Resources Planning and Management* **131**, 2 (2005) 1–11.
10. N. Mao and L. W. Mays, Gold programming models for determining freshwater inflows to estauries, *Journal of Water Resources Planning and Management* **120**, 3 (1994) 316–329.
11. P. Prabnarong and J. Kaewrat, The uthokawiphatprasit water gate: a man-made change in pak phanang river basin, *Walailak J Sci & Tech.* **3**, 2 (2006) 131–143.
12. A. Prempree, *Optimal design of water distribution piping system using differential evolution algorithm*, MS Thesis, Kasetsart University, 2005.
13. K. Price and R. Storn, Differential evolution a simple and efficient adaptive scheme for global optimization over continuous spaces, Technical Report TR-95-012, ICSI, 1998.
14. Royal Irrigation Department (RID), The study of the Operations of Uthokawiphatprasit water gate, ROG, PALCON, and SDCON Consultant Ltd, Bangkok, Thailand, 2004.
15. S. Thasaduak and S. Chittaladakorn, Efficiency evaluation of differential evolution algorithms for resesrvoir operation, *The 10th Nat. Conf. on Civil Eng.* Thammasart University, Chon Buri, Thailand, 2005, pp. WRE108–WRE113.
16. Y. Tung and L. W. Mays, Optimization of freshwater inflow to estuaries, *Journal of Water Resources Planning and Management* **116**, 4 (1990) 567–584.
17. B. Zhao and L. W. Mays, Estuary management by stochastic linear quadratic optimal control, *Journal of Water Resources Planning and Management* **121**, 5 (1995) 382–391.

www.ingramcontent.com/pod-product-compliance
Ingram Content Group UK Ltd.
Pitfield, Milton Keynes, MK11 3LW, UK
UKHW020403150926
8624IPUK00004B/318

9 789812 838117